Advanced Research in Parasitology

Advanced Research in Parasitology

Editor: Cherilyn Jose

www.callistoreference.com

Callisto Reference,
118-35 Queens Blvd., Suite 400,
Forest Hills, NY 11375, USA

Visit us on the World Wide Web at:
www.callistoreference.com

ISBN: 978-1-64116-148-0 (Hardback)

Trademark Notice: Registered trademark of products or corporate names are used only for explanation and identification without intent to infringe.

Cataloging-in-Publication Data

Advanced research in parasitology / edited by Cherilyn Jose.
   p. cm.
Includes bibliographical references and index.
ISBN 978-1-64116-148-0
1. Parasitology. 2. Parasitology--Research. I. Jose, Cherilyn.
QL757 .A38 2019
591.785 7--dc23

Table of Contents

Permissions

List of Contributors

Index

Preface

The study of parasites, their hosts and the dynamics between them is under the scope of parasitology. It integrates the techniques of cell biology, molecular biology, bioinformatics, immunology, biochemistry and genetics. Some of the sub-fields of parasitology are medical parasitology, veterinary parasitology, structural parasitology and quantitative parasitology. Study in parasitology also encompasses the study of taxonomy, phylogenetics, conservation biology and ecology of parasites. This book discusses the fundamentals as well as modern approaches of parasitology. From theories to research to practical applications, case studies related to all contemporary topics of relevance to this field have been included herein. This book, with its detailed analysis and data, will prove immensely beneficial to professionals and students involved in this area at various levels.

Various studies have approached the subject by analyzing it with a single perspective, but the present book provides diverse methodologies and techniques to address this field. This book contains theories and applications needed for understanding the subject from different perspectives. The aim is to keep the readers informed about the progresses in the field; therefore, the contributions were carefully examined to compile novel researches by specialists from across the globe.

Indeed, the job of the editor is the most crucial and challenging in compiling all chapters into a single book. In the end, I would extend my sincere thanks to the chapter authors for their profound work. I am also thankful for the support provided by my family and colleagues during the compilation of this book.

Editor

The Seroprevalence of Human Cystic Echinococcosis in Iran

Reza Shafiei,[1] Saeed Hosseini Teshnizi,[2] Kurosh Kalantar,[3] Maryam Gholami,[1] Golnush Mirzaee,[4] and Fatemeh Mirzaee[5]

[1]Vector-Borne Diseases Research Center, North Khorasan University of Medical Sciences, Bojnurd, Iran
[2]Clinical Research Development Center of Children Hospital, Hormozgan University of Medical Sciences, Bandar Abbas, Iran
[3]Department of Immunology, School of Medicine, Shiraz University of Medical Sciences, Shiraz, Iran
[4]Rehabilitation Management, University of Social Welfare and Rehabilitation Sciences, Tehran, Iran
[5]School of Nursing and Midwifery, Shahrekord University of Medical Sciences, Shahrekord, Iran

Correspondence should be addressed to Saeed Hosseini Teshnizi; saeed.teshnizi@gmail.com

Academic Editor: José F. Silveira

Human cystic echinococcosis (HCE), a zoonotic infection of the larval stage of *Echinococcus granulosus*, has high effect on public health in human population all around the world. Iran is one of the most important endemic areas in the Middle East. This systematic review and meta-analysis was performed to evaluate the seroprevalence of HCE in Iranian population. An electronic search for articles from 1985 until April 2015 was performed using data bases PubMed, Scopus, Google Scholar, Magiran, IranMedex, Iran Doc, and Scientific Information Database (SID) both in English and in Persian. A random-effects meta-analysis was used to combine results from individual studies. The information was analyzed by STATA version 11.1. A total of 33 articles met our eligibility criteria and were included in a meta-analysis. The pooled estimate of the prevalence of HCE based on random-effects model was estimated 6.0% (95% CI: 4.0%, 7.0%). The prevalence of the disease significantly increased with age and prevalence rate in males was significantly lower than females ($p < 0.001$). The using of CIE or CCIEP method was also significantly greater than the other methods ($p < 0.001$). There was a publication bias in prevalence of studies. HCE is highly prevalent in Iran. Public education for preventive strategies and finally reducing transmission of the parasite and infection in population is needed.

1. Introduction

Human hydatidosis or human cystic echinococcosis (HCE) is a chronic parasitic infection disease caused by the larval stage of *Echinococcus granulosus* which has an important effect on public health in human populations [1, 2]. This zoonotic disease is initiated by accidental ingestion of the parasite's egg. It can be transmitted by infected feces of dogs via soil, vegetables, contact with dog, water, food, and so forth [3, 4].

The disease was widely distributed mostly in regions where sheep-rearing is a major industry [5, 6]. This multihost disease is one of the most important public health infection diseases in Iran [6, 7]. Echinococcosis is one problem not only in humans but also in traps. It causes a huge economic burden for governments. That is why Ministry of Health and veterinary organizations pay more attention to it every year and have regular instruction programs for prevention of this infection. For example, they kill wild dogs; they also prescribe antihelminthes drugs in pet dogs and train the people to wash vegetables. The prevalence of hydatidosis in dogs has been recorded to be 5–45% in Iran [8]. The infection rate with different strains of *E. granulosus* sensu lato in various domestic livestock has been reported to be 24.41%, 8.51%, 18.89%, 35.76%, and 35.21% in sheep, goat, cattle, buffalo, and camels, respectively [8–10]. The incidence of surgical cases of HCE is estimated to be 1.18–3 per 100,000 in different medical centers of Iran provinces and territories [3].

Early diagnosis of HCE is difficult due to being asymptomatic in early stages while using physical imaging; particularly ultrasound (US) examination is helpful not only at late stages but also for early diagnosis (cysts under 1 cm in diameter) [2, 11]. The early diagnosis of HCE is based on available immunodiagnostic techniques with specific immunodominant antigens such as Ag B and US imaging. Methods for detecting specific antibodies can provide opportunities for early treatment of the disease [11–13]. Immunodiagnostic techniques have been used for total screening of population in endemic regions but the sensitivity and specificity of the diagnostic antigen are important [11, 12]. Several immunochemical tests such as ELISA and IFA are developed for determining anti-echinococcus IgG in serum for the diagnosis of HCE in Iran.

Given the high prevalence and the greater importance of this parasite, as well as the economic losses and the significant mortality and morbidity that HCE has in Iran, and due to the fact that the prevalence of this disease has not been identified in Iran yet, the need for a comprehensive study with the aim of serological monitoring of this disease in Iran seemed necessary. In this systematic review and meta-analysis study, we provide some insights into seroprevalence of HCE in different provinces of Iran from 1985 to 2015 where noncoordinated mass screenings have been performed in the past.

2. Methods

2.1. Search Method. The medical publications in English and Persian electronic databases were searched including PubMed Medical Subject Headings (MeSH/mh), Google Scholar, Magiran, Iran Medex, Iran Doc, and Scientific Information Database (SID) both English and Persian databases from 1985 to April 2015. Publication searches were applied using the keywords: "Seroprevalence", "Serological prevalence", "Cystic Echinococcosis", "Hydatid cyst", "Hydatidosis", "Human", "IgG antibody", and "Iran" in combination or alone. To reduce the possibility of selection bias in this study, criteria were clearly defined and studied.

2.2. Data Extraction. The information was extracted from the included studies using a standard form by the two independently reviewers (RSH, STH). Any disagreement was resolved by discussion between the two reviewers. If consensus could not be reached, a third reviewer was consulted (MGH). Kappa index showed an agreement of 89% between the two reviewers.

The standard form consisted of the following variables: first author; year of publication; location, sample size, positive subjects, age and sex of participants, and lab methods. The outcome was the prevalence of seroprevalence and this was obtained for each study by dividing the number of positive cases to the total sample size.

2.3. Statistical Analysis. In this meta-analysis study outcomes were the prevalence of the seroprevalence of HCE. Forest plot was used to visualize the heterogeneity among studies. The results for each study and pooled outcome were revealed as a forest plot [reported as effect size (ES) with a 95% confidence interval (CI)]. Cochran's heterogeneity statistic or Q-test ($p < 0.1$ indicated heterogeneity) and I^2 statistic were used to examine the difference in study variability due

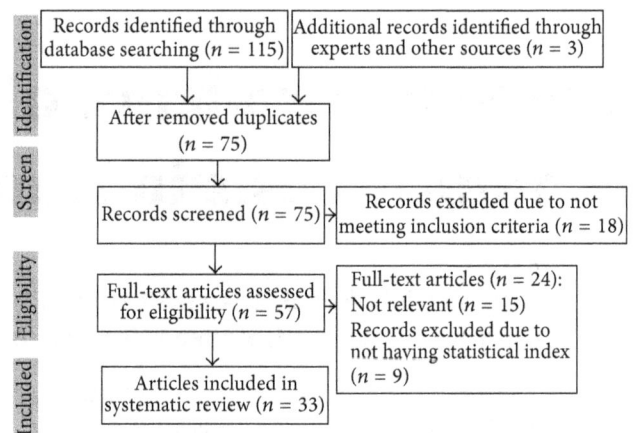

FIGURE 1: PRISMA flowchart describing the study design process.

FIGURE 2: Graphic representation of prevalence (%) of HCE in provinces in Iran. The asterisk (∗) means the minimum and maximum of prevalence in province.

to heterogeneity rather than chance, with a range from 0 to 100 percent (values of 25%, 50%, and 75% are considered to represent low, medium, and high heterogeneity, resp.). Subgroup meta-analysis analysis was used to compare the prevalence of hydatidosis among age, sex, and lab methods groups. Egger's test was used to evaluate the publication bias [14, 15]. Statistical analyses were conducted using the Statistical Software Package (STATA) version 11.1.

3. Results

Our initial database searches identified 115 articles and an additional 3 studies through hand searches and expert suggestions, giving a total of 118 articles that were screened. Out of these, 57 were chosen for reading of full text and 33 were included in this meta-analysis. Figure 1 shows the diagram of article selection according to the PRISMA statement.

Generally, the most prevalence of this disease took place in the western and southwestern areas of Iran and the highest prevalence was related to Lorestan, Fars, and Khuzestan provinces and the lowest rate was related to Tehran (Figure 2).

TABLE 1: General characteristics of studies included in the present systematic review and meta-analysis.

Author [reference]	Year	Province	sample	Positive (n)	Positive % (95% CI)	Lab method (Ag)	Target group
Jamali [37]	1995	Uromieh	300	5	1.67 (0.22, 3.12)	IFA	Villagers
Arbabi [18]	1998	Hamedan	1530	46	3.01 (2.15, 3.86)	IFA	Healthy volunteers
Saberi-Firouzi [21]	1998	Fars	1000	137	13.70 (11.57, 15.83)	CIE	Healthy volunteers
Mohamadi [38]	1998	Tehran	700	68	9.71 (7.52, 11.91)	IFA	Healthy volunteers
Zariffard [39]	1999	Western Iran	4138	230	5.56 (4.86, 6.26)	ELISA	Healthy volunteers
Nilfroshan [40]	1998	Esfehan	1000	36	3.60 (2.45, 4.75)	IFA	Healthy volunteers
Sadjjadi [13]	2001	Shiraz	1227	76	6.19 (4.85, 7.54)	CCIEP	Healthy volunteers
Sedaghat Gohar [27]	2001	Tehran	1052	62	5.89 (4.47, 7.32)	IFA	Healthy volunteers
Yousefi Darani [28]	2003	Chaharmahal	2524	120	4.75 (3.92, 5.58)	CIE	Surgical patients
Amiri [41]	2001	Kermanshah	1072	86	8.02 (6.4, 9.65)	IFA	Patients and blood donors
Farrokhzad [17]	2004	Tehran	437	1	0.23 (−0.22, 0.68)	IFA	Healthy volunteers
Haniloo [29]	2002	Zanjan	2367	71	3.00 (2.31, 3.69)	ELISA	Healthy volunteers
Aflaki [22]	2005	Eilam	3000	37	1.23 (0.84, 1.63)	Dot-ELISA	Healthy volunteers
Rafiei [42]	2005	Khuzestan	4596	437	9.51 (8.66, 10.36)	ELISA	Healthy volunteers
Akhlaghi [43]	2005	Kordestan	1114	37	3.32 (2.27, 4.37)	IFA	Healthy volunteers
Rafiei [20]	2007	Khuzestan	3446	475	13.78 (12.63, 14.94)	ELISA	Healthy volunteers
Baharsefat [34]	2007	Golestan	1024	46	4.49 (3.22, 5.76)	ELISA, IFA	Healthy volunteers
Mirzanejadasl [30]	2008	Ardabil	1003	111	11.07 (9.13, 13.01)	ELISA	Healthy volunteers
Hadadian [23]	2008	Kordestan	1979	22	1.11 (0.65, 1.57)	ELISA	Healthy volunteers
Moazezi [31]	2009	Kerman	451	37	8.20 (5.67, 10.74)	ELISA	Blood donors
Akhlaghi [44]	2009	Tehran	1100	18	1.64 (0.01, 0.02)	Dot-ELISA	Healthy volunteers
Esmaeili [45]	2010	Kashan	361	11	3.05 (0.01, 0.05)	ELISA, IFA	Healthy volunteers
Srakari [24]	2010	Yasuj	500	36	7.20 (0.05, 0.09)	ELISA	Patients referred to lab
Dadkhah [46]	2011	East Azarbaijan	250	8	3.20 (0.01, 0.05)	IFA	Healthy volunteers
Harandi [35]	2011	Kerman	1140	34	2.98 (0.02, 0.04)	ELISA	Healthy volunteers
Kavous [25]	2010	Jahrom	1096	69	6.30 (0.05, 0.08)	ELISA	Patients referred to lab
Garedaghi [47]	2011	East Azarbaijan	1500	11	0.73 (0, 0.01)	ELISA	Healthy volunteers
Heidari [26]	2011	Ardabil	670	12	1.79 (0.01, 0.03)	ELISA	Healthy volunteers
Rakhshanpour [36]	2012	Qom	1564	25	1.60 (0.01, 0.02)	ELISA	Healthy volunteers
Zibaei [16]	2013	Khorram abad	617	95	15.40 (0.13, 0.18)	ELISA	Patients referred to lab
Asgari [19]	2013	Arak	578	20	3.46 (0.02, 0.05)	ELISA	Healthy volunteers
Shahrokhabadi [48]	2014	Kerman	486	9	1.85 (0.01, 0.03)	ELISA	Patients referred to HC
Ilbeigi [33]	2015	Isfahan	635	7	1.10 (0, 0.02)	ELISA	Patients referred to HC

N: number of positive; HC: health center.

From 42706 people 2551 were positive for anti-echinococcosis. Indirect fluorescent antibody (IFA), enzyme-linked immunosorbent assay (ELISA), counterimmunoelectrophoresis (CIE) or counter-current immunoelectrophoresis (CCIEP) were used in all of the studies. ELISA was the mostly used test in this study, 17 (51.52%). Target group of most of studies was healthy volunteers 24 (72.72%) (Table 1).

There was a strong heterogeneity in the prevalence of the studies ($I^2 = 98\%$, $p < 0.0001$). The pooled estimate of the prevalence of HCE based on random-effects meta-analysis was obtained 5.0% (95% CI: 4.0%, 6.0%). The highest prevalence was related to study of Zibaei et al. (2013) with prevalence 15.4% in Khorramabad province [16] study carried out by ELISA and the lowest prevalence was 0.2% in Farrokhzad et al. (2004) in Tehran study carried out by IFA [17]. Also the pooled prevalence significantly was higher than zero line (ES = 0: $z = 10.03$, $p < 0.001$) (Figure 3).

The subgroup analysis indicated that the prevalence of HCE was significantly increased with the increase of age ($p < 0.001$). Also prevalence of CE among males, 2.1 (95% CI: 1.8%, 2.4%), was significantly lower than females, 3.6% (95% CI: 3.2%, 3.9%) ($p < 0.001$). Also the prevalence for CIE method

Study ID	ES (95% CI)	% weight
Jamali et al. (1985)	0.02 (0.00, 0.03)	3.02
Arbabi et al. (1998)	0.03 (0.02, 0.04)	3.14
Saberi et al. (1998)	0.14 (0.12, 0.16)	2.83
Mohammadi (1998)	0.10 (0.08, 0.12)	2.81
Zariffard et al. (1999)	0.06 (0.05, 0.06)	3.16
Nilfroshan et al. (1998)	0.04 (0.02, 0.05)	3.08
Sadjhadi et al. (2001)	0.06 (0.05, 0.08)	3.04
Sedaghat Gohar et al. (2001)	0.06 (0.04, 0.07)	3.02
Yousefi Darani et al. (2003)	0.05 (0.04, 0.06)	3.14
Amiri et al. (2001)	0.08 (0.06, 0.10)	2.97
Farokhzad et al. (2004)	0.00 (−0.00, 0.01)	3.19
Haniloo et al. (2002)	0.03 (0.02, 0.04)	3.16
Aflaki et al. (2005)	0.01 (0.01, 0.02)	3.19
Rafiei and Hamzeloee (2005)	0.10 (0.09, 0.10)	3.14
Akhlaghi et al. (2005)	0.03 (0.02, 0.04)	3.10
Rafiei et al. (2007)	0.14 (0.13, 0.15)	3.08
Baharsefat et al. (2007)	0.04 (0.03, 0.06)	3.06
Mirzanejadasl et al. (2008)	0.11 (0.09, 0.13)	2.88
Hadadian et al. (2008)	0.01 (0.01, 0.02)	3.19
Moazezi et al. (2009)	0.08 (0.06, 0.11)	2.69
Akhlaghi et al. (2005)	0.02 (0.01, 0.02)	3.15
Esmaeili and Arbabi (2008)	0.03 (0.01, 0.05)	2.93
Sarkari et al. (2010)	0.07 (0.05, 0.09)	2.78
Dadkhah et al. (2011)	0.03 (0.01, 0.05)	2.81
Harandi et al. (2011)	0.03 (0.02, 0.04)	3.12
Solhjoo et al. (2011)	0.06 (0.05, 0.08)	3.02
Garedaghi et al. (2011)	0.01 (0.00, 0.01)	3.19
Heidari et al. (2011)	0.02 (0.01, 0.03)	3.11
Rakhshanpour et al. (2012)	0.02 (0.01, 0.02)	3.17
Zibaei et al. (2013)	0.15 (0.13, 0.18)	2.59
Asgari et al. (2013)	0.03 (0.02, 0.05)	3.01
Shahrokhabadi (2014)	0.02 (0.01, 0.03)	3.07
Ilbeigi (2015)	0.01 (0.00, 0.02)	3.14
Overall (I^2 = 97.8%, p = 0.000)	0.05 (0.04, 0.06)	100.00

Note: weights are from random-effects analysis

−0.182   0   0.182

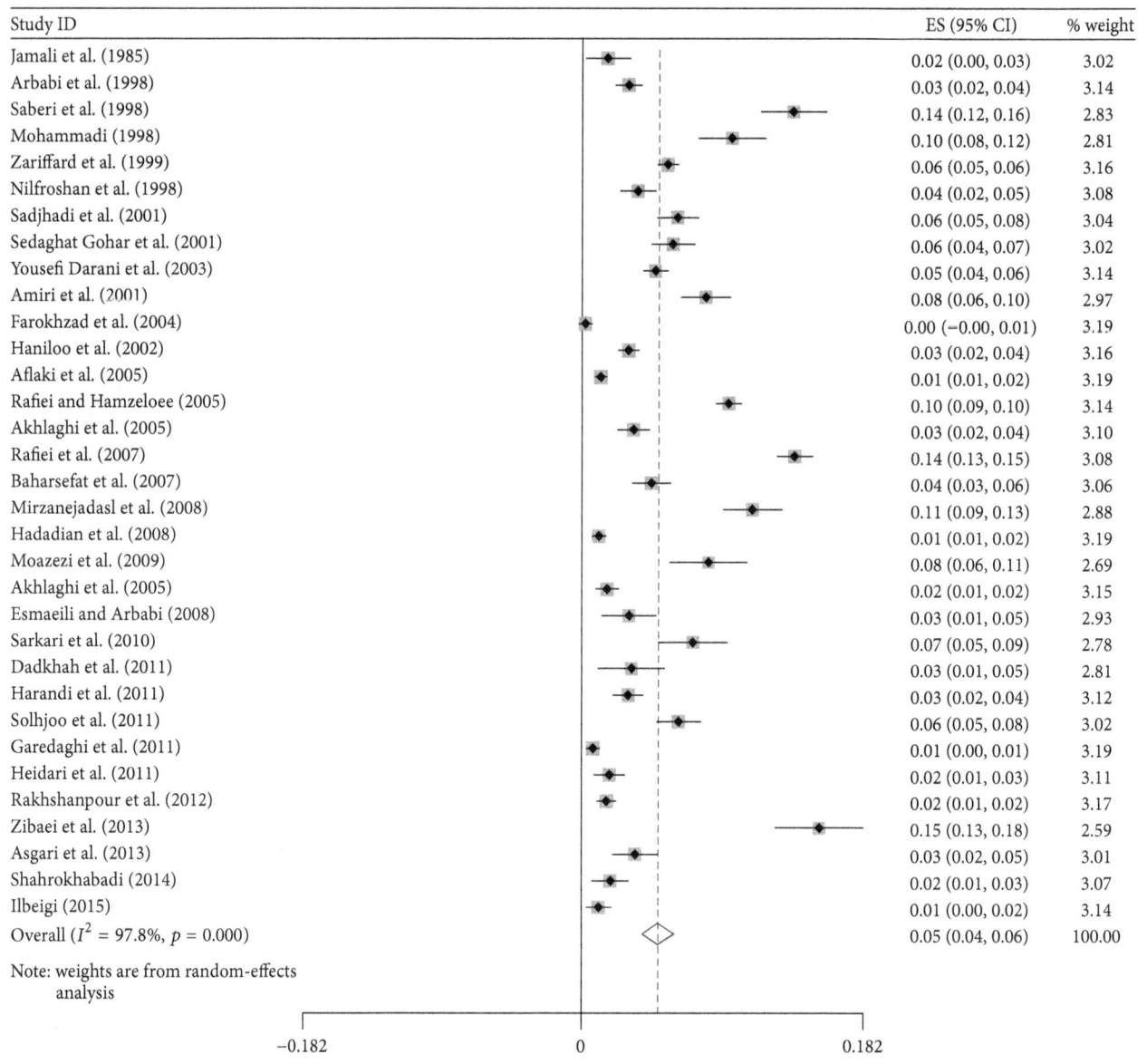

FIGURE 3: Forest plot of meta-analysis results for prevalence of CE. The middle point of each line indicates the prevalence rate and the length of line indicates 95% confidence interval of each study. Rhombus shape indicates 95% confidence interval for all studies.

was significantly greater than other methods ($p < 0.001$) (Table 2). The results of Egger's test showed there was not a publication bias among studies (coef. = −0.011, SE = 0.008, $p = 0.326$). Therefore, this meta-analysis included two types of studies, high and low prevalence of HCE.

4. Discussion

HCE as an emerging neglected disease is a major public health problem in many countries which results in substantial economic resource loss [5]. The mentioned disease has a global distribution with an annual occurrence ranging from 1 to 200 per 100,000 individuals [1]. The prevalence rate of HCE based on hospital cases is different in Iran with rate of >1% of total population [7, 18] and this is the most commonly

used index of HCE [16]. Most of the infected people with the larva of E. granulosus have a delay in showing related symptoms such as cyst-like mass which grows gradually among various groups [5]. So serological examination alone is useful for giving an approximate evaluation of the infection pressure and might be useful (used on already collected serum samples, such as those found in blood banks) to provide data on the level of presence of E. granulosus in a given area to identify asymptomatic cyst carriers generally. We must also have plans to control this disease. To achieve a more accurate diagnosis, mass screening should include both ultrasound examination and serology [19].

The result of our investigation showed that the range of HCE is 0.2% in Tehran with IFA by Farrokhzad et al. (2004) [17] while Zibaei et al. recorded a higher seroprevalence

TABLE 2: Subgroup meta-analysis of cystic echinococcosis for age, sex, and lab methods.

Variable	Number	Prevalence (%)	95% CI		I^2	Q-test	p
			Lower	Upper			
Age (year)							
<30	7	2.2	1.9	3.3	93.2%		
30–40	9	2.8	2.1	4.2	94.5%	862.8	$p < 0.001$
>40	6	4.1	3.0	4.8	98.7%		
Sex							
Male	21	2.1	1.8	2.4	94.9%		
Female	10	3.6	3.2	3.9	96.2%	398.1	$p < 0.001$
Lab methods							
CIE	3	6.0	5.3	6.7	96.6%		
ELISA	15	3.2	3.0	3.4	98.6%		
IFA	9	2.2	1.8	2.5	97.4%	720.9	$p < 0.001$
ELISA & Others	4	1.8	1.5	2.1	91.8%		

(15.4%) in Khorramabad in the southwest of Iran with ELISA [16]. The prevalence of HCE infection in this survey was higher in southwest and south of Iran with 13.7% by Rafiei et al. (2007) in Khouzestan [20] and Saberi-Firouzi et al. (1998) in Fars [21]. The higher prevalence found by the studies which used counterimmunoelectrophoresis (CIE) or counter-current immunoelectrophoresis (CCIEP) may seem paradoxical since this test is usually considered to be more specific than sensitive (it is used as a confirmation test in some countries; it is time-consuming and is rarely used as a screening test) [18]. Besides this test, ELISA with B antigen was used in most of the studies (20 studies) because this method is acceptable, easy, efficient, and affordable and has a high level of sensitivity and specificity. In addition, since preparing the B antigen is easy, using local area antigen shows highly accurate test results. Applying of different kinds of serological methods could be the reason for controversy of obtained results. In addition, it seems that using the serological methods with high sensitivity and specificity would be helpful and something must be noticed that for each experiment in each part of world it is highly recommended to use the antigen which is prepared from that area. Thus, using the antigen which is from a specific district should be used in ELISA test. This strategy could have made the ELISA test more reliable in comparison to those tests which are using the universal antigen. So, we recommend that ELISA test by means of local antigens could help us to get rid of the controversy results obtained from different groups [9].

Table 1 and Figure 2 show the model studies in the country and the seropositivity rate in patients. As it can be seen, no studies have been performed in the eastern and central parts of Iran and the reason is the low prevalence of the disease especially in southeastern areas of the country. The eastern and central parts of Iran do not have enough pasture. That is why it is not a good place for animal husbandry, and consequently there are not enough intermediate hosts for this kind of helminthes. Furthermore, the parasite egg is sensitive to high temperature and low humidity. So, on one hand lack of intermediate host for this worm and on

the other hand existing industrial abattoir make the life cycle of such helminthes unfinished because the infected meat of intermediate hosts is not available for the final host. Based on the mentioned reasons we have less infected people with HCE. That is why we have a small number of infected people who do surgery to remove cysts. This issue makes the seroepidemiological study difficult because of the lack of samples [21]. On the other hand, the highest prevalence rate of this parasite belongs to the western parts of the country especially in the provinces of Lorestan, Kohgiluyeh and Boyer-Ahmad, Khuzestan, Fars, and Ardabil which have a mild climate and more rainfall and humidity. This fact consequently increases the survival rate of the eggs and the transmission cycle of the parasite. Therefore, the different climate conditions of the country have an effective role in the infection prevalence rate. On the other hand, due to the good weather conditions, the rate of livestock raising and grazing is very high in these areas which can cause parasite infection in the intermediate hosts and eventually in the final hosts. It must be noted that the stray dogs which are infected with the parasite play a very big role in spreading the disease and increasing the prevalence rate in these areas [11].

On average, the hydatidosis rate in stray dogs is 5–94% in different regions of Iran [8] and the median rate of infection in the stray dogs is 20% in the western areas [9] which shows a high rate of infection in the final hosts. One more thing to be kept in our mind is about cultural and religious issues existing in Iran. Here in Iran in the rural parts, most of the people have dogs and these dogs are vaccinated and cannot consider them as a source of infections [3]. These infected stray dogs can easily spread large numbers of parasite eggs on the agricultural fields where vegetables grow. This is considered as one of the main factors in increasing the prevalence of the disease in intermediate hosts including humans in these regions. Also the human behavior in contact with dogs plays an important role in the transmission of infection in humans. This behavior is closely related to peoples' cultural and economic conditions. Because people who keep dogs in these areas are more sensitive to the dogs' infections, these people are also inferior to other people in terms of educational and economic conditions [8].

This study identified that the prevalence of HCE infection was usually higher in rural inhabitants rather than in urban ones [16, 19, 22–26]; also most of the studies showed that females were the main subject of HCE [9, 20, 22, 25, 27–31]. One evidence of this result is that most farmers and housewives that come in contact with infection source are females that live in rural areas [8]. In addition, the presence of an unlimited dog population in rural communities contributes to the exposure to *E. granulosus* and high seroprevalence rate. Slaughtering practices (noncontrolled family slaughtering in rural areas versus more publicly or privately organized and controlled slaughtering in cities, especially Tehran) and permanent presence of intermediate hosts (sheep) to maintain the cycle are important too [5].

The results of our investigation showed that prevalence of HCE is high among subjects in the middle-aged people in most of the studies. They reported the age range of 10–19 as the highest infected age group in Zanjan [29], 20–40 age range in Kurdistan [32], 60–69, 60–80, and 60–90 age range in Isfahan, Hamadan, and Ardabil [18, 26, 33], 20–29 years old in Khorramabad [16], 30–39 years old in Yasuj [24], and 29–39, 40–49, and 30–60 years old in Kerman, Arak, Golestan, and Qom, respectively [19, 34–36].

These heterogeneity differences seen in the studies performed in different regions of the country are due to the classification of different age groups, the difference in the type of the studied people, and the geographical location of each study. Also long patent period of this disease is one of the problems [18]. The issue of age groups also includes the population's access to care: if diagnosis is made rather early because of easy availability of US examination and if young people are operated on because of the proximity of well-equipped hospitals and of surgeons ready to work (and earn their living!), there will be higher prevalence in older age groups. As different studies show different results and with most cases being diagnosed many years after their infection, the detection of true age group in the context of infection with HCE is more difficult. Mass screening has well shown that the majority of cysts will remain asymptomatic and even spontaneously degenerate in the majority of subjects; serology does not distinguish between rapidly progressing cases and stable or even aborted cases. Several publications support this [11].

4.1. Strengths and Weaknesses of Study Designs. These studies demonstrate the importance of serologic HCE in various groups of people in Iran. The control of the disease in Iran where most dogs are stray and are infected with adult worms all around the cities is difficult. Therefore, public education that highlights the importance of washed vegetables and inhabitation of exposure to the source of infection could reduce the transmission of the parasite and the consequences of infection in humans [6, 13].

In general, performing such systematic studies on the rate of the human hydatid cysts seroepidemiology in different parts of the world especially in parasite endemic countries like Iran which is located in the Middle East can show the general trend of the disease in its different parts. Studying the general pattern of the disease in different regions can help a lot in planning health and disease prevention programs. Considering the very high economic losses in terms of mutilating the infected organs of the domestic animals as well as the various surgeries performed on the infected humans existing in the country, these control programs can decrease the general trend of the disease.

5. Conclusion

HCE is highly prevalent in Iran and could be a cause of considerable health problems in the country. Educational programs, serological screening, and the continuation of the treatment of the patients when possible could help reduce the national impacts of the disease. Further studies are needed to describe the exact epidemiology of the disease at a national level in other parts of Iran.

Competing Interests

The authors declare that there are no competing interests.

References

[1] Z. Pawlowski and D. Vuitton, "Echinococcosis in humans: clinical aspects, diagnpsis and treatment," in *WHO/OIE Manual on Echinococcosis in Humans and Animals: A Public Health Problem of Global Concern*, J. Eckert, M. A. Gemmel, F.-X. Meslin, and Z. S. Pawłowski, Eds., pp. 20–66, World Health Organization and World Organization for Animal Health, Paris, France, 2001.

[2] W. Zhang, H. Wen, J. Li, R. Lin, and D. P. McManus, "Immunology and immunodiagnosis of cystic echinococcosis: an update," *Clinical and Developmental Immunology*, vol. 2012, Article ID 101895, 10 pages, 2012.

[3] M. Fasihi Harandi, C. M. Budke, and S. Rostami, "The monetary burden of cystic echinococcosis in Iran," *PLoS Neglected Tropical Diseases*, vol. 6, no. 11, Article ID e1915, 2012.

[4] P. R. Torgerson and P. Deplazes, "Echinococcosis: diagnosis and diagnostic interpretation in population studies," *Trends in Parasitology*, vol. 25, no. 4, pp. 164–170, 2009.

[5] P. Moro and P. M. Schantz, "Echinococcosis: a review," *International Journal of Infectious Diseases*, vol. 13, no. 2, pp. 125–133, 2009.

[6] S. M. Sadjjadi, "Present situation of echinococcosis in the Middle East and Arabic North Africa," *Parasitology International*, vol. 55, supplement, pp. S197–S202, 2006.

[7] M. B. Rokni, "The present status of human helminthic diseases in Iran," *Annals of Tropical Medicine and Parasitology*, vol. 102, no. 4, pp. 283–295, 2008.

[8] M. B. Rokni, "Echinococcosis/hydatidosis in Iran," *Iranian Journal of Parasitology*, vol. 4, pp. 1–16, 2009.

[9] A. Dalimi, G. Motamedi, M. Hosseini et al., "Echinococcosis/hydatidosis in western Iran," *Veterinary Parasitology*, vol. 105, no. 2, pp. 161–171, 2002.

[10] M. F. Harandi, R. P. Hobbs, P. J. Adams, I. Mobedi, U. M. Morgan-Ryan, and R. C. A. Thompson, "Molecular and morphological characterization of Echinococcus granulosus of human and animal origin in Iran," *Parasitology*, vol. 125, no. 4, pp. 367–373, 2002.

[11] B. Sarkari and Z. Rezaei, "Immunodiagnosis of human hydatid disease: where do we stand?" *World Journal of Methodology*, vol. 5, no. 4, pp. 185–195, 2015.

[12] M. B. Rokni and B. Aminian, "Evaluation of the Enzyme-linked Immuno-electro Transfer Blot (EITB) technique using hydatid cyst antigens B/5 and total IgG antibodies in laboratory diagnosis of human hydatidosis," *Pakistan Journal of Medical Sciences*, vol. 22, no. 2, pp. 127–131, 2006.

[13] S. M. Sadjjadi, S. Ardehali, B. Noman-Pour, V. Kumar, and A. Izadpanah, "Diagnosis of cystic echinococcosis: ultrasound imaging or countercurrent immunoelectrophoresis?" *Eastern Mediterranean Health Journal*, vol. 7, no. 6, pp. 907–911, 2001.

[14] H. Z. Hezarjaribi, M. Fakhar, A. Shokri, S. Hosseini Teshnizi, A. Sadough, and M. Taghavi, "*Trichomonas vaginalis* infection among Iranian general population of women: a systematic review and meta-analysis," *Parasitology Research*, vol. 114, no. 4, pp. 1291–1300, 2015.

[15] M. T. Rahimi, A. Daryani, S. Sarvi et al., "Cats and *Toxoplasma gondii*: a systematic review and meta-analysis in Iran," *Onderstepoort Journal of Veterinary Research*, vol. 82, no. 1, article 823, pp. 1–10, 2015.

[16] M. Zibaei, A. Azargoon, M. Ataie-Khorasgani, K. Ghanadi, and S. M. Sadjjadi, "The serological study of cystic echinococcosis and assessment of surgical cases during 5 years (2007–2011) in Khorram Abad, Iran," *Nigerian Journal of Clinical Practice*, vol. 16, no. 2, pp. 221–225, 2013.

[17] G. L. B. Farrokhzad, M. Nariman, and M. R. Nazari Poya, "Investigation of the prevalence of hydatid cysts in rural areas shemiranat of Tehran and reviews of IFA test," *Research in Medicine*, vol. 30, no. 3, pp. 241–244, 2004.

[18] M. Arbabi, J. Masoud, A. Dalimi Asl, and M. Sajadi, "Seroepidemiologic prevalence of Hydatid cyst in Hamadan," *Feyz*, vol. 2, no. 2, pp. 43–50, 1998.

[19] M. Asgari, M. Mohebali, E. B. Kia et al., "Seroepidemiology of human hydatidosis using AgB-ELISA test in Arak, central Iran," *Iranian Journal of Public Health*, vol. 42, no. 4, pp. 391–396, 2013.

[20] A. Rafiei, A. Hemadi, S. Maraghi, B. Kaikhaei, and P. S. Craig, "Human cystic echinococcosis in nomads of south-west Islamic Republic of Iran," *Eastern Mediterranean Health Journal*, vol. 13, no. 1, pp. 41–48, 2007.

[21] M. Saberi-Firouzi, F. Kaffashian, E. Hayati et al., "Prevalence of hydatidosis in nomadic tribes of southern Iran," *Medical Journal of the Islamic Republic of Iran*, vol. 12, no. 2, pp. 113–118, 1998.

[22] A. Aflaki, F. Ghaffarifar, and A. Dalimi Asl, "Seroepidemiological survey of hydatidosis by Dot-ELISA in Ilam province," *Modarres Journal of Medical Sciences*, vol. 8, pp. 1–6, 2005.

[23] M. Hadadian, F. Ghaffarifar, A. Dalimi Asl, and S. Roudbar Mohammadi, "Seroepidemiological survey of hydatid cyst by ELISA in Kordestan province," *Modares Journal of Medical Sciences: Pathobiology*, vol. 10, pp. 13–18, 2008.

[24] B. Sarkari, S. M. Sadjjadi, M. M. Beheshtian, M. Aghaee, and F. Sedaghat, "Human cystic echinococcosis in Yasuj district in Southwest of Iran: an epidemiological study of seroprevalence and surgical cases over a ten-year period," *Zoonoses and Public Health*, vol. 57, no. 2, pp. 146–150, 2010.

[25] S. Kavous, A. Kazemi, and S. Jelodari, "Seroepidemiology of human hydatid cyst in jahrom," *Journal of Jahrom Uiversity of Medical Sciences*, vol. 8, no. 3, pp. 18–24, 2010.

[26] Z. Heidari, M. Mohebali, Z. Zarei et al., "Seroepidemiological study of human hydatidosis in Meshkinshahr district, Ardabil province, Iran," *Iranian Journal of Parasitology*, vol. 6, no. 3, pp. 19–25, 2011.

[27] M. Sedaghat Gohar, J. Massoud, M. Rokni, and E. Kia, "Seroepidemiologic survey of human hydatidosis in Shahriar region," *Journal of Kerman University of Medical Sciences*, vol. 1, pp. 44–49, 2001.

[28] H. Yousefi Darani, M. Avijgan, K. Karimi, K. Manouchehri, and J. Masood, "Seroepidemiology of hydatid cyst in Chaharmahal va Bakhtiari Province," *Iranian Journal of Public Health*, vol. 32, no. 2, pp. 31–33, 2003.

[29] A. Haniloo, H. Badali, and A. Esmaeil Zadeh, "Seroepidemiological study of Hydatidosis in Zanjan, Islam-Abad, 2002," *Journal of Zanjan University of Medical Sciences & Health Services*, vol. 12, pp. 41–46, 2002.

[30] H. Mirzanejadasl, M. F. Harandi, and P. Deplazes, "Serological survey of human cystic echinococcosis with ELISA method and CHF Ag, in moghan plain, Ardabil Province, Iran," *Research Journal of Biological Sciences*, vol. 3, no. 1, pp. 64–67, 2008.

[31] S. S. Moazezi, M. Fasihi Harandi, M. Saba, H. Kamyabi, and F. Sheikhzadeh, "Sonographic and serological survey of hydatid disease in rural regions of Shahdad and Chatroud, Kerman province, 2006-2007," *Journal of Kerman University of Medical Sciences*, vol. 16, no. 1, pp. 25–34, 2009.

[32] S. A. Hosseini and J. Masood, *Seroepidemiological study of hydati-dosis in Divandarreh, Kurdistan [M.S. thesis]*, Medical Parasitology, School of Public Health, Tehran University of Medical Sciences, Tehran, Iran, 1997.

[33] P. Ilbeigi, M. Mohebali, E. B. Kia et al., "Seroepidemiology of human hydatidosis using AgB-ELISA test in Isfahan City and Suburb Areas, Isfahan Province, Central Iran," *Iranian Journal of Public Health*, vol. 44, no. 9, pp. 1219–1224, 2015.

[34] M. Baharsefat, J. Massoud, I. Mobedi, A. Farahnak, and M. Rokni, "Seroepidemiology of cystic echinococcosis in referred patients to health centers in Golestan Province using ELISA and IFA," *Iranian Journal of Parasitology*, vol. 2, pp. 20–24, 2007.

[35] M. F. Harandi, S. S. Moazezi, M. Saba et al., "Sonographical and serological survey of human cystic echinococcosis and analysis of risk factors associated with seroconversion in rural communities of Kerman, Iran," *Zoonoses and Public Health*, vol. 58, no. 8, pp. 582–588, 2011.

[36] A. Rakhshanpour, M. Fasihi Harandi, S. S. Moazezi et al., "Seroprevalence of human hydatidosis using ELISA method in Qom province, central Iran," *Iranian Journal of Parasitology*, vol. 7, no. 3, pp. 10–15, 2012.

[37] R. Jamali, B. Naghili, and Sh. Mozafari, "Seroepidemiological study of Hydatid cyst prevalence in Uromieh villagers," *Medical Journal of Tabriz University of Medical Sciences & Health Service*, vol. 29, no. 27, pp. 23–30, 1995.

[38] H. Mohamadi, *Seroepidemiological study of hydatidosis in man in Varamin area south of Tehran [M.S. dissertation]*, Tehran University of Medical Sciences, Irantion SoPH, Tehran, Iran, 1998.

[39] M. R. Zariffard, N. Abshar, M. A. Akhavizadegan, and G. R. Motamedi, "Seroepidemiological survey of human hydatidosis in western parts of Iran," *Archives of Razi Institute*, vol. 50, pp. 71–75, 1999.

[40] M. R. Nilfroshan, A. Deylami, and H. Niazi, "Epidmiology of Hydatid cyst in Fardin dirstrict," *Pajouhesh Va Sazandgi*, vol. 36, pp. 80–83, 1998.

[41] Z. Amiri, *Prevalence of hydatid cyst in Kermanshah using IFA and ELISA method in the year 2001 [M.S. thesis]*, Medical Parasitology, School of Public Health, Tehran University of Medical Sciences, Tehran, Iran, 2001.

[42] A. Rafiei, P. S. Craig, and S. A. Maraghi, "Seroepidemiological servey of human cysitic ecinicoccnsis in Iran," in *Proceedings of the 20th International Congress of Hydatidology*, vol. 193, Kusadasi, Turkey, June 2001.

[43] L. Akhlaghi, J. Massoud, and A. Housaini, "Observation on hydatid cyst infection in Kordestan province (West of Iran) using epidemiological and seroepidemiological criteria," *Iranian Journal of Public Health*, vol. 34, no. 4, pp. 73–75, 2005.

[44] L. Akhlaghi, H. Ourmazdi, S. H. Sarvi et al., "Using Dot-ELISA method to study the prevalence of human hydatidosis in people referred to blood transfusion center in Tehran, 2005-2006," *Journal of Iran University of Medical Sceinces*, vol. 16, no. 67, pp. 52–58, 2009.

[45] N. Esmaeili and M. Arbabi, "Seroepidemiology of hydatidosis among adult human at Kashan region, Iran in 2008," *Journal of Kashan University of Medical Sciences*, vol. 13, no. 4, pp. 321–326, 2010.

[46] M. A. Dadkhah, M. Yeganehzad, and B. Nadery, "Survey on hydatid cyst infestation in Sarab city (Northwest of Iran) using epidemiological and seroepidemiological," *Journal of Animal and Veterinary Advances*, vol. 10, no. 16, pp. 2099–2101, 2011.

[47] Y. Garedaghi and S. R. Bahavarnia, "Seroepidemiology of human hydatidosis by ELISA method in East-Azarbaijan province in Iran in year 2009," *Iranian Journal of Epidemiology*, vol. 7, no. 2, pp. 25–29, 2011.

[48] R. Shahrokhabadi, E. Rahimi, and R. Poursahebi, "Seroepidemiological study of human hydatidosis in Rafsanjan, Kerman," *Zahedan Journal of Research in Medical Sciences*, vol. 16, no. 4, p. 46, 2014.

Immune Profile of Honduran Schoolchildren with Intestinal Parasites: The Skewed Response against Geohelminths

José Antonio Gabrie,[1] María Mercedes Rueda,[2] Carol Anahelka Rodríguez,[2] Maritza Canales,[2] and Ana Lourdes Sanchez[1,3]

[1]Department of Health Sciences, Brock University, St. Catharines, ON, Canada
[2]School of Microbiology, National Autonomous University of Honduras (UNAH), Tegucigalpa, Honduras
[3]Microbiology Research Institute, National Autonomous University of Honduras (UNAH), Tegucigalpa, Honduras

Correspondence should be addressed to José Antonio Gabrie; jgabrie@brocku.ca

Academic Editor: Emmanuel Serrano Ferron

Soil-transmitted helminth infections typically induce a type-2 immune response (Th2), but no immunoepidemiological studies have been undertaken in Honduras, an endemic country where the main control strategy is children's annual deworming. We aimed to characterize the immune profile of Honduran schoolchildren harbouring these parasitoses. Demographic and epidemiological data were obtained through a survey; nutritional status was assessed through anthropometry; intestinal parasites were diagnosed by formol-ether and Kato-Katz; and blood samples were collected to determine immunological markers including Th1/Th2 cytokines, IgE, and eosinophil levels. A total of 225 children participated in the study, all of whom had received deworming during the national campaign five months prior to the study. Trichuriasis and ascariasis prevalence were 22.2% and 20.4%, respectively. Stunting was associated with both age and trichuriasis, whereas ascariasis was associated with sex and household conditions. Helminth infections were strongly associated with eosinophilia and hyper-IgE as well as with a Th2-polarized response (increased levels of IL-13, IL-10, and IL4/IFN-γ ratios and decreased levels of IFN-γ). Pathogenic protozoa infections were associated with a Th1 response characterized by elevated levels of IFN-γ and decreased IL10/IFN-γ ratios. Even at low prevalence levels, STH infections affect children's nutrition and play a polarizing role in their immune system.

1. Introduction

Soil-transmitted helminths (STH) or geohelminths infect more than 2 billion people worldwide, especially in developing countries located in tropical and subtropical regions. The most prevalent species are *Ascaris lumbricoides* (roundworm), *Trichuris trichiura* (whipworm), and hookworms (*Necator americanus* and *Ancylostoma duodenale*). STH infections account for about 40% of global morbidity due to infectious diseases [1]. Children and women of childbearing age are among the high-risk groups for these parasitic diseases. The morbidity caused by STH infections is highly associated with the size of the worm population residing in the intestines (also known as worm burden). Infections of heavy intensity are more likely to cause impaired physical growth and cognitive development as well as micronutrient deficiencies, including iron-deficiency anaemia [1, 2]. Further, due to their chronic and insidious nature, even light intensity infections may compromise health outcomes. To reduce the morbidity caused by these parasitoses, the World Health Organization (WHO) recommends cyclical (annual or biannual) benzimidazole-based mass drug administration to groups at high-risk of infection, especially pre- and schoolchildren [1]. The basic premise of this strategy is that even though this intervention may not decrease transmission and prevalence, it will in time reduce hosts' worm burden with the ensuing reduction in health impact.

In addition to health impact, STH infections affect the human host in more subtle manner: as extracellular metazoan parasites, helminths trigger type-2 immune response mediated by T-helper type 2 (Th2) cells. This response is typically characterized by expansion of mast cells, eosinophils, basophils, group 2 innate lymphoid cells (ILC-2), and alternatively activated macrophages (AAMs), as well as increased production of Th2 cytokines (e.g., IL-4, IL-5, IL-9, and IL-13) and IgE [3–5].

On the other hand, intracellular pathogens such as viruses, bacteria, protozoa, and fungi elicit a type-1 response, characterized by increased numbers of phagocytic neutrophils and macrophages, as well as cytotoxic CD8$^+$ T cells and Th1 cells. Antigen-presenting dendritic cells (DCs) secrete interleukin- (IL-) 12 promoting differentiation of naive CD4$^+$ cells, first into null T-helper (Th0) and then into Th1. Via production of interferon-gamma (IFN-γ) and IL-2 to a lesser degree, Th1 cells direct, maintain, and enhance the antimicrobial effect of type-1 response [6]. Both subsets of T-helper cells antagonize and negatively regulate each other. Naturally, a healthy resolution of infections relies on a dynamic combination of the two [5, 7].

Chronic helminth infections such as the ones caused by STH exhibit a modified type-2 response, in which a superimposition of regulatory mechanisms exerted over the basic response pattern takes place. This regulation is mainly achieved by the expansion and induction of regulatory T cells (Treg). Increased levels of anti-inflammatory cytokines IL-10 and transforming growth factor beta (TGF-β) are hallmarks of this response [4, 8, 9]. It has been suggested that, during chronic helminthiases, this anti-inflammatory network may play a key role in the lower prevalence of allergic disease observed in Th2-skewed populations [7, 10–12]. Conversely, helminth-induced IL-10-mediated immunosuppression has raised concerns for populations in which STH infections coexist with other morbidities such as HIV/AIDS, malaria, and tuberculosis, as clearance for the latter depends on an active and timely type-1 response [8, 13–16].

In this context, successful deworming treatment has been proposed as a practical and inexpensive way to rebalance the immune response [17, 18]. Recent studies have observed that after administration of anthelminthic medication, eosinophil counts [19], IL-10 levels [19–21], and IgE concentrations [17] show a significant decrease from pretreatment values. Moreover, in HIV-1 infected individuals, deworming was shown to promote a significant decrease in plasma HIV RNA levels [18, 22] and CD8$^+$ T cells counts [18]. Also, a randomized control trial in Kenya demonstrated a significant increase in CD4$^+$ T cell counts after deworming [22]. Other studies, however, found no evidence of this effect [18, 19].

The principal aim of the present study was to characterize the immune profile in schoolchildren infected with geohelminths. Secondary aims involved assessing the health impact and epidemiological determinants of such parasitoses.

2. Materials and Methods

2.1. Study Population and Ethical Approval. The study was conducted in April 2014 and data were collected from schoolchildren living in Linaca, a rural community of the Municipality of Tatumbla, Department of Francisco Morazán in Honduras. The study was approved by Brock University's Bioscience Research Ethics Board (file number BU 12-262,

dated June 27, 2013) as well as by the Research Ethics Board of the Master's Program in Infectious and Zoonotic Diseases, School of Microbiology, National Autonomous University of Honduras, UNAH (file number CEI-MEIZ 01-2013, dated September 23, 2013). Linaca school authorities also approved the study. The school of Linaca participates in the Honduras National Deworming Program, which provides enrolled children with a 400 mg single-dose albendazole annually. At the time of the study, the last deworming round occurred in November 2013. Children who had not received deworming treatment from other sources in the past 3 months were included in the study. Written informed consent was provided by children's parents. Children's oral assents were individually obtained and documented through a child assent form.

2.2. Data Collection. Basic demographic and epidemiological data of participant children were obtained using a structured questionnaire during a 5-minute face-to-face interview conducted in Spanish. Body weight and height of participants were measured twice by different researchers and their average was used to calculate the following anthropometric indicators: (a) height-for-age Z-score (HAZ), (b) weight-for-age Z-score (WAZ), and (c) body-mass-index-for-age Z-score (BAZ). These indicators were used to ascertain children's nutritional status as per international parameters: stunted growth (chronic malnutrition), thinness, and underweight, respectively [23].

2.3. Stool Collection and Parasite Determination. A same-day single stool sample was provided by each participant the morning of the interview. Samples were kept in portable coolers and examined early afternoon at the laboratory facilities of UNAH's School of Microbiology. The presence of intestinal parasites was assessed using both Kato-Katz and formol-ethyl acetate concentration methods. For the latter, an aliquot was preserved in 10% formalin. Microscopic examination of Kato-Katz smears was done between 30 and 60 min of preparation. Helminth eggs were identified and counted and the number of eggs per gram (epg) was calculated. Infection intensities were classified as light, moderate, or heavy, based on the epg calculations, according to WHO criteria [1]. Diagnostic accuracy was ensured by having 100% of negative and 10% of positive smears read again by a different researcher immediately after the first reading. Formol-ethyl acetate was done one month after sample collection and sediments were observed with dry and immersion oil objective lenses in order to identify helminth and protozoa stages, respectively.

2.4. Blood Collection and Hematological Analyses. Blood samples obtained from the cubital vein were collected in 3 mL tubes of each K$_2$EDTA and serum separator and clot activator (BD Vacutainer, NJ, USA). The latter were centrifuged within 4 hours and serum was stored at $-21°$C until immunological analysis. Hematological automated analyses were performed within 4 hours of sample collection and were done with an ABX Pentra 120 (HORIBA-ABX-SAS, Montpellier, France). According to the parameters established in the most recent (2011-2012) Demographic and Health Survey in Honduras

(DHS-HN) [24], anaemia was defined as Hb concentration < 12 g/dL. Eosinophilia was defined according to international standards as eosinophil count ≥ 500/μL in peripheral blood. Mild, moderate, or severe eosinophilia were defined if cell counts were 500–1500/μL, >1500–5000/μL, or >5000/μL, respectively [25].

2.5. Immunological Biomarkers. Cytokine concentrations in serum samples were measured using a MagPix magnetic beads platform (Luminex xMAP, Austin, TX, USA). Multiplex panels containing magnetic beads covered with fluorescent dyed conjugated to a monoclonal antibody specific for each target molecule were used to quantify Th1/Th2 cytokines (IL-2, IL-6, IL-8, IL-12p70, IFN-γ, GM-CSF, TNF-α, IL-4, IL-5, and IL-13) and the regulatory cytokine IL-10.

Serum concentrations of total IgE were also quantified using magnetic beads platform but with singleplex kits, Bio-Plex Pro™ Human IgE Isotyping (Bio-Rad laboratories, Inc. Hercules, CA, USA). Hyper-IgE was defined as a serum concentration exceeding 100 IU/mL and was classified as mild (>100–399 IU/mL), moderate (>399–999 IU/mL), or severe (≥1000 IU/mL) [26]. Cytokine and IgE tests were performed according to the manufacturer's instructions (Bio-Rad laboratories, Inc. Hercules, CA, USA).

2.6. Statistical Analyses. Descriptive statistics were used to characterize the study population. Point prevalence with 95% confidence intervals (95% CI) was calculated for overall STH infections and for each parasite species, as well as for mixed STH infections (i.e., infected with two or more species). Anthropometric indicators were calculated using the WHO AnthroPlus software ver. 1.04 (WHO). Associations between nutritional indicators and parasite infections were explored using both univariate and multivariable logistic regression models. Unadjusted and adjusted odds ratio (OR and adj. OR) with 95% CI were determined. Due to the non-Gaussian distribution of the immunological markers, nonparametric methods were used as follows. Geometric means and 95% CI were calculated for each marker. Relationships were evaluated through Spearman's rank correlation coefficient. Kruskal-Wallis test was used to assess differences among groups, followed by Dunn's test to investigate individual group differences. All statistical analyses were conducted using Stata 13 (StataCorp LP, TX, USA) and the level of significance was defined as $p < 0.05$.

3. Results

3.1. Characteristics of the Study Population. Table 1 summarizes the characteristics and parasitological findings in the study population. A total of 225 schoolchildren attending grades 1 to 6 (age 6–13) participated in the study. All provided stool samples and all but one child agreed to provide blood samples. Most of the children lived in households with piped water (88%) and flushing toilet and/or latrines (98%) but 40% lived in households with earthen floor.

3.2. Parasitological Findings. Qualitative Kato-Katz results were 100% correlated with those of formol-ethyl acetate. No

TABLE 1: Characteristics and parasitological findings of the study population ($n = 225$).

Characteristics	n (%)
Age—mean (SD)	8.96 (1.8)
Girls	104 (46.2)
Household conditions	
Earthen floor (complete or partial)	91 (40.4)
Type of sanitary facility available	
None	5 (2.2)
Latrine	115 (51.1)
Toilet	105 (46.7)
Access to piped water	198 (88.0)
Nutritional indicators	
Height-for-age Z-score (HAZ)—mean (SD)	−0.76 (1.03)
Weight-for-age Z-score (WAZ)—mean (SD) ($n = 155$)[a]	−0.19 (1.05)
Body mass index-for-age Z-score (BAZ)—mean (SD)	0.22 (0.93)
Stunting (< −2 SD HAZ)	22 (9.8)
Underweight (< −2 SD WAZ)[a]	3 (1.9)
Thinness (< −2 SD BAZ)	1 (0.4)
Overweight (>1 SD & ≤2 SD BAZ)	26 (11.6)
Obesity (>2 SD BAZ)	11 (4.9)
Hemoglobin (g/dL)—mean (SD) ($n = 224$)[b]	13.7 (0.8)
Presence of anemia[b]	8 (3.6)
Parasitic profile	
Overall prevalence of STH infections	67 (29.8)
Overall prevalence of *Ascaris lumbricoides*	46 (20.4)
Overall prevalence of *Trichuris trichiura*	50 (22.2)
Overall prevalence of Hookworms	0 (0.0)
Single *Ascaris lumbricoides* infections ($n = 67$)	17 (25.4)
Single *Trichuris trichiura* infections ($n = 67$)	21 (31.3)
Mixed infections ($n = 67$)	29 (43.3)
Moderate-to-heavy infections by *Ascaris lumbricoides* ($n = 46$)	15 (32.6)
Moderate-to-heavy infections by *Trichuris trichiura* ($n = 50$)	8 (16.0)

STH: soil-transmitted helminth.
[a]Not calculated in children older than 10 years of age.
[b]One child did not agree on providing blood sample.

cases of *S. stercoralis* or hookworm infections were found with either technique but one case of *Taenia solium* was identified by the presence of gravid proglottids in the stool sample. Among pathogenic protozoa, *Giardia intestinalis* and *Entamoeba histolytica/dispar* were present in 4% and 9% of the samples, respectively. Additionally, one case of *Cyclospora cayetanensis* was identified.

The overall STH prevalence was 30% (95% CI = 24.1–36.1). Prevalence for *T. trichiura* and *A. lumbricoides* was 22.2% (95% CI = 17.2–28.2) and 20.4% (95% CI = 15.6–26.2), respectively. Mixed STH infections represented 43% of all infections

TABLE 2: Geometric means (95%CI) of immunological markers determined in schoolchildren by STH infection status ($n = 224$)[a].

Parameter	Nonparasitized	Ascariasis	Trichuriasis	Mix infections
	$n = 178$	$n = 17$	$n = 21$	$n = 29$
Eosinophils (cells/μL)	273.5 (240.7–310.8)	320.8 (229.7–448.0)	410.1 (277.1–606.9)[*]	448.9 (292.2–689.5)[**]
IgE (IU/mL)	722.9 (573.0–912.1)	443.9 (159.1–1238.2)	726.1 (352.4–1495.9)	1463.7 (805.9–2658.4)[*]
IL-2 (pg/mL)	26.1 (21.0–32.4)	34.7 (19.7–61.2)	24.1 (16.0–36.3)	21.3 (14.4–31.6)
IL-4 (pg/mL)	1.2 (0.8–1.7)	0.7 (0.1–5.4)	0.9 (0.1–5.0)	0.8 (0.2–2.4)
IL-5 (pg/mL)	10.9 (4.6–25.8)	17.1 (4.4–66.6)	11.8 (9.4–14.9)	5.1 (1.2–21.7)
IL-6 (pg/mL)	13.7 (11.2–16.8)	16.1 (7.9–32.8)	10.8 (5.9–19.7)	9.8 (6.4–14.8)
IL-8 (pg/mL)	107.2 (91.8–125.2)	105.8 (65.5–170.9)	91.8 (61.1–138.1)	99.3 (71.1–138.7)
IL-10 (pg/mL)	8.5 (7.1–10.1)	15.3 (6.8–34.6)[*]	7.1 (4.8–10.5)	6.6 (4.9–8.8)
IL-12p70 (pg/mL)	9.5 (7.5–12.1)	15.3 (5.7–41.1)	7.4 (5.2–10.5)	6.3 (4.1–9.8)
IL-13 (pg/mL)	1.2 (0.9–1.5)	4.7 (1.5–15.0)[**]	1.3 (0.6–2.7)	1.9 (0.9–3.9)
IFN-γ (pg/mL)	131.5 (98.4–175.8)	162.9 (49.7–533.2)	70.1 (35.6–138.0)[*]	95.5 (38.7–235.7)
GM-CSF (pg/mL)	55.1 (43.9–69.0)	72.8 (38.8–136.5)	45.5 (22.1–93.4)	45.0 (23.0–87.9)
TNF-α (pg/mL)	8.2 (6.7–10.2)	10.4 (3.7–28.9)	9.1 (5.1–16.0)	6.4 (3.8–10.7)

STH: soil-transmitted helminth.
[a]One child did not agree on providing blood sample.
[*]$p < 0.05$.
[**]$p < 0.01$.

observed. While the majority (84%) of trichuriasis cases were light intensity, about one-third (32.6%) of ascariasis cases were moderate-to-heavy intensity (Table 1). Only 20% of moderate-to-heavy infections with both *A. lumbricoides* and *T. trichiura* occurred as single infections. In the multivariable analysis, age was not associated with parasitism. Ascariasis was associated with sex and some household conditions of participating children. Almost three-quarters (71.7%) of ascariasis cases occurred in boys and they had almost three times the odds of having *A. lumbricoides* infections compared to girls (adj. OR = 2.82, 95% CI = 1.34–5.94, and $p = 0.006$). Children living in households with earthen floor had twice the odds of helminthic infections (adj. OR = 2.29, 95% CI = 1.15–4.55, and $p = 0.018$). Conversely, children with access to piped water in their households showed 64% reduced odds of ascariasis (adj. OR = 0.36, 95% CI = 0.14–0.93, and $p = 0.035$). *T. trichiura* infections were not significantly associated with any of the abovementioned risk factors.

3.3. Nutritional Status. About 12% of the children were overweight, and this condition was significantly higher in boys than girls (73% versus 27%, $p = 0.036$); only one of these children was stunted. Stunted growth was observed in 9.8% of the studied children. Multivariable logistic models controlling for age and sex were constructed, and adjusted OR were calculated. In this population, stunting was significantly associated with the age of participants as well as with trichuriasis. Per every year of age, the odds of stunting increased in 34% (adj. OR = 1.34, 95% CI = 1.03–1.73, and $p = 0.025$). Similarly, children harbouring *T. trichiura* infections had almost four times the odds of being stunted (adj. OR = 3.93, 95% CI = 1.03–14.93, and $p = 0.045$) when compared with children without STH infection. Further, children harbouring moderate-to-heavy trichuriasis had an additional 70% increased odds of being stunted (adj. OR = 6.64, 95% CI = 1.19–37.09, and $p = 0.031$).

3.4. Eosinophilia and IgE Levels. Figures 1 and 2 depict the variation observed in eosinophils count (cells/μL) and serum levels of IgE by STH infection and infection intensity. The geometric means and 95% CI of these parameters by STH infection are shown in Table 2. The overall prevalence of eosinophilia was 31.7% (95% CI = 25.9–38.1), and mild eosinophilia accounted for the vast majority (91.5%) of cases. No severe cases of eosinophilia were identified.

Multivariable logistic models found no association between eosinophilia and sex or age of the studied children. Significantly higher mean counts of eosinophils were found in children infected with *T. trichiura* alone or in those harbouring mixed infections compared to nonparasitized children. A significant association was found between eosinophilia and STH infection and infection intensity. Children harbouring mixed infections had 2.5-fold increased odds of having eosinophilia when compared to nonparasitized children (adj. OR = 2.59, 95% CI = 1.13–5.90, and $p = 0.023$). Similar increased odds were observed in children with single infection by *T. trichiura*, although this effect was only marginally significant (adj. OR = 2.48, 95% CI = 0.98–6.26, $p = 0.054$). Eosinophilia was positively correlated with intensity of infection by both *A. lumbricoides* and *T. trichiura* ($r_s = 0.26$, $p < 0.001$ and $r_s = 0.33$, $p < 0.001$, resp.). Children with moderate-to-heavy ascariasis had five times the odds of presenting eosinophilia (adj. OR = 5.23, 95% CI = 1.69–16.12, and $p = 0.004$) compared to their nonparasitized counterparts. Likewise, moderate-to-heavy trichuriasis was associated with a 7-fold increased odds of eosinophilia (adj. OR = 7.19, 95% CI = 1.41–36.70, and $p = 0.018$).

The overall prevalence of hyper-IgE was 90.2% (95% CI = 85.5–93.5). More than half of these cases (51.0%) were severe, whereas mild and moderate hyper-IgE accounted for 24.3% and 24.7% of the cases, respectively. No significant correlation was found between IgE levels and sex or age of participant

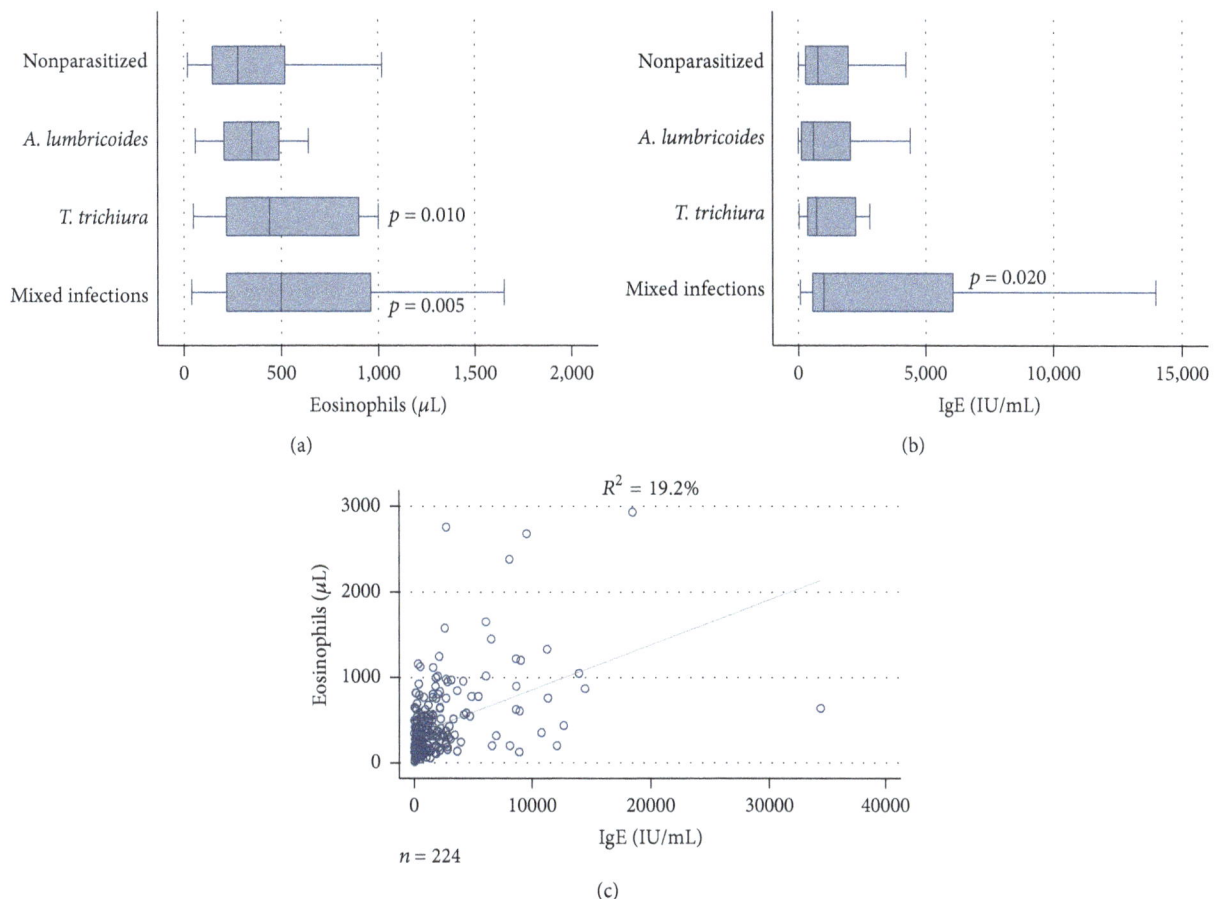

FIGURE 1: Association between STH infections and eosinophils count and serum levels of IgE in studied children. Higher counts in eosinophils were significantly associated with trichuriasis and mixed infections (a). Higher total IgE levels were associated with mixed infections (b). Moderate positive correlation of eosinophils count and total IgE levels (c).

children. Compared to those without STH infections, children with mixed STH infections had significantly higher mean levels of IgE (Table 2). There was a moderate positive correlation between IgE serum levels and eosinophils count ($r_s = 0.43$, $p < 0.001$). Similarly, IgE levels were positively correlated with intensity of infection of both A. lumbricoides and T. trichiura. Children with moderate-to-heavy ascariasis or trichuriasis had significantly higher mean values of IgE compared to those uninfected or with only light infections ($p < 0.001$) (Figure 2).

3.5. Serum Cytokines Levels. Th1 and Th2 representative cytokines were determined in serum samples. Table 2 summarizes the geometric mean values and 95% CI of these cytokines by STH infection. In general, cytokine mean concentrations did not differ significantly when comparing the parasitized with the nonparasitized group. However, children with single infections by A. lumbricoides had significantly higher levels of IL-10 and IL-13 compared to children without infection, with single T. trichiura infections or with mixed infections ($p = 0.018$ and $p = 0.004$, resp.). Spearman's rank correlation coefficients showed a significant, although weak, negative correlation between IL-10 levels and age

($r_s = -0.23$, $p < 0.001$). IL-10 levels were not significantly correlated with infection intensity, whereas for IL-13 a weak but significant positive correlation with A. lumbricoides infection intensity was found ($r_s = 0.22$, $p = 0.012$). Significantly lower values of IFN-γ were found in children with single infections by T. trichiura ($p = 0.043$) and these values showed a weak, negative correlation with T. trichiura infection intensity, although this correlation was not statistically significant ($r_s = -0.17$, $p = 0.069$) (Figure 3).

On the other hand, IFN-γ values were significantly negatively associated with A. lumbricoides infection intensity ($p = 0.007$). Children with moderate-to-heavy ascariasis had lower IFN-γ mean values (37.6 [15.7–90.0] pg/mL) compared with those having no or light infections (131.5 [98.4–175.8] pg/mL and 214.4 [99.0–464.4] pg/mL, resp.) (Figure 3). Th2/Th1 ratios, as well as IL-4/IFN-γ and IL-10/IFN-γ, were also calculated; significantly higher IL-4/IFN-γ ratios were obtained in children with moderate-to-heavy infections by A. lumbricoides (0.020 [0.010–0.042]), compared to those without (0.011 [0.08–0.014]) or light infections (0.007 [0.002–0.022]) ($p = 0.046$ and $p = 0.021$, resp.) (Figure 4). A very similar pattern in IL-4/IFN-γ ratio was found for T. trichiura infection intensity, although differences were only marginally

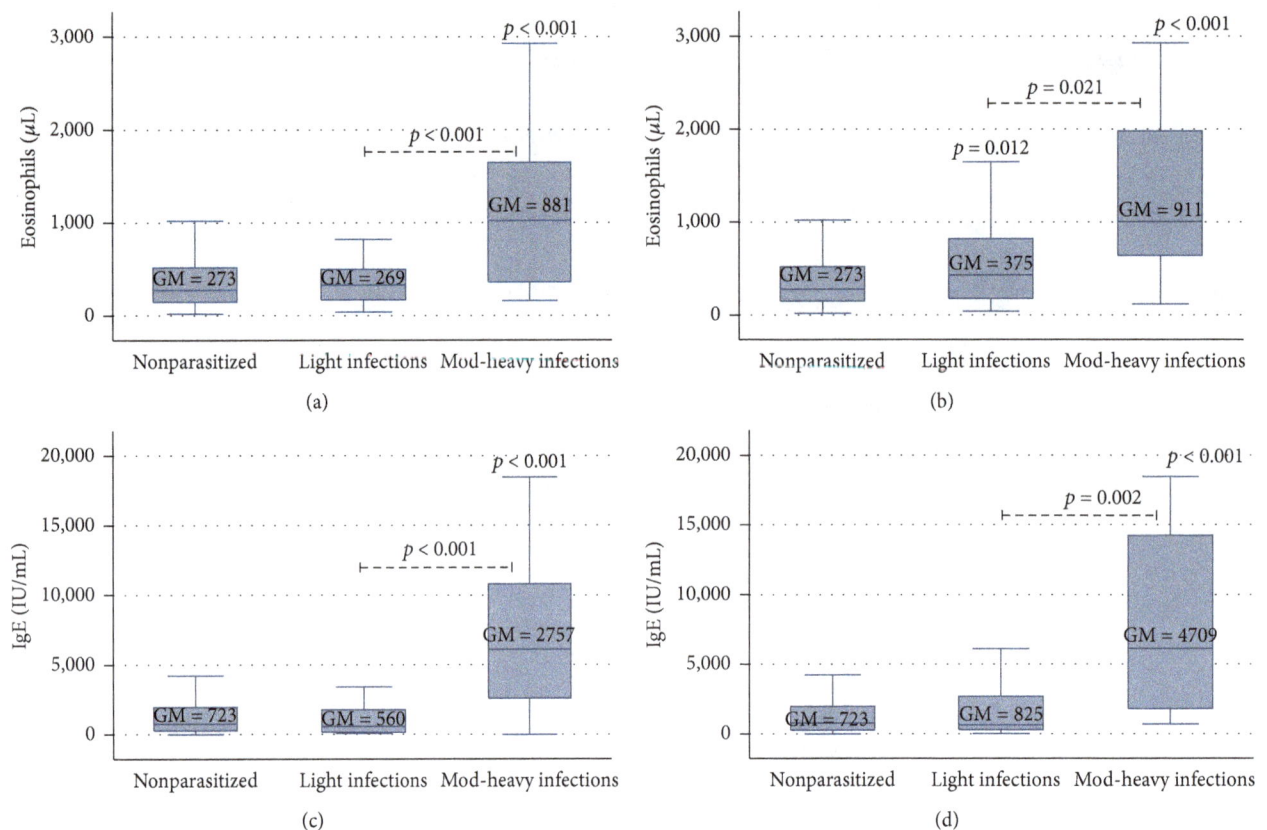

FIGURE 2: Association between STH infection intensity and eosinophils count and serum levels of IgE in studied children. Moderate-to-heavy infections of *A. lumbricoides* (a, c) and *T. trichiura* (b, d) were associated with higher eosinophils count and total IgE levels. Geometric means (GM) are depicted in their respective boxes.

significant ($p = 0.057$ and $p = 0.059$, resp.). Conversely, no significant differences in IL-10/IFN-γ ratios were observed.

Finally, the potential effect of pathogenic protozoa in children's immune response was assessed but this assessment was done only in children without STH infections. It was found that IFN-γ mean values were significantly higher in children harbouring pathogenic protozoa (i.e., *G. intestinalis* and/or *E. histolytica/dispar*) compared to those infected with commensals only or to those with no protozoa at all (332.9 [146.7–755.8] pg/mL, 93.7 [60.9–144.2] pg/mL, and 145.6 [93.8–225] pg/mL, resp.). IL-10/IFN-γ ratios were significantly lower in children infected with pathogenic protozoa when compared to those uninfected or infected with commensals only (0.044 [0.019–0.105], 0.095 [0.062–0.144], and 0.124 [0.072–0.214], resp.) (Figure 4). No statistically significant associations could be established between protozoa infection and eosinophil counts, IgE levels, or the remaining cytokines studied.

4. Discussion

The parasitological findings of this study demonstrate a moderately high (30%) overall STH prevalence and specific prevalence for *T. trichiura* and *A. lumbricoides* of about

20% among schoolchildren living in Linaca. School teachers reported that the school is reached by the national deworming program and thus receives a single-dose 400 mg albendazole tablet per child on a yearly basis. No records are kept at the school in terms of deworming tablets intake and no parasitological baseline data have been obtained during governmental surveys (data reviewed in [27]). To our knowledge, prior to the present study, two parasitological surveys had been undertaken in the study community. In 1985, a study investigating causes of diarrhea in Honduras included a small number of preschoolchildren and found prevalence of 48% and 30% for *A. lumbricoides* and *T. trichiura*, respectively [28, 29]. More recently, a study involving schoolchildren determined prevalence of 64% for *A. lumbricoides* and 46% for *T. trichiura* (Canales et al. 2008, unpublished). Compared to the STH prevalence found in the present study, it appears that there has been an important decrease in STH prevalence among Linaca's children. Due to insufficient data on deworming, we did not attempt to identify if this strategy has served an important function in decreasing STH prevalence or at least worm burden. Thus, whether or not deworming has contributed to STH prevalence reduction remains an open question. Our findings, however, show that improved household conditions were strong protective factors for ascariasis. In this case, there might be a synergy between the

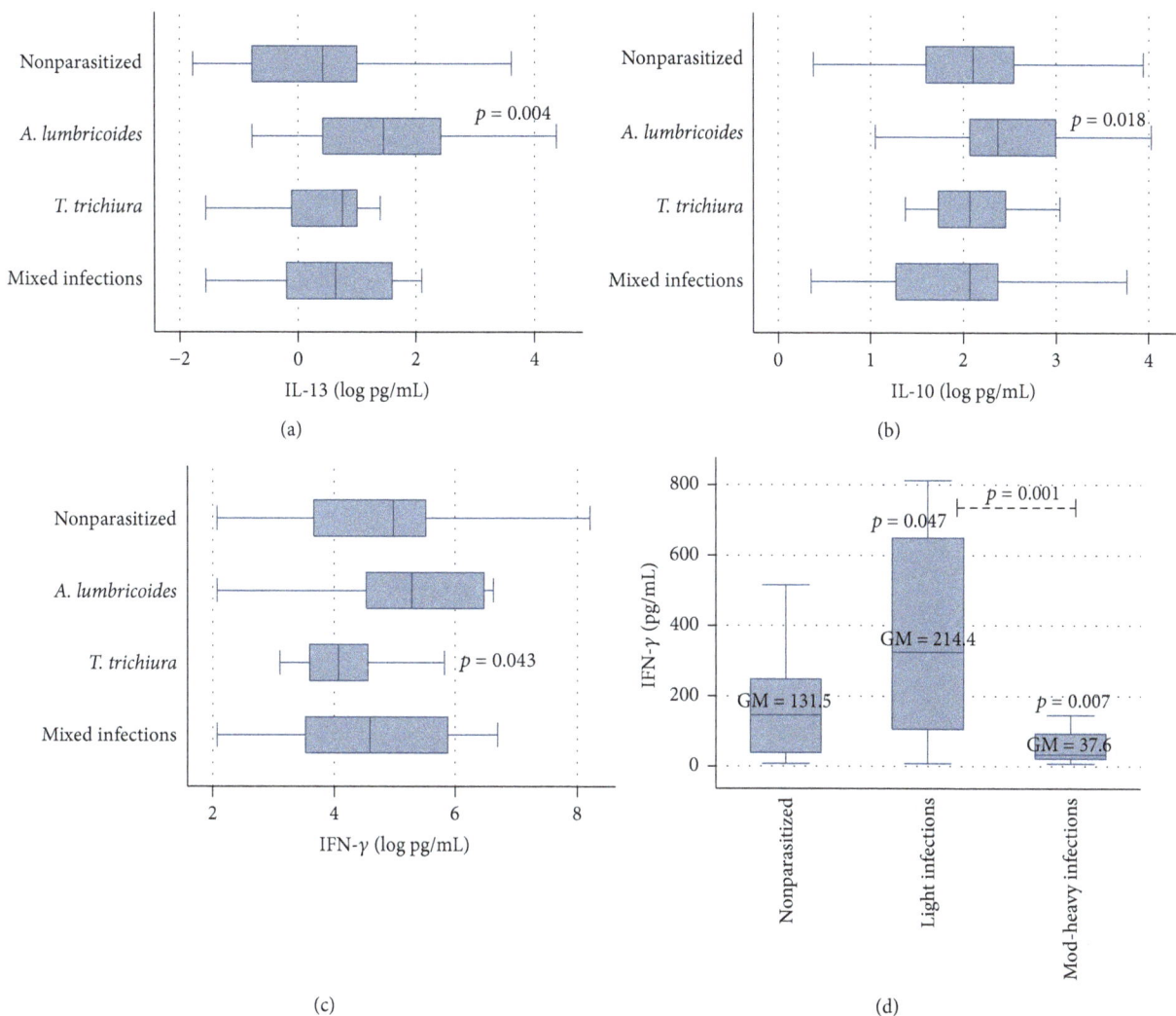

(a)

(b)

(c)

(d)

FIGURE 3: Association between STH infection and cytokines in studied children. *A. lumbricoides* infections were associated with higher levels of IL-13 (a) and IL-10 (b), whereas lower levels of IFN-γ were associated with *T. trichiura* infections (c) and moderate-to-heavy ascariasis (d). Geometric means (GM) are depicted in their respective boxes.

built environment and anthelminthic medication, which may not be sufficient to protect against *T. trichiura* infection as one dose of ALB treatment is largely ineffective to clear this parasite.

Even though a similar proportion of children were found with either excess weight (12%) or chronic malnutrition (stunted growth, ~10%), the latter was strongly associated with parasitism. Similarly, despite the fact that trichuriasis prevalence was at a moderate level and that the majority of infections were light, a significant association was documented between stunting and both *T. trichiura* infection and infection intensity. This finding aligns with observations from other studies conducted in Honduras [30], Ethiopia [31], Mexico [32], and Brazil [33, 34].

The concerning association between trichuriasis and chronic malnutrition, along with the fact that *T. trichiura* has been found consistently as the most prevalent geohelminth in Honduras [27, 35, 36], suggests that (i) a reexamination of the current deworming guidelines are necessary; (ii)

improvements in sanitary conditions and health education are indispensable components of STH control initiatives; and (iii) it is important to monitor for potential benzimidazole resistance in this parasite [37].

The prevalence of eosinophilia was high among the studied children. It has been shown that, in developing countries, eosinophilia is most commonly induced by tissue-invasive parasites, particularly helminths [25, 38]. STH infections have been largely associated with eosinophilia, especially in early stages of infection, when larval migration occurs. The association between eosinophilia and STH in the present study is consistent with results from previous work conducted in Honduras [39, 40] and other countries such as Brazil [41, 42], Philippines [43], Indonesia [44], and Spain [26].

Whereas other causes of eosinophilia cannot be ruled out, it is likely that, in this particular group of children, such eosinophilia was reactive (i.e., secondary to an external stimulus) and most likely associated to STH infection. This inference is supported by data contained in Linaca's Health Centre

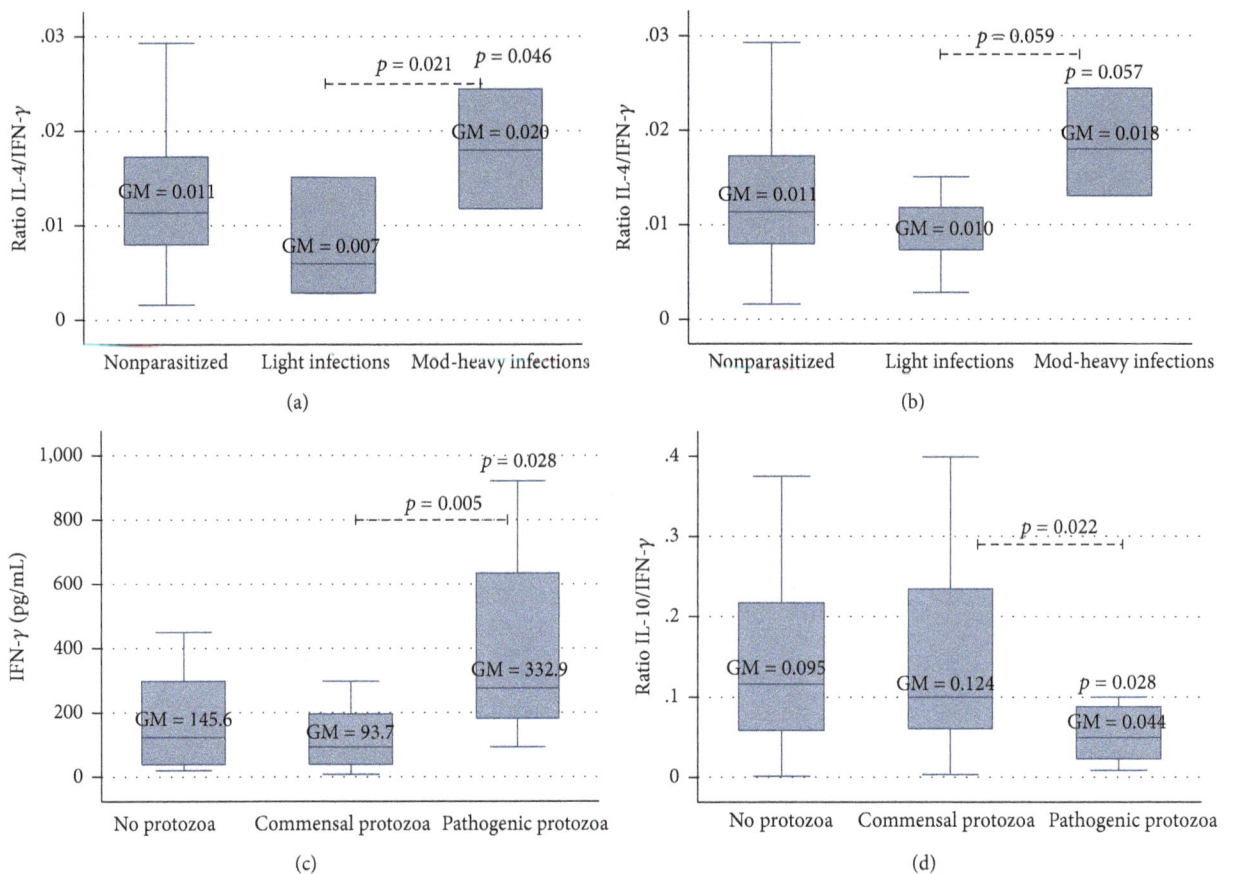

FIGURE 4: Association between STH infection intensity and Th2/Th1 cytokines ratio in studied children. Higher IL-4/IFN-γ ratios were associated with moderate-to-heavy ascariasis (a) and trichuriasis (b), although marginally significant in the latter. In children without STH infections, pathogenic protozoa were associated with higher levels of IFN-γ (c) and lower IL-10/IFN-γ (d). Geometric means (GM) are depicted in their respective boxes.

morbidity report for 2014, which shows that <2% of children's visits were due to asthma or allergic dermatitis (Linaca Health Care Centre 2014 report, unpublished). Moreover, a comprehensive socioeconomic and life conditions study conducted in the municipality of Tatumbla, where Linaca is located, contained no reference to allergic conditions being a health issue among inhabitants. Rather, it revealed that childhood acute respiratory infections accounted for >50% of visits to the health centre [45].

The use of eosinophilia as a biomarker for helminthiases remains without consensus. As a predictor of current helminthic infections, some authors consider eosinophilia's predictive value very limited [38], while others regard it as a suitable indicator of helminthiases in people from tropical and subtropical areas [26, 42, 44, 46]. Our data support the latter since eosinophilia correlated with both STH infection and intensity. Future studies should investigate eosinophilia at the individual level in order to better understand this dynamic relationship.

Our study revealed strikingly high prevalence of hyper-IgE among the studied children; it also highlighted a significant association between STH infections and increased total IgE levels. Such findings are consistent with observations

in studies conducted in Nigeria [47, 48], Ecuador [49], Venezuela [50], Spain [26], and Brazil [41, 51, 52]. Helminth-induced IgE is mostly characterized by a nonspecific polyclonal stimulation with only a small fraction of parasite-specific IgE. However, it is the parasite-specific IgE which plays an important role in helminth clearance as well as in preventing reinfection [48, 50, 53–55]. In a study of Venezuelan children, Hagel and collaborators demonstrated that total IgE levels were inversely correlated with parasite-specific IgE levels. They also showed that reinfection with A. lumbricoides was significantly associated with high pretreatment total IgE but low parasite-specific IgE levels [50].

Another interesting finding in the present investigation was the significant association observed between high total IgE levels and intensity of infection (i.e., worm burden). Other researchers have made the same observation [49, 51, 56]. Some studies have shown that specific IgE levels reflect the parasite infection intensity; therefore, higher specific antibody levels would be expected in lightly infected individuals when compared with those heavily infected [57]. Total IgE levels in the studied children were significantly positively correlated with eosinophils counts; this finding is not surprising due to the ability of IgE to induce a cytotoxic response against helminthic parasites mediated by mast cells and eosinophils [55].

It has been demonstrated that helminth infections strongly induce an immune response involving elevated Th2 cytokines (i.e., IL-4, IL-5, IL-9, and IL-13), IgE, IgA, eosinophilia, and mucus secretion [58–60]. Helminth-caused tissue damage triggers the response by releasing danger-associated molecular patterns (DAMPs) and cytokine alarmins, particularly IL-25 and IL-33. These alarmins promote the activation of basophils and innate lymphoid cells (ILC-2) to support a type-2 innate immune response needed to signal the type-2 adaptive immunity. In this context, large quantities of IL-5 and IL-13 are primarily produced by ILC-2 cells. IL-5 promotes the eosinophils differentiation and activation which promote IL-4 production, the main primer of $CD4^+$ Th2 cells activation and expansion, with further release of IL-4, IL-5, and IL-13 by this type of cells [3, 60–62]. In this environment, IL-4 and IL-13 induce AAMs, which mainly promote tissue repair and fibrosis by expression of different markers such as arginase-1, Yml, Ym2, and RELM-α. These AAMs also express regulatory IL-10 and TGF-β able to downregulate inflammatory response [3–5, 9, 61–63].

Our study could not find a significant correlation between IL-5 and eosinophilia. Instead, the majority of children showed either low or nondetectable values of this cytokine. This was an unexpected finding since IL-5 is essential for eosinophilic differentiation, proliferation, and activation. There are a few potential explanations for this finding. Since the participant children have probably experienced repeated STH infections since they were much younger, their immune response may have already changed to a modified Type 2 response, where inflammatory Th2 shifted to an attenuated phenotype characterized by the shutdown of effector cytokines such as IL-5, retention of IL-4, and reinforcement of anti-inflammatory cytokines such as IL-10 and/or TGF-β [4, 9]. Instead, as recently determined by Fulkerson and colleagues, an alternative IL-5-independent pathway for promoting eosinophilia might be in place. This pathway would contribute to persistent eosinophil differentiation and survival even after a complete IL-5 withdrawal [64]. Finally, periodic deworming followed by reinfection along several years is likely disturbing the antiparasitic immune responses as described in the literature.

We found that IL-13 levels were significantly increased in children infected with A. lumbricoides and that these levels were positively correlated with worm burden. IL-13 is a Th2 pleiotropic cytokine with an important function in the intestinal epithelia, controlling the rate of transit of epithelial cells toward the outmost layer and promoting a harsh environment that might help in gut parasite expulsion (i.e., increased luminal fluids, mucus, from goblet cells, increased edema, and muscle contractility) [7, 60, 61]. Gallo and colleagues demonstrated that IL-13 may also be produced by Th1 and Th17 cells, suggesting that this cytokine can have either a pro- or anti-inflammatory effect, depending on the environment [65]. Nevertheless, demonstrating an association between STH and IL-13 overproduction has not been achieved by other research groups in India [66] and Brazil [67]. Some studies in Cameroon [68, 69] found high

levels of IL-13 associated with lower intensity of infection, but the present study did not establish such association.

IL-10 is the anti-inflammatory cytokine by excellence. It can be produced by diverse type of cells such as macrophages, dendritic cells (DC), B cells, and various subsets of $CD4^+$ and $CD8^+$ T cells. Regardless of the type of infection, IL-10 limits Th1 and Th2 effector responses by suppressing function of macrophages and DCs. The timing, site of its production, and strength may favor (a) simultaneous pathogen clearance and suppression of downstream pathologies; (b) potential benefit to both the host and the pathogen (limiting pathology and allowing persistent infection); (c) severe tissue damage; or (d) overwhelming infection [70, 71]. In our study, statistically significant higher IL-10 values were found in children with ascariasis, although they did not correlate with infection intensity of this parasite. Similar findings have been frequently reported by other researchers [66, 69, 72–74], with some exceptions [47]. We also found a negative correlation between children's age and IL-10 levels. This is congruent with previous studies conducted in a different community in Honduras [75] as well as with other studies in Nigeria [73] and Cameroon [68, 76]. This might be due to altered immune modulation provoked by deworming treatment. Although its efficacy is species-dependent, albendazole generally reduces the duration and/or intensity of helminth infections [75].

In our study, IFN-γ levels were significantly inversely associated with intensity of infection by A. lumbricoides. It is well known that IFN-γ is downregulated by type-2 cytokines making our finding consistent with the literature. IFN-γ is the most important type-1 cytokine and is produced by a diverse type of cells from both innate and adaptive immune arms such as natural killer (NK), $CD4^+$ Th1, $CD8^+$ Tc, and eosinophils. It interacts with macrophages to activate direct antimicrobial and antitumor mechanisms as well as upregulating antigen processing and presentation pathways [77].

The overall interpretation of the different cytokines, their levels, and relationships with helminthic infections is an interesting challenge. In moderate-to-heavy ascariasis, we observed a distinctive type-2 immune response. This response was characterized by upregulation of IL-13 and downregulation of IFN-γ, as well as by significantly higher IL-4/IFN-γ ratio and IgE levels. Results from T. trichiura infections were not as well-defined as those from ascariasis. Nonetheless and similarly to A. lumbricoides, infections of high intensity were associated with lower values of IFN-γ and higher IL-4/IFN-γ ratios and IgE levels. These findings confirm that a type-2 response against this parasite was also displayed in our study population.

As mentioned earlier, a type-1 immune response is needed for the control and clearance of intestinal protozoa such as Entamoeba histolytica, Cryptosporidium spp., and Giardia intestinalis. Although, particularly in the case of G. intestinalis, mixed Th1-Th2 responses have been reported, human and animal studies have shown that increased levels of IFN-γ play a crucial role in orchestrating the immune response and appear to be one of the main signatures in individuals with these parasitoses [78–81]. This description is in agreement with our findings of significantly higher (up to 3-fold) levels of IFN-γ found in children harbouring *G. intestinalis* and/or *E. histolytica/dispar* without the immune influence of STH infections (i.e., negative for helminthic infections). Moreover, the significantly lower IL10/IFN-γ ratios observed in this group of children indicate a Th1-biased response. Additionally, in a 6-year longitudinal study in Bolivia, Blackwell and colleagues described an antagonist relationship between *G. intestinalis* and geohelminths where *giardiasis* was less likely to be present in helminth-infected individuals. Likewise, they also found that infection with helminths was less likely for individuals with giardiasis [82]. Our data, though cross-sectional, support these observations: none of the children infected with *Giardia intestinalis* had ascariasis and only 10% of them had concomitant trichuriasis. Since clearance and protective immunity against *G. intestinalis* require also Th2 cytokines and antibodies, IgA, IgG, and IgE [78, 79, 81], a plausible scenario has been suggested: helminth-induced Th2 response might provide cross-immunity against this protozoan. We believe that the STH/*G. intestinalis* interaction is worth investigating.

A proper interpretation of our findings must be done considering the limitations of the present study. Firstly, its cross-sectional nature only allows establishing associations but no causality and, therefore, strong inferences cannot be made. Secondly, we did not obtain a complete medical history from participating children or an accurate deworming history. This is important as a variety of unmeasured confounding variables (e.g., bacterial and viral infections, autoimmune diseases, asthma, and other allergic diseases, as well as environmental and genetic factors) may have contributed to skew the Th1/Th2 balance among research participants [6]. Thirdly, we measured circulating levels of cytokines, which may not be an accurate representation of local immune responses in the gut milieu. Notwithstanding, the study has several strengths: an adequate sample size, determination of both helminthic and protozoal infections, and the use of highly sensitive and accurate methodology for biomarkers analysis.

5. Conclusion

This is the first study providing a comprehensive immune profile in Honduran children infected with geohelminths or pathogenic protozoa. Our study shows that higher intensities of STH infections are associated with a polarized Th2 response, characterized by reduced proinflammatory and increased regulatory cytokines, eosinophilia, and hyper-IgE. It also demonstrates that an enhanced Th1 response is elicited by pathogenic protozoa. The interpretation of the host's immune response against helminth represents challenging undertaking due to the high complexity of multiple host-parasite interactions and potential unmeasured confounders. Pivotal to characterizing and understanding the immune response is obtaining a meticulous clinical history from study participants, their deworming history, and very importantly the existence of multiple infections including pathogenic protozoa. Further, the role of the intestinal microbiota and its interplay with the host and its parasites can no longer be

ignored. Longitudinal studies are required to better characterize such immune responses.

Competing Interests

All authors declare that there is no conflict of interests regarding the publication of this paper.

Acknowledgments

The authors thank all the children who participated in this study, as well as their parents and school authorities of Linaca. They also thank all volunteers from the School of Microbiology (UNAH) for their cooperation and enthusiasm which ensured the successful completion of this study. This study received partial financial support from UNAH through an internal grant.

References

[1] WHO, "Soil-transmitted helminthiases: eliminating soil-transmitted helminthiases as a public health problem in children," in *Progress Report 2001–010 and Strategic Plan 2011–2020*, p. 79, World Health Organization, Geneva, Switzerland, 2012.

[2] P. J. Hotez, M. E. Bottazzi, C. Franco-Paredes, S. K. Ault, and M. R. Periago, "The neglected tropical diseases of Latin America and the Caribbean: a review of disease burden and distribution and a roadmap for control and elimination," *PLoS Neglected Tropical Diseases*, vol. 2, no. 9, article e300, 2008.

[3] P. K. Mishra, M. Palma, D. Bleich, P. Loke, and W. C. Gause, "Systemic impact of intestinal helminth infections," *Mucosal Immunology*, vol. 7, no. 4, pp. 753–762, 2014.

[4] A. Díaz and J. E. Allen, "Mapping immune response profiles: the emerging scenario from helminth immunology," *European Journal of Immunology*, vol. 37, no. 12, pp. 3319–3326, 2007.

[5] W. C. Gause, T. A. Wynn, and J. E. Allen, "Type 2 immunity and wound healing: evolutionary refinement of adaptive immunity by helminths," *Nature Reviews Immunology*, vol. 13, no. 8, pp. 607–614, 2013.

[6] P. Kidd, "Th1/Th2 balance: the hypothesis, its limitations, and implications for health and disease," *Alternative Medicine Review*, vol. 8, no. 3, pp. 223–246, 2003.

[7] J. Hopkin, "Immune and genetic aspects of asthma, allergy and parasitic worm infections: evolutionary links," *Parasite Immunology*, vol. 31, no. 5, pp. 267–273, 2009.

[8] H. J. McSorley and R. M. Maizels, "Helminth infections and host immune regulation," *Clinical Microbiology Reviews*, vol. 25, no. 4, pp. 585–608, 2012.

[9] M. D. Taylor, N. van der Werf, and R. M. Maizels, "T cells in helminth infection: the regulators and the regulated," *Trends in Immunology*, vol. 33, no. 4, pp. 181–189, 2012.

[10] P. Endara, M. Vaca, M. E. Chico et al., "Long-term periodic anthelmintic treatments are associated with increased allergen skin reactivity," *Clinical & Experimental Allergy*, vol. 40, no. 11, pp. 1669–1677, 2010.

[11] A. M. J. Van Den Biggelaar, L. C. Rodrigues, R. Van Ree et al., "Long-term treatment of intestinal helminths increases mite skin-test reactivity in Gabonese schoolchildren," *The Journal of Infectious Diseases*, vol. 189, no. 5, pp. 892–900, 2004.

[12] M. Yazdanbakhsh, A. van den Biggelaar, and R. M. Maizels, "Th2 responses without atopy: immunoregulation in chronic helminth infections and reduced allergic disease," *Trends in Immunology*, vol. 22, no. 7, pp. 372–377, 2001.

[13] G. Borkow and Z. Bentwich, "Chronic immune activation associated with chronic helminthic and human immunodeficiency virus infections: role of hyporesponsiveness and anergy," *Clinical Microbiology Reviews*, vol. 17, no. 4, pp. 1012–1030, 2004.

[14] J. E. Fincham, M. B. Markus, V. J. Adams et al., "Association of deworming with reduced eosinophilia: implications for HIV/AIDS and co-endemic diseases," *South African Journal of Science*, vol. 99, no. 3-4, pp. 182–184, 2003.

[15] A. Mulu, M. Legesse, B. Erko et al., "Epidemiological and clinical correlates of malaria-helminth co-infections in Southern Ethiopia," *Malaria Journal*, vol. 12, article 227, 2013.

[16] V. H. Salazar-Castañon, M. Legorreta-Herrera, and M. Rodriguez-Sosa, "Helminth parasites alter protection against *Plasmodium* infection," *BioMed Research International*, vol. 2014, Article ID 913696, 19 pages, 2014.

[17] A. Mulu, B. Anagaw, A. Gelaw, F. Ota, A. Kassu, and S. Yifru, "Effect of deworming on Th2 immune response during HIV-helminths co-infection," *Journal of Translational Medicine*, vol. 13, article 236, 2015.

[18] A. Mulu, M. Maier, and U. G. Liebert, "Deworming of intestinal helminths reduces HIV-1 subtype C viremia in chronically co-infected individuals," *International Journal of Infectious Diseases*, vol. 17, no. 10, pp. e897–e901, 2013.

[19] E. Abate, D. Elias, A. Getachew et al., "Effects of albendazole on the clinical outcome and immunological responses in helminth co-infected tuberculosis patients: a double blind randomised clinical trial," *International Journal for Parasitology*, vol. 45, no. 2-3, pp. 133–140, 2015.

[20] R. Anuradha, S. Munisankar, Y. Bhootra et al., "IL-10- and TGFβ-mediated Th9 responses in a human helminth infection," *PLoS Neglected Tropical Diseases*, vol. 10, no. 1, Article ID e0004317, 2016.

[21] C. A. Blish, L. Sangaré, B. R. Herrin, B. A. Richardson, G. John-Stewart, and J. L. Walson, "Changes in plasma cytokines after treatment of Ascaris lumbrkoides infection in individuals with HIV-1 infection," *The Journal of Infectious Diseases*, vol. 201, no. 12, pp. 1816–1821, 2010.

[22] J. L. Walson, P. A. Otieno, M. Mbuchi et al., "Albendazole treatment of HIV-1 and helminth co-infection: a randomized, double-blind, placebo-controlled trial," *AIDS*, vol. 22, no. 13, pp. 1601–1609, 2008.

[23] M. De Onis, A. W. Onyango, E. Borghi, A. Siyam, C. Nishida, and J. Siekmann, "Development of a WHO growth reference for school-aged children and adolescents," *Bulletin of the World Health Organization*, vol. 85, no. 9, pp. 660–667, 2007.

[24] Honduras-Ministry-of-Health(SS), National-Institute-of-Statistics(INE), and ICF-International, *National Demographic and Health Survey 2011-2012*, SS, INE and ICF International, 2013 (Spanish).

[25] J. Gotlib, "World Health Organization-defined eosinophilic disorders: 2014 update on diagnosis, risk stratification, and management," *American Journal of Hematology*, vol. 89, no. 3, pp. 325–337, 2014.

[26] M. Belhassen-García, J. Pardo-Lledías, L. Pérez Del Villar et al., "Relevance of eosinophilia and hyper-IgE in immigrant children," *Medicine*, vol. 93, no. 6, article e43, 2014.

[27] A. L. Sanchez, J. A. Gabrie, M. M. Rueda, R. E. Mejia, M. E. Bottazzi, and M. Canales, "A scoping review and prevalence analysis of soil-transmitted helminth infections in Honduras," *PLoS Neglected Tropical Diseases*, vol. 8, no. 1, article e2653, 2014.

[28] M. Figueroa, E. Poujol, H. Cosenza, and R. G. Kaminsky, "Etiology of childhood diarrhea in three Honduran communities," *Revista Médica Hondureña*, vol. 58, no. 4, p. 9, 1990 (Spanish).

[29] R. G. Kaminsky, "Parasitism and diarrhoea in children from two rural communities and marginal barrio in Honduras," *Transactions of the Royal Society of Tropical Medicine and Hygiene*, vol. 85, no. 1, pp. 70–73, 1991.

[30] A. L. Sanchez, J. A. Gabrie, M.-T. Usuanlele, M. M. Rueda, M. Canales, and T. W. Gyorkos, "Soil-transmitted helminth infections and nutritional status in school-age children from rural communities in honduras," *PLoS Neglected Tropical Diseases*, vol. 7, no. 8, article e2378, 2013.

[31] M. Wolde, Y. Berhan, and A. Chala, "Determinants of underweight, stunting and wasting among schoolchildren," *BMC Public Health*, vol. 15, no. 1, article 8, 2015.

[32] L. Quihui-Cota, M. E. Valencia, D. W. T. Crompton et al., "Prevalence and intensity of intestinal parasiticinfections in relation to nutritional status in Mexican schoolchildren," *Transactions of the Royal Society of Tropical Medicine and Hygiene*, vol. 98, no. 11, pp. 653–659, 2004.

[33] A. Beltrame, C. Scolari, C. Torti, and C. Urbani, "Soil Transmitted Helminth (STH) infections in an indigenous community in Ortigueira, Paraná, Brazil and relationship with its nutritional status," *Parassitologia*, vol. 44, no. 3-4, pp. 137–139, 2002.

[34] R. D. C. R. Silva and A. M. O. Assis, "Association between geohelminth infections and physical growth in schoolchildren," *Revista de Nutricao*, vol. 21, no. 4, pp. 393–399, 2008.

[35] J. A. Gabrie, M. M. Rueda, M. Canales, T. W. Gyorkos, and A. L. Sanchez, "School hygiene and deworming are key protective factors for reduced transmission of soil-transmitted helminths among schoolchildren in Honduras," *Parasites and Vectors*, vol. 7, no. 1, article 354, 2014.

[36] A. Sanchez, J. Gabrie, M. Canales et al., "Soil-transmitted helminths, poverty, and malnutrition in honduran children living in remote rural communities," *Human Parasitic Diseases*, vol. 8, p. 27, 2016.

[37] A. Diawara, L. J. Drake, R. R. Suswillo et al., "Assays to detect β-tubulin codon 200 polymorphism in *Trichuris trichiura* and *Ascaris lumbricoides*," *PLoS Neglected Tropical Diseases*, vol. 3, no. 3, article e397, 2009.

[38] V. Khanna, K. Tilak, C. Mukhopadhyay, and R. Khanna, "Significance of diagnosing parasitic infestation in evaluation of unexplained eosinophilia," *Journal of Clinical and Diagnostic Research*, vol. 9, no. 7, pp. DC22–DC24, 2015.

[39] L. M. Espinoza, R. J. Soto, and J. Alger, "Eosinophilia associated to helminthiases in children attending Hospital Escuela, Honduras," *Revista Mexicana de Patología Clínica*, vol. 46, no. 2, pp. 79–85, 1999 (Spanish).

[40] R. G. Kaminsky, R. J. Soto, A. Campa, and M. K. Baum, "Intestinal parasitic infections and eosinophilia in an human immunedeficiency virus positive population in Honduras," *Memorias do Instituto Oswaldo Cruz*, vol. 99, no. 7, pp. 773–778, 2004.

[41] N. M. Alcântara-Neves, G. S. G. Britto, R. V. Veiga et al., "Effects of helminth co-infections on atopy, asthma and cytokine production in children living in a poor urban area in Latin America," *BMC Research Notes*, vol. 7, article 817, 2014.

[42] C. C. V. Silva, R. R. N. Ferraz, J. V. Fornari, and A. S. Barnabe, "Epidemiological analysis of eosinophilia and elevation

of immunoglobulin E as a predictable and relative risk of enteroparasitosis," *Revista Cubana de Medicina Tropical*, vol. 64, no. 1, pp. 22–26, 2012.

[43] J. B. Sumagaysay and F. M. Emverda, "Eosinophilia and incidence of soil-transmitted helminthic infections of secondary students of an indigenous school," *Asian Journal of Health*, vol. 1, no. 1, pp. 172–182, 2012.

[44] S. Jiero, M. Ali, S. Pasaribu, and A. P. Pasaribu, "Correlation between eosinophil count and soil-transmitted helminth infection in children," *Asian Pacific Journal of Tropical Disease*, vol. 5, no. 10, pp. 813–816, 2015.

[45] R. Diaz, *Socio-Economic Study and Baseline Indicators of Municipality of Tatumbla, Francisco Morazan*, Secretaría del Interior y Población. Gobierno de Honduras, 2012 (Spanish).

[46] O. G. Sara, H. C. J. Manuel, and V. G. Patricia, "Relationship of eosinophilia with parasitoses and alergies in children," *Bioquimia*, vol. 34, p. 78, 2009 (Spanish).

[47] G. O. Arinola, O. A. Morenikeji, K. S. Akinwande et al., "Serum levels of cytokines and IgE in helminth-infected nigerian pregnant women and children," *Annals of Global Health*, vol. 81, no. 5, pp. 689–693, 2015.

[48] O. G. Arinola, A. S. Yaqub, and K. S. Rahamon, "Reduced serum IgE level in Nigerian children with helminthiasis compared with protozoan infection: implication on hygiene hypothesis," *Annals of Biological Research*, vol. 3, pp. 5754–5757, 2012.

[49] P. J. Cooper, N. Alexander, A.-L. Moncayo et al., "Environmental determinants of total IgE among school children living in the rural tropics: importance of geohelminth infections and effect of anthelmintic treatment," *BMC Immunology*, vol. 9, article 33, 2008.

[50] I. Hagel, N. R. Lynch, M. C. Di Prisco, E. Rojas, M. Perez, and N. Alvarez, "Ascaris reinfection of slum children: relation with the IgE response," *Clinical and Experimental Immunology*, vol. 94, no. 1, pp. 80–83, 1993.

[51] J. P. Figueiredo, R. R. Oliveira, L. S. Cardoso et al., "Adult worm-specific IgE/IgG4 balance is associated with low infection levels of Schistosoma mansoni in an endemic area," *Parasite Immunology*, vol. 34, no. 12, pp. 604–610, 2012.

[52] N. Rosário Filho, M. Carneiro Filho, E. Ferreira, M. Baranski, and I. Cat, "Niveis de IgE total no soro e contagens de eosinofilos em criancas com enteroparasitoses: efeito do tratamento anti-helmintico," *Jornal de Pediatria*, vol. 52, no. 4, pp. 209–215, 1982.

[53] M. Amarasekera, "Immunoglobulin E in health and disease," *Asia Pacific Allergy*, vol. 1, no. 1, pp. 12–15, 2011.

[54] F. Hamid, A. S. Amoah, R. Van Ree, and M. Yazdanbakhsh, "Helminth-induced ige and protection against allergic disorders," *Current Topics in Microbiology and Immunology*, vol. 388, pp. 91–108, 2015.

[55] W. E. Winter, N. S. Hardt, and S. Fuhrman, "Immunoglobulin E: importance in parasitic infections and hypersensitivity responses," *Archives of Pathology and Laboratory Medicine*, vol. 124, no. 9, pp. 1382–1385, 2000.

[56] A. Mulu, A. Kassu, M. Legesse et al., "Helminths and malaria co-infections are associated with elevated serum IgE," *Parasites & Vectors*, vol. 7, article 240, 2014.

[57] D. R. Palmer, A. Hall, R. Haque, and K. S. Anwar, "Antibody isotype responses to antigens of Ascaris lumbricoides in a case-control study of persistently heavily infected Bangladeshi children," *Parasitology*, vol. 111, no. 3, pp. 385–393, 1995.

[58] J. E. Bradley and J. A. Jackson, "Immunity, immunoregulation and the ecology of trichuriasis and ascariasis," *Parasite Immunology*, vol. 26, no. 11-12, pp. 429–441, 2004.

[59] C. M. Fitzsimmons, F. H. Falcone, and D. W. Dunne, "Helminth allergens, parasite-specific IgE, and its protective role in human immunity," *Frontiers in Immunology*, vol. 5, article 61, 2014.

[60] T. A. Wynn, "IL-13 effector functions," *Annual Review of Immunology*, vol. 21, pp. 425–456, 2003.

[61] R. M. Anthony, L. I. Rutitzky, J. F. Urban, M. J. Stadecker, and W. C. Gause, "Protective immune mechanisms in helminth infection," *Nature Reviews Immunology*, vol. 7, no. 12, pp. 975–987, 2007.

[62] T. B. Nutman, "Looking beyond the induction of Th2 responses to explain immunomodulation by helminths," *Parasite Immunology*, vol. 37, no. 6, pp. 304–313, 2015.

[63] K. J. Mylonas, M. G. Nair, L. Prieto-Lafuente, D. Paape, and J. E. Allen, "Alternatively activated macrophages elicited by helminth infection can be reprogrammed to enable microbial killing," *Journal of Immunology*, vol. 182, no. 5, pp. 3084–3094, 2009.

[64] P. C. Fulkerson, K. L. Schollaert, C. Bouffi, and M. E. Rothenberg, "IL-5 triggers a cooperative cytokine network that promotes eosinophil precursor maturation," *Journal of Immunology*, vol. 193, no. 8, pp. 4043–4052, 2014.

[65] E. Gallo, S. Katzman, and A. V. Villarino, "IL-13-producing Th1 and Th17 cells characterize adaptive responses to both self and foreign antigens," *European Journal of Immunology*, vol. 42, no. 9, pp. 2322–2328, 2012.

[66] N. Malla, B. A. Fomda, and M. A. Thokar, "Serum cytokine levels in human ascariasis and toxocariasis," *Parasitology Research*, vol. 98, no. 4, pp. 345–348, 2006.

[67] S. Geiger, C. L. Massara, J. Bethony, P. T. Soboslay, O. S. Carvalho, and R. Corrêa-Oliveira, "Cellular responses and cytokine profiles in Ascaris lumbricoides and Trichuris trichiura infected patients," *Parasite Immunology*, vol. 24, no. 11-12, pp. 499–509, 2002.

[68] J. A. Jackson, J. D. Turner, L. Rentoul et al., "Cytokine response profiles predict species-specific infection patterns in human GI nematodes," *International Journal for Parasitology*, vol. 34, no. 11, pp. 1237–1244, 2004.

[69] J. D. Turner, H. Faulkner, J. Kamgno et al., "Th2 cytokines are associated with reduced worm burdens in a human intestinal helminth infection," *The Journal of Infectious Diseases*, vol. 188, no. 11, pp. 1768–1775, 2003.

[70] K. N. Couper, D. G. Blount, and E. M. Riley, "IL-10: the master regulator of immunity to infection," *Journal of Immunology*, vol. 180, no. 9, pp. 5771–5777, 2008.

[71] S. A. Redpath, N. M. Fonseca, and G. Perona-Wright, "Protection and pathology during parasite infection: IL-10 strikes the balance," *Parasite Immunology*, vol. 36, no. 6, pp. 233–252, 2014.

[72] P. J. Cooper, L. D. Amorim, C. A. Figueiredo et al., "Effects of environment on human cytokine responses during childhood in the tropics: role of urban versus rural residence," *World Allergy Organization Journal*, vol. 8, no. 1, article 71, 2015.

[73] N. O. P. Goddey, I. D. Osagie, and A. Maliki, "Serum cytokines profiles in Nigerian children with Ascaris lumbricoides infection," *Asian Pacific Journal of Tropical Medicine*, vol. 3, no. 4, pp. 288–291, 2010.

[74] M. Reina Ortiz, F. Schreiber, S. Benitez et al., "Effects of chronic ascariasis and trichuriasis on cytokine production and gene expression in human blood: a cross-sectional study," *PLoS Neglected Tropical Diseases*, vol. 5, no. 6, article e1157, 2011.

[75] A. L. Sanchez, D. L. Mahoney, and J. A. Gabrie, "Interleukin-10 and soil-transmitted helminth infections in Honduran children," *BMC Research Notes*, vol. 8, no. 1, article 55, 2015.

[76] H. Faulkner, J. Turner, J. Kamgno, S. D. Pion, M. Boussinesq, and J. E. Bradley, "Age- and infection intensity-dependent cytokine and antibody production in human trichuriasis: the importance of IgE," *The Journal of Infectious Diseases*, vol. 185, no. 5, pp. 665–672, 2002.

[77] K. Schroder, P. J. Hertzog, T. Ravasi, and D. A. Hume, "Interferon-γ: an overview of signals, mechanisms and functions," *Journal of Leukocyte Biology*, vol. 75, no. 2, pp. 163–189, 2004.

[78] J. C. Jiménez, J. Fontaine, C. Creusy et al., "Antibody and cytokine responses to Giardia excretory/secretory proteins in Giardia intestinalis-infected BALB/c mice," *Parasitology Research*, vol. 113, no. 7, pp. 2709–2718, 2014.

[79] J. Matowicka-Karna, V. Dymicka-Piekarska, and H. Kemona, "IFN-gamma, IL-5, IL-6 and IgE in patients infected with Giardia intestinalis," *Folia Histochemica et Cytobiologica*, vol. 47, no. 1, pp. 93–97, 2009.

[80] K. Nakada-Tsukui and T. Nozaki, "Immune response of amebiasis and immune evasion by Entamoeba histolytica," *Frontiers in Immunology*, vol. 7, article 175, 2016.

[81] K. Roxström-Lindquist, D. Palm, D. Reiner, E. Ringqvist, and S. G. Svärd, "Giardia immunity—an update," *Trends in Parasitology*, vol. 22, no. 1, pp. 26–31, 2006.

[82] A. D. Blackwell, M. Martin, H. Kaplan, and M. Gurven, "Antagonism between two intestinal parasites in humans: the importance of co-infection for infection risk and recovery dynamics," *Proceedings of the Royal Society B: Biological Sciences*, vol. 280, no. 1769, p. 1671, 2013.

3

Intestinal Schistosomiasis among Primary Schoolchildren in Two On-Shore Communities in Rorya District, Northwestern Tanzania: Prevalence, Intensity of Infection and Associated Risk Factors

David Z. Munisi,[1,2] Joram Buza,[1] Emmanuel A. Mpolya,[1] and Safari M. Kinung'hi[3]

[1]*Department of Global Health and Biomedical Sciences, School of Life Sciences and Bioengineering,*
 Nelson Mandela African Institution of Science and Technology, P.O. Box 447, Arusha, Tanzania
[2]*Department of Biomedical Sciences, School of Medicine and Dentistry, College of Health Sciences, University of Dodoma,*
 P.O. Box 259, Dodoma, Tanzania
[3]*National Institute for Medical Research (NIMR), Mwanza Research Centre, Isamilo Road, P.O. Box 1462, Mwanza, Tanzania*

Correspondence should be addressed to David Z. Munisi; munisid@nm-aist.ac.tz

Academic Editor: Emmanuel Serrano Ferron

In Tanzania, *Schistosoma mansoni* is of great public health importance. Understanding the prevalence and infection intensity is important for targeted, evidence-based control strategies. This study aimed at studying the prevalence, intensity, and risk factors of *S. mansoni* among schoolchildren in the study area. A cross-sectional study was conducted in Busanga and Kibuyi villages. Sampled 513 schoolchildren provided stool specimens which were examined using kato-katz method. Pretested questionnaire was used to collect sociodemographic data and associated risk factors. The prevalence of *S. mansoni* infection was 84.01%, with geometric mean egg intensity of 167.13 (95% CI: 147.19–189.79) eggs per gram of stool (epg). Other parasites detected were *Ascaris lumbricoides* (1.4%) and hookworms (1.4%). The geometric mean infection intensity in Busanga and Kibuyi were 203.70 (95% CI: 169.67–244.56) and 135.98 (95% CI: 114.33–161.73) epg, respectively. Light, moderate, and heavy infection intensities were 34.11%, 39.91%, and 25.99%, respectively. Village of residence, parent's level of education, toilet use, and treatment history were predictors of infection. The high prevalence and infection intensity in this study were associated with village, parent's level of education, inconsistent toilet use, and treatment history. To control the disease among at-risk groups, these factors need to be considered in designing integrated schistosomiasis control interventions.

1. Background

Schistosomiasis is a chronic and debilitating disease caused by a waterborne digenetic trematode of the genus *Schistosoma* [1]. The disease is one of the most widespread parasitic infections in tropical and subtropical countries where it ranks second to malaria in terms of its socioeconomic and public health significance [2].

In Sub-Saharan Africa (SSA) two *Schistosoma* species are the main cause of schistosomiasis. These are *S. mansoni* and *S. haematobium* that cause intestinal and urinary schistosomiasis, respectively. The region harbours 93% of the world's 207 million estimated cases of schistosomiasis [3, 4]. The disease causes high morbidity and considerable mortality in many endemic areas where children tends to be mostly affected [5].

Schistosomiasis owes its clinical significance from its tendency to slowly damage host organs due to granuloma formation around eggs trapped in tissues, resulting in development of chronic inflammation and fibrosis in the liver and spleen causing hepatosplenomegaly that leads to severe portal hypertension, ascites, gastroesophageal varices, gastrointestinal bleeding, cancer, and death [6, 7]. Despite the serious health impact resulting from these infections and their predominance in areas of poverty, their geographical distribution especially in rural areas of SSA remains incompletely studied [8, 9].

In Tanzania, both *S. mansoni* and *S. haematobium* are highly endemic, and the country ranks second to Nigeria in terms of disease burden in Africa [4, 10, 11]. Intestinal schistosomiasis is of great public health significance along the shores of Lake Victoria [11]. High exposure to infested water bodies makes schoolchildren in this region the most affected group and thereby besides its clinical implication, it contributes to their growth retardation and poor school performance [12]. A number of factors that range from political, demographic, social, economic, environmental, climatic, and cultural trends are known to determine the transmission of schistosomiasis, directly or indirectly [13, 14]. High infection prevalence has been correlated to coming into contact with infested water bodies in various ways [15].

Underlying any sound and effective control strategy for schistosomiasis is a thorough understanding of the prevalence, intensity, and local transmission pattern of the parasite, of which in Mara region many parts have not been well studied making epidemiological data sparse and very incomplete [16]. Although several studies have been conducted on the prevalence of *S. mansoni* and their risk factors in Tanzania, there is still a lack of epidemiological information in some localities of Northwestern Tanzania. This study therefore aimed at studying the prevalence and intensity of *S. mansoni* and its associated risk factors among primary schoolchildren in the study area. This information is important for strengthening the understanding of local schistosomiasis transmission patterns which in turn will be used in developing sound, targeted, and evidence-based control interventions.

2. Methods

2.1. Study Area. This study was conducted in Rorya District, Mara region, Northwestern Tanzania. The district is bordered by Tarime District to the East, Butiama District to the South, Lake Victoria to the West, and the Republic of Kenya to the North (Figure 1) [17]. The majority of inhabitants of Rorya District are from the Luo tribe. Other ethnic groups are Kurya, Kine, Simbiti, Sweta, and Suba. The district is situated in Northwestern Tanzania and lies between latitudes $1°00''$–$1°45''$ south of the Equator and longitudes $33°30''$–$35°0''$ east of Greenwich meridian. Rorya District has two agroecological zones, namely, the midlands and the lowlands with temperature varying from 14°C to 30°C. The annual rainfall ranges from 700 mm to 1200 mm. The district has a total area of 9,345 square kilometers. In the study district five most commonly reported causes of morbidity and mortality are malaria, acute respiratory infections/upper respiratory tract infections, diarrhoea, intestinal worms, and pneumonia [18].

2.2. Study Design. This was a cross-sectional study which was part of a longitudinal randomized intervention trial. This cross-sectional baseline survey assessed the prevalence and intensity of *Schistosoma mansoni* infection among primary schoolchildren in the selected schools.

2.3. Study Population and Inclusion and Exclusion Criteria. The study population consisted of primary schoolchildren

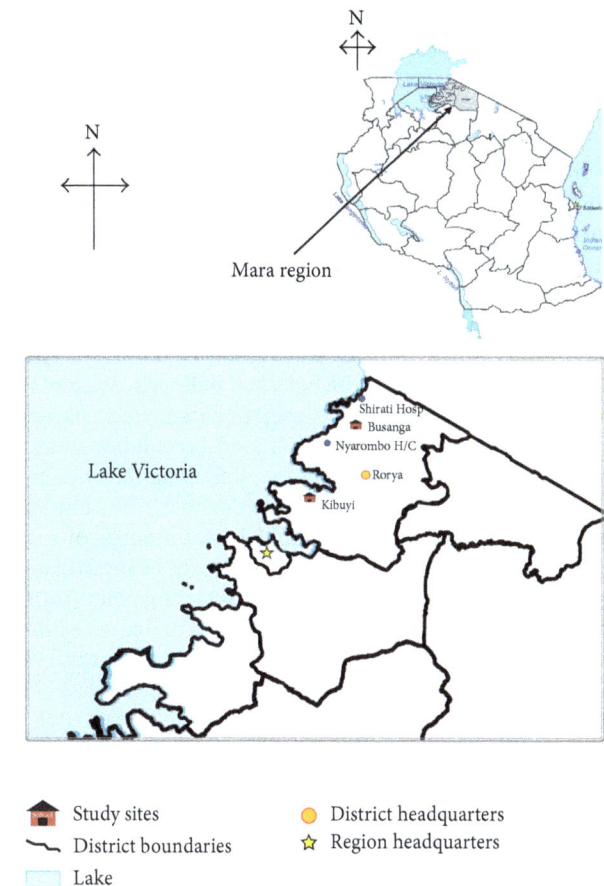

FIGURE 1: The study sites in Rorya District, Tanzania.

aged 6–16 attending pre-grade one to grade six in Busanga and Kibuyi primary schools in two villages of Busanga and Kibuyi, respectively. All schoolchildren between 6 and 16 years of age who agreed to participate in the study and whose parents gave a written informed consent were eligible for the study. Schoolchildren who had a history of being clinically ill during the time of recruitment and those who used antischistosomal drugs within a period of six months before the study and those whose parents refused to sign written informed consent forms were excluded from the study.

2.4. Sample Size Determination and Sampling Procedures. This study was part of a longitudinal interventional study, which aimed at comparing cure rates for two different treatment regimens. Therefore the sample size was calculated using a formula used for comparing two rates [19]. In the calculations we used cure rates reported from a study of communities living along the shores of Lake Albert in Uganda, which reported cure rates of 41.9% and 69.1% for single dose and two-dose treatment regimen, respectively [20]. We set the level of significance at 95% and power of 90%. Adding 30% annual loss to follow-up, a total sample size of 257 per treatment group was required, but we managed to recruit a total of 513 study participants for the entire study.

Conveniently two schools along Lake Victoria shores were selected from two villages, namely, Busanga and Kibuyi.

A total of 246 and 267 schoolchildren were recruited from Busanga and Kibuyi primary schools, respectively. We sampled children from pre-grade one to grade six. Children in grade seven were excluded because they were about to do their final national examinations and they would not be around during the follow-up surveys. The number of schoolchildren selected from each class was determined by the probability proportional to number of children in the class. An attempt was made to sample equal numbers of boys and girls from each class. The total number of schoolchildren selected from each class was determined by the probability proportional to the number of children in the class. Then half of this number was to be boys and half girls. Systematic random sampling method was used to obtain study participants for each sex from each class. The schoolchildren in each class were requested to stand in two lines, one for boys and the other one for girls, and they were counted. The sampling interval was obtained by dividing the total number of each sex in the class with the number of each sex to be investigated from that class (N/n). After obtaining a starting point from a table of random numbers, children were sampled according to the sampling interval. The same interval was kept until the required number of children for each sex in each class was obtained.

2.5. Data Collection

2.5.1. Assessment of Sociodemographic Information and Risk Factors. A pretested Kiswahili translated semistructured questionnaire was used to gather demographic information and risk factors for *S. mansoni* infection. Variables such as age, sex, socioeconomic activities of parents/guardians, sanitary practices, and water contact behaviour were assessed as potential risk factors for the disease. The questionnaire was initially developed in English and then translated to Kiswahili and backtranslated by a different person who was blinded to the original questionnaires.

2.5.2. Stool Sample Collection, Processing, and Examination. A day before stool sample collection, the study objectives were explained to the school teachers and children. Then schoolchildren were provided with informed consent forms to take home to their parents/guardians. They were instructed to tell their parents/guardians to read and understand and then sign if they agree that their children participate in the study. The next morning children with signed written informed consent forms were provided with labelled, small, clean, dried, and leak proof stool containers and clean wooden applicator sticks. Then, they were informed to bring a sizeable stool sample of their own. A single stool sample was collected from all study participants. Each of the specimens was checked for its label, quantity, and procedure of collection. Four Kato-Katz thick smears were prepared from different parts of the single stool sample using a template of 41.7 mg (Vestergaard Frandsen, Lausanne, Switzerland) following a standard protocol [21–23]. Examination of slides for hookworm eggs was performed within 1 hour of slide preparation. Then the slides were transported to the National Institute for Medical Research (NIMR), Mwanza laboratory where they were examined for *S. mansoni* eggs by two experienced laboratory technicians. A Kato-Katz slide prepared for each child was used to determine egg per gram of stool sample (EPG) for *S. mansoni*. For quality assurance, a random sample of 10% of the negative and positive Kato-Katz thick smears was reexamined by a third technician. Since a template delivering 41.7 mg of stool was used to prepare the slides, the eggs of each parasite in the slide were counted and the number of eggs was multiplied by 24 to calculate EPG for each helminth species [21–23]. The intensity of *S. mansoni* infection was calculated based on the intensity classes set by WHO as light (1–99 epg), moderate (100–399 epg), and heavy (epg ≥ 400) [23].

2.6. Data Analysis.
The collected data were entered into a database using EpiData version 3.1. Data analysis was done using STATA version 12.1 (StataCorp, Texas, USA). The chi-square test was used to compare proportions and to test for association between *S. mansoni* infection prevalence and different exposure groups. Parasite counts were normalized by log transformation, averaged, and then back transformed to the original scale. *S. mansoni* infection intensities were calculated as geometric mean of eggs per gram of faeces. The student's *t*-test and one-way analysis of variance (ANOVA) were used to compare geometric mean parasite counts where two or more than two groups were compared, respectively. Logistic regression analysis was performed to determine the independent effect of the independent variables with dependent variable by calculating the strength of the association between intestinal parasites infection and determinant factors using odds ratio (OR) and 95% confidence interval (CI). Crude and adjusted OR were estimated by bivariate and multivariate logistic regression analysis with respective 95% CIs, respectively. A *p* value less than or equal to 0.05 was considered as statistically significant.

2.7. Ethical Statement.
The study was approved by the Medical Research Coordination Committee (MRCC) of the National Institute for Medical Research (NIMR), Tanzania (Reference number NIMR/HQ/R.8a/Vol. IX/1990). The study received further approval from the District Executive Director, District Education Officer, and Medical Officer of Rorya District Council. Before commencement of the study, the research team conducted meetings with the village executive officers, teachers, and students of selected villages and schools, respectively. During these meetings, the objectives of the study, the study procedures to be followed, samples to be taken, study benefits, and potential risks and discomforts were explained. Informed consent for all children who participated in the study was sought from parents and legal guardians by signing an informed consent form. Assent was sought from children who were also informed of their rights to refuse to participate in the study and to withdraw from the study at any time during the study. At baseline, all children were given a standard dose of praziquantel (40 mg/kg) and albendazole (400 mg) as a single dose after stool sample collection. Treatment with praziquantel was given after a meal which was prepared and offered at school to minimize potential side effects. Treatment was performed under direct observation (DOT) of a qualified nurse.

3. Results

3.1. Sociodemographic Characteristics of the Study Participants.

A total of 513 schoolchildren from the two primary schools were enrolled into the study. Of these children, 49.71% ($n = 255$) were boys and 50.29% ($n = 258$) were girls. Of all the study participants 246 (47.95%) and 267 (52.05%) were from Busanga and Kibuyi primary schools, respectively. The numbers of girls and boys in Busanga primary school were 125 (50.81%) and 121 (49.19%), respectively, whereas the numbers of girls and boys in Kibuyi primary school were 133 (49.81%) and 134 (50.19%), respectively. The age of the schoolchildren ranged from 6 to 16 years with the mean of 10.9 (±2.4) years. The number of children at Busanga and Kibuyi primary schools in the age categories was as follows: 6–9 years 87 (56.13%) and 68 (43.87%), respectively; 10–12 years 97 (46.19%) and 113 (53.81%), respectively; and 13–16 years 62 (41.89%) and 86 (58.11%), respectively.

3.2. Prevalence of S. mansoni and Other Soil-Transmitted Helminths (STH) among Primary Schoolchildren at Busanga and Kibuyi Primary Schools.

Overall, 84.01% (431/513) of all the study participants were infected with *S. mansoni*. Other parasites found on Kato-katz technique were hookworms 1.4% (7/513) and *Ascaris lumbricoides* 1.4% (7/513). All children who were positive for *Ascaris lumbricoides* were also positive for *S. mansoni* while six of those with hookworms were also positive for *S. mansoni*. None had both *Ascaris lumbricoides* and hookworm infections. The prevalence of soil-transmitted helminths in this study was too low for any valid statistical analysis to be done.

3.3. Prevalence of S. mansoni Stratified by Demographic Characteristics.

Girls had slightly higher prevalence of *S. mansoni* than boys but the difference was not statistically significant ($p = 0.31$). However the prevalence of infection varied significantly between age groups ($p = 0.004$) with those aged 10–12 having the highest prevalence and those aged 6–9 having the lowest prevalence. There was also a very strong association between infection prevalence and children's village, where children at Busanga village had a significantly higher prevalence of infection as compared to those at Kibuyi village ($p = 0.001$). *S. mansoni* infection seemed to vary significantly with parent's level of education ($p = 0.036$). Toilet use was also associated with *S. mansoni* infection whereby those who reported to use a toilet at home only sometimes had a significantly higher prevalence of infection ($p = 0.01$). Those who reported to visit the lake had a significantly higher prevalence of infection as compared to those who reported not to ($p = 0.018$). Children who reported to have ever had a person with intestinal schistosomiasis at home had a significantly higher prevalence than those who had no history of having a person with intestinal schistosomiasis at home ($p = 0.005$). Children who spent most of their time on the shoreline when at the lake had a significantly higher prevalence of *S. mansoni* infection as compared to those who spent most of their time when at the lake on the inner (deeper) parts of the lake ($p = 0.022$). Table 1 shows prevalence of *S. mansoni* stratified by sociodemographic characteristics of the study participants.

3.4. Intensity of Schistosoma mansoni Infection among Study Participants.

The overall geometrical mean egg per gram of faeces (GM-epg) for individuals with *S. mansoni* infection was 167.13 (95% CI: 147.19–189.79). The GM-epg intensity for Busanga was 203.69 (95% CI: 169.67–244.56) epg and for Kibuyi was 135.98 (95% CI: 114.33–161.73) epg. The distribution of light, moderate, and heavy intensity infection as categorized by WHO was 34.11%, 39.91%, and 25.99%, respectively. Boys had slightly higher GM-epg than girls but the difference was not statistically significant ($p > 0.05$). The geometric mean egg counts per gram of stool seemed to increase across age group with those between 6 and 9 years having the lowest mean epg and those between 13 and 16 years having the highest mean epg, but the observed difference was not statistically significant ($p > 0.05$). Parent's level of education was significantly associated with geometric mean epg whereby children who reported their parents not having any formal education had the highest mean epg than other categories ($p = 0.005$) (Table 2). Children who reported that their parents are fishing had a significantly higher intensity of infection as compared to those whose parents were not involved in fishing ($p < 0.001$). Again parent employment status was significantly associated with intensity of *S. mansoni* infection whereby those children whose parents were not employed had higher intensity as compared to those whose parents were employed ($p = 0.018$). Children who reported to have had a person with intestinal schistosomiasis in their household had significantly higher intensity of infection as compared to those who reported otherwise ($p < 0.001$). The intensity of infection seemed to vary significantly between villages with children at Busanga village bearing higher intensity than those at Kibuyi village ($p = 0.002$). Again children who reported to use the toilet at home only sometimes had a slightly higher intensity of infection as compared to those who use the toilet always, but their difference was not statistically significant. No statistical significant difference in the mean egg intensity between those who reported to visit the lake and those who reported not to visit was observed, though those who visited the lake had a slightly higher mean egg counts (Table 2).

3.5. Prevalence and Intensity of S. mansoni by History of Clinical Morbidity and Treatment History among Study Participants.

S. mansoni infection was more common among children who reported to experience stomach pain in the past two weeks as compared to those who reported not to have stomach pain and the difference was statistically significant ($p = 0.002$). These children also had significantly higher egg intensity than children who reported not to have stomach pain in the past two weeks. History of ever being treated for intestinal schistosomiasis was associated with significantly higher prevalence of *S. mansoni* ($p < 0.001$) (Table 3).

3.6. Determinants of S. mansoni Infection among Study Participants.

On bivariate analysis, children's age, village of residence, parent's level of education, parent reporting fishing,

TABLE 1: Prevalence of *S. mansoni* stratified by sociodemographic characteristics of study participants.

Variable	No examined	Prevalence (%)	P value
Sex (n = 513)			
Male	255	210 (82.35)	
Female	258	221 (85.66)	0.31
Age (in years) (n = 513)			
6–9	155	122 (78.71)	0.004
10–12	210	190 (90.48)	
13–16	148	119 (80.41)	
Village (n = 513)			
Busanga	246	220 (89.43)	0.001
Kibuyi	267	211 (79.03)	
Parent's level of education (n = 488)			
No formal education	48	45 (93.75)	0.036
Primary education	337	290 (86.050)	
Secondary education	58	45 (77.59)	
Collage education	5	5 (100.00)	
University education	1	1 (100.00)	
Do not know	39	28 (71.79)	
Parent is a farmer/livestock keeper (n = 488)			
Yes	221	187 (84.62)	0.90
No	267	227 (85.02)	
Parent is fishing (n = 488)			
Yes	241	212 (87.97)	0.06
No	247	202 (81.78)	
Parent is doing small businesses (n = 488)			
Yes	70	58 (82.86)	0.62
No	418	356 (85.17)	
Parent is employed (n = 488)			
Yes	32	29 (90.63)	0.35
No	456	385 (84.43)	
Use of toilet at home (n = 414)			
Always	229	183 (79.91)	0.01
Only sometimes	185	165 (89.19)	
Visit the lake (n = 488)			
Yes	471	403 (85.56)	0.018
No	17	11 (64.71)	
Part of the lake (n = 370)			
On the shoreline	350	307 (87.71)	0.022
On deeper part of the lake	120	95 (79.17)	
Ever had a person with intestinal schistosomiasis in household (n = 488)			
Yes	251	224 (89.24)	0.005
No	237	190 (80.17)	

p values calculated based on chi-square statistic.

using toilet only sometimes, visiting the lake, spending most of the time along the shoreline when at the lake, history of ever having a patient of intestinal schistosomiasis at home, and history of ever being treated for intestinal schistosomiasis were significantly associated with higher odds of having *Schistosoma mansoni* infection ($p < 0.05$). On multivariate analysis, village of residence, parent level of education, use of toilet at home, and history of ever being treated for intestinal

schistosomiasis remained significant predictors of *S. mansoni* infection after adjusting for age and sex (Table 4).

4. Discussion

Efforts have been made to document the distribution of *Schistosomiasis mansoni* in different parts of Tanzania [16, 24–26]. However there are still many areas whose prevalence and

TABLE 2: Intensity of *Schistosoma mansoni* infection by sociodemographic characteristics among study participants.

Variable	Number	GM-epg	95% CI	P value
Sex (n = 431)				
Male	210	171.23	142.55–205.67	0.716[*]
Female	221	163.34	136.74–195.11	
Age (in years) (n = 431)				
6–9	122	156.67	122.73–198.34	0.769[**]
10–12	190	167.70	138.38–204.38	
13–16	119	177.62	141.17–223.63	
Parent's level of education (n = 414)				
No formal education	45	295.95	164.02–428.38	0.005[**]
Primary education	290	172.94	149.90–200.33	
Secondary education	45	105.30	67.36–164.02	
Collage/university education	6	94.66	89.98–99.34	
Don't know	28	185.56	106.70–323.76	
Parent is a farmer/livestock keeper (n = 414)				
Yes	187	162.80	136.19–194.62	0.402[*]
No	227	181.93	151.13–219.01	
Parent is fishing (n = 414)				
Yes	212	228.53	192.93–270.71	<0.001[*]
No	202	129.21	106.86–156.24	
Parent is doing small businesses (n = 414)				
Yes	58	131.03	87.41–196.43	0.088[*]
No	356	181.05	158.10–207.32	
My parent is employed (n = 414)				
Yes	29	98.39	61.12–158.40	0.0184[*]
No	385	180.54	157.90–206.43	
Ever had a person with intestinal schistosomiasis (n = 414)				
Yes	224	216.41	182.64–256.43	<0.001[*]
No	190	132.91	109.52–161.30	
Use of toilet at home (n = 348)				
Always	183	158.85	131.33–192.14	0.257[*]
Only sometimes	165	187.94	150.19–235.18	
Visit the lake (n = 414)				
Yes	403	174.91	153.43–199.41	0.32[*]
No	11	116.33	46.38–291.74	
Village (n = 431)				
Busanga	220	203.70	169.67–244.56	0.002[*]
Kibuyi	211	135.98	114.33–161.73	

p values = [*]*t*-test and [**]ANOVA.

intensities of infection are yet to be documented. This study attempted to document the prevalence, intensity of infection, and factors associated with intestinal schistosomiasis among primary schoolchildren in two communities in Rorya District that lies along the shores of Lake Victoria, Northwestern Tanzania.

The findings from this study have shown that schistosomiasis due to *Schistosoma mansoni* is highly endemic in the study area. The prevalence of *Schistosoma mansoni* observed among Schoolchildren in the present study was slightly higher compared to what has been reported around Lake Victoria basin, 64.3% [25] and 63.91% [25] in Tanzania, Mbita Island in Western Kenya (60.5%) [27], and Ssese Islands in

Lake Victoria in Uganda (58.1%) [28]. The high prevalence of *Schistosoma mansoni* in the present study is likely to be due to high dependency of the surveyed community on the lake water for different domestic and economic activities and the inadequacy of portable water supply in the area. In addition, the absence of any major control interventions which have been implemented in the study area could further explain the observed high prevalence and intensities of infection. Contrasting findings have been reported on the prevalence of schistosomiasis among boys and girls with some studies reporting boys being more affected by intestinal schistosomiasis than girls [29–32]. In these cases, higher frequency of boys coming into contact with cercaria infested water than

TABLE 3: Prevalence and intensity of *S. mansoni* by history of clinical morbidity and treatment history.

Variable	No examined	Prevalence	P value	GM-epg (95% CI)	P value
Had blood in stool in the past two weeks					
Yes	59	51 (86.44)	0.714[†]	172.12 (149.69–197.91)	0.8318[*]
No	429	363 (84.62)		179.61 (126.41–255.22)	
Stomach pain in the past two weeks (488)					
Yes	286	255 (89.16)	0.002[†]	129.92 (103.46–163.14)	<0.001[*]
No	202	159 (78.71)		206.88 (177.72–240.82)	
Had bloody diarrhoea in the past two weeks					
Yes	51	40 (78.43)	0.178[†]	169.16 (147.42–194.10)	0.2936[*]
No	437	374 (85.58)		213.80 (145.78–313.56)	
Had blood in stool, stomach pain, and bloody diarrhoea in the past two weeks					
Yes	8	6 (75.00)	0.436[†]	171.68 (150.67–195.61)	0.7436[*]
No	479	407 (84.97)		205.55 (57.16–739.1%)	
Ever been treated for intestinal schistosomiasis					
Yes	217	197 (90.78)	<0.001[†]	159.92 (133.77–191.20)	0.4924[**]
No	251	206 (82.07)		187.22 (154.25–227.23)	
I do not know	20	11 (55.00)		162.04 (73.77–355.94)	

p values = χ^2-test[†], *t*-test[*], and ANOVA[**].

TABLE 4: Multivariate logistic regression for factors associated with *Schistosoma mansoni* infection.

Independent variable	Categories	Adjusted OR (95% CI)	P value
Age (in years)	6–9	1	
	10–12	2.24 (0.90–5.55)	0.083
	13–16	0.80 (0.33–1.92)	0.616
Sex	Boys	1	
	Females	0.92 (0.50–1.70)	0.783
Village	Kibuyi	1	
	Busanga	3.30 (1.60–6.89)	0.001
Parent's level of education (*n* = 488)	No formal education	12.52 (1.33–117.80)	0.027
	Primary education	2.76 (1.16–6.61)	0.022
	Secondary education	1	
	Collage/university education	—	—
	Do not know	1.19 (0.34–4.16)	0.782
Parent is fishing	No	1	
	Yes	1.82 (0.94–3.53)	0.076
Use of toilet at home (*n* = 414)	Always	1	
	Only sometimes	2.15 (1.04–4.48)	0.040
Part of the lake	On deeper part of the lake	1	
	On the shoreline	1.45 (0.69–3.06)	0.325
Ever had a patient at home	No	1	
	Yes	1.31 (0.67–2.56)	0.436
Ever been treated for intestinal schistosomiasis	No	1	
	Yes	2.46 (1.190–5.08)	0.015
	Do not know	0.57 (0.13–2.55)	0.466

girls was noted to be the likely cause of the observed difference. Other studies have suggested hormonal differences being the reason for the observed higher prevalence in boys than girls [15] while other studies have also reported the opposite [33–35]. However, our study found a nonsignificant difference in the infection prevalence and intensity between sexes suggesting equal exposure pattern to cercarial infested water among boys and girls in the study area. This contrasting observation calls for further studies to elucidate sex predispositions to *Schistosoma mansoni* infections in endemic areas.

Although age was not retained on multivariate analysis in our study, it has been reported to be a significant predictor of schistosomiasis. Haftu et al. reported that children in the age group 10–14 had relatively higher infection intensities than children below 9 years of age [36]. In our study, this was shown on bivariate analysis where children in the age group of 10–12 had the highest infection prevalence when compared to children in the age group of 6–9. This observation is in liaison with a common theory that in endemic areas infection may start at an early age, increasing and reaching peak at 19 years, after which it starts to decline gradually with an increase in age [37–39].

This study found that the prevalence and infection intensity varied significantly by village with children at Busanga village having significantly higher prevalence and infection intensities as compared to children at Kibuyi village. The variation in infection prevalence and intensities of *S. mansoni* by geographical area has been reported elsewhere, citing variation on intensity of parasite transmission and frequency of exposure to cercariae contaminated water bodies [40]. This observation in our study is likely to be due to a relatively higher dependency of people at Busanga on lake water for domestic and economic uses as compared to Kibuyi and also to differences in the numbers and infection levels in the snails.

It has been reported that one of the primary presenting symptoms for intestinal schistosomiasis is abdominal pain [41], and the key determinants for morbidity progression are repeated infection, intensity of infection, and duration of infection [42–44]. In line with this knowledge, our study found both prevalence and infection intensities to be significantly higher among children who reported to have had stomach pain within a period of two weeks preceding this study as compared to those who did not. It was further noted that children with a history of ever being treated for intestinal schistosomiasis had higher prevalence of infection than those who reported otherwise. This observation is likely to be due to the fact that *S. mansoni* and other intestinal helminths infections in communities tend to be aggregately distributed, with only a few number of individuals harbouring most of the infection in the community, the kind of distribution which is due to host heterogeneities in exposure and susceptibility to infection [45]. These individuals are likely to be reinfected following treatment if there has not been a change in the behaviour thereby altering their exposure pattern.

The findings in this study have shown that almost 26% of the *S. mansoni* infections are heavy intensity infections and close to 40% are of moderate intensity. This pattern of infection has been reported elsewhere [46]. These observed rates of moderate and heavy intensity infections in the study area

are of significant concern owing to the fact that clinical manifestations and other complications related to intestinal schistosomiasis are highly related to the intensity of infections [42, 44]. Though not statistically significant, we found that the intensity of infection increased with age suggesting that the observed infection level is cumulative over a long time period and that there has been no major control intervention in the area.

The present study has further demonstrated that *S. mansoni* geometric mean egg count varies with parent's level of education, whereby children who reported their parents to have no formal education bearing the highest mean egg count per gram of faeces. This observation is comparable to what has been reported elsewhere that father's level of education was significantly associated with infection with *S. mansoni*. Children from illiterate parents have higher chances of being infected as compared to children form literate parents [36, 46]. This observation in our study may be due to the fact that as schistosomiasis is a disease of poverty, it is likely that parents with no formal education are poor and therefore children under their households are living in poverty and therefore more likely to involve themselves in activities that expose them to infections by schistosomiasis, for example, fishing and gardening along the lake shore.

Another study elsewhere in Tanzania, reported a nonsignificant higher *S. mansoni* geometric mean egg count per gram of faeces among children who reported their parents to be involved in fishing activities than those who reported not to [16]. In contrast our study has shown that schoolchildren who reported their parents to be involved in fishing activities had significantly higher mean egg intensity per gram of faeces as compared to those children whose parents do not fish. This observation may be because children of fishing parents are likely to start visiting lakes early in their life and have more frequent visits to the lake as compared to children of nonfishing parents. Further, parent employment status was associated with intensity of infection. Children who reported their parents not to be employed had higher mean parasite egg count per gram of stool compared to children whose parents were employed. This observation is similar to what was reported in Bamako Mali, where parent's occupation was seen to be a significant factor associated with intestinal schistosomiasis with children of nonofficials having higher infection prevalence than officials [47].

The present study investigated important risk factors associated with intestinal schistosomiasis. We found a significant relationship between *Schistosoma mansoni* infection and village where participants lived, parent's level of education, use of toilet at home, and history of ever being treated for intestinal schistosomiasis.

This study demonstrated that parent's level of education was a significant predictor of schistosomiasis whereby children of parents with no any formal education have the highest infection prevalence as compared to children whose parents had secondary education. This observation is similar to what was reported in western Africa where lower education level of the head of household was a significant predictor of schistosomiasis [48]. The present study has further shown that inconsistent use of toilet at home is a significant predictor of schistosomiasis. This observation has been reported by other studies [23, 49]. On visual examination, indiscriminate defecation practice was common in the study area as there were many faecal materials along the lake shoreline. It is apparent that children are more likely to clean themselves in the lake soon after defecation, a practice that could be responsible for the observed higher rates of infection among children who do not always use toilets at home.

Schistosomiasis is a water associated infection. Surprisingly, visiting the lake was not retained in the multivariate logistic regression analysis model as a significant predictor for intestinal schistosomiasis although it was demonstrated to be a significant factor on bivariate analysis. Coming into contact with infested water has also been reported as a significant predictor of *Schistosoma mansoni* infection in other studies [16, 50].

5. Conclusion and Recommendations

The present study has demonstrated that the prevalence and infection intensity of *Schistosoma mansoni* among schoolchildren in the study area are alarmingly high. We found that the village in which the study participant lived, parent's level of education, use of toilet at home, and history of ever being treated for intestinal schistosomiasis were significantly associated with *S. mansoni* infection. We recommend that public health interventions to control the disease should take into consideration the associated risk factors demonstrated by this study.

Abbreviations

NIMR: National Institute for Medical Research
SSA: Sub-Saharan Africa
EPG: Eggs per gram of faeces
GM: Geometric mean.

Additional Points

The datasets during and/or analysed during the current study are available from the corresponding author upon reasonable request.

Ethical Approval

The study was approved by the Medical Research Coordination Committee (MRCC) of the National Institute for Medical Research (NIMR), Tanzania (Reference No. NIMR/HQ/R.8a/Vol. IX/1990). The study received further approval from the District Executive Director, District Education Officer, and Medical Officer of Rorya District Council. Before commencement of the study, the research team conducted meetings with the village executive officers, teachers, and students of selected villages and schools, respectively. During these meetings, the objectives of the study, the study procedures to be followed, samples to be taken, study benefits, and potential risks and discomforts were explained. Informed consent for all children who participated in the study was

sought from parents and legal guardians by signing an informed consent form. Assent was sought from children who were also informed of their rights to refuse to participate in the study and to withdraw from the study at any time during the study. At baseline, all children were given a standard dose of praziquantel (40 mg/kg) and albendazole (400 mg) as a single dose. Treatment with praziquantel was given after a meal, which was prepared and offered at school to minimize potential side effects. Treatment was performed under direct observation (DOT) of a qualified nurse.

Competing Interests

The authors declare that they have no competing interests.

Authors' Contributions

The study was designed by David Z. Munisi and Safari M. Kinung'hi. Data collection was done by David Z. Munisi and Safari M. Kinung'hi. Safari M. Kinung'hi provided scientific guidance in data collection, planning, and implementation of day-to-day field and laboratory activities. David Z. Munisi, Safari M. Kinung'hi, and Emmanuel A. Mpolya participated in data analysis and manuscript preparation. Safari M. Kinung'hi, Joram Buza, and Emmanuel A. Mpolya critically reviewed the manuscript and the interpretation of the results. All authors read and approved the final manuscript.

Acknowledgments

The authors thank the management of the National Institute for Medical Research (NIMR), Mwanza Research Centre, for providing a conducive environment to do laboratory activities. They also thank the District Executive Director, District Medical Officer, and District Education Officer for Rorya District for granting permission to undertake this study. They also thank the head teachers, teachers, and pupils in each of the schools that participated in this study and all the research team members for their assistance in the field work. This study received financial support from the Government of Tanzania through the Nelson Mandela African Institution of Science and Technology and formed part of Ph.D. training programme for David Z. Munisi.

References

[1] B. Senghor, A. Diallo, S. N. Sylla et al., "Prevalence and intensity of urinary schistosomiasis among school children in the district of Niakhar, region of Fatick, Senegal," *Parasites & Vectors*, vol. 7, article 5, 2014.

[2] P. Jordan, "From Katayama to the Dakhla Oasis: the beginning of epidemiology and control of bilharzia," *Acta Tropica*, vol. 77, no. 1, pp. 9–40, 2000.

[3] M. J. van der Werf, S. J. de Vlas, S. Brooker et al., "Quantification of clinical morbidity associated with schistosome infection in sub-Saharan Africa," *Acta Tropica*, vol. 86, no. 2-3, pp. 125–139, 2003.

[4] A. G. P. Ross, P. B. Bartley, A. C. Sleigh et al., "Schistosomiasis," *The New England Journal of Medicine*, vol. 346, no. 16, pp. 1212–1220, 2002.

[5] C. H. King, "Parasites and poverty: the case of schistosomiasis," *Acta Tropica*, vol. 113, no. 2, pp. 95–104, 2010.

[6] T. Harrison, "Schistosomiasis and other trematode infection," in *Harrrison's Principles of Internal Medicine*, pp. 1266–1271, 2005.

[7] B. J. Vennervald, L. Kenty, A. E. Butterworth et al., "Detailed clinical and ultrasound examination of children and adolescents in a *Schistosoma mansoni* endemic area in Kenya: hepatosplenic disease in the absence of portal fibrosis," *Tropical Medicine & International Health*, vol. 9, no. 4, pp. 461–470, 2004.

[8] P. J. Hotez and A. Kamath, "Neglected tropical diseases in sub-Saharan Africa: review of their prevalence, distribution, and disease burden," *PLoS Neglected Tropical Diseases*, vol. 3, no. 8, article e412, 2009.

[9] N. McCreesh and M. Booth, "Challenges in predicting the effects of climate change on *Schistosoma mansoni* and *Schistosoma haematobium* transmission potential," *Trends in Parasitology*, vol. 29, no. 11, pp. 548–555, 2013.

[10] P. Steinmann, J. Keiser, R. Bos, M. Tanner, and J. Utzinger, "Schistosomiasis and water resources development: systematic review, meta-analysis, and estimates of people at risk," *The Lancet Infectious Diseases*, vol. 6, no. 7, pp. 411–425, 2006.

[11] H. D. Mazigo, F. Nuwaha, S. M. Kinung'hi et al., "Epidemiology and control of human schistosomiasis in Tanzania," *Parasites and Vectors*, vol. 5, no. 1, article 274, 2012.

[12] A. Assefa, T. Dejenie, and Z. Tomass, "Infection prevalence of Schistosoma mansoni and associated risk factors among schoolchildren in suburbs of Mekelle city, Tigray, Northern Ethiopia," *Momona Ethiopian Journal of Science*, vol. 5, pp. 174–188, 2013.

[13] M. Beniston, "Climatic change: possible impacts on human health," *Swiss Medical Weekly*, vol. 132, no. 25-26, pp. 332–337, 2002.

[14] E. Cox, *A Text Book of Parasitology*, Blackwell Science Ltd., London, UK, 1993.

[15] N. B. Kabatereine, S. Brooker, E. M. Tukahebwa, F. Kazibwe, and A. W. Onapa, "Epidemiology and geography of *Schistosoma mansoni* in Uganda: implications for planning control," *Tropical Medicine and International Health*, vol. 9, no. 3, pp. 372–380, 2004.

[16] M. Mugono, E. Konje, S. Kuhn, F. J. Mpogoro, D. Morona, and H. D. Mazigo, "Intestinal schistosomiasis and geohelminths of Ukara Island, North-Western Tanzania: prevalence, intensity of infection and associated risk factors among school children," *Parasites & Vectors*, vol. 7, article 612, 2014.

[17] G. C. Webber and B. Chirangi, "Women's health in women's hands: a pilot study assessing the feasibility of providing women with medications to reduce postpartum hemorrhage and sepsis in rural Tanzania," *Health Care for Women International*, vol. 35, no. 7-9, pp. 758–770, 2014.

[18] TDS, "Mara Regional Profile," 2013.

[19] A. Hardon, P. Boonmongkon, P. Streefland et al., *Applied Health Research Manual Anthropology of Health and Health Care*, CIP-Data Konkinklijke Bibliotheek, The Hague, Netherlands, 1994.

[20] N. B. Kabatereine, J. Kemijumbi, J. H. Ouma et al., "Efficacy and side effects of praziquantel treatment in a highly endemic Schistosoma mansoni focus at Lake Albert, Uganda," *Transactions of the Royal Society of Tropical Medicine and Hygiene*, vol. 97, no. 5, pp. 599–603, 2003.

[21] N. Katz, A. Chaves, and J. Pellegrino, "A simple device for quantitative stool thick-smear technique in *Schistosomiasis mansoni*," *Revista do Instituto de Medicina Tropical de Sao Paulo*, vol. 14, no. 6, pp. 397–400, 1972.

[22] WHO, *Basic Laboratory Methods in Medical Parasitology*, World Health Organization, Geneva, Switzerland, 1991.

[23] WHO, *Prevention and Control of Schistosomiasis and Soil-Transmitted Helminthiasis: Report of a WHO Expert Committee*, WHO, 2002.

[24] S. M. Kinung'hi, P. Magnussen, G. M. Kaatano, C. Kishamawe, and B. J. Vennervald, "Malaria and helminth co-infections in school and preschool children: a cross-sectional study in Magu district, North-Western Tanzania," *PLoS ONE*, vol. 9, no. 1, article e86510, 2014.

[25] H. D. Mazigo, R. Waihenya, G. M. Mkoji et al., "Intestinal schistosomiasis: prevalence, knowledge, attitude and practices among school children in an endemic area of north western Tanzania," *Journal of Rural and Tropical Public Health*, vol. 9, pp. 53–60, 2010.

[26] N. J. S. Lwambo, J. E. Siza, S. Brooker, D. A. P. Bundy, and H. Guyatt, "Patterns of concurrent hookworm infection and schistosomiasis in schoolchildren in Tanzania," *Transactions of the Royal Society of Tropical Medicine and Hygiene*, vol. 93, no. 5, pp. 497–502, 1999.

[27] M. R. Odiere, F. O. Rawago, M. Ombok et al., "High prevalence of schistosomiasis in Mbita and its adjacent islands of Lake Victoria, western Kenya," *Parasites and Vectors*, vol. 5, no. 1, article 278, 2012.

[28] C. J. Standley, M. Adriko, F. Besigye, N. B. Kabatereine, and R. J. Stothard, "Confirmed local endemicity and putative high transmission of *Schistosoma mansoni* in the Sesse Islands, Lake Victoria, Uganda," *Parasites and Vectors*, vol. 4, no. 1, article 29, 2011.

[29] W. Tilahun, E. Tekola, S. Teshome et al., "Intestinal parasitic infections in Western Abaya with special reference to Schistosomiasis mansoni," *Ethiopian Journal of Health Development*, vol. 13, pp. 21–26, 1999.

[30] A. Tsehai, W. Tilahun, and D. Amare, "Intestinal parasitism among students in three localities in south Wello, Ethiopia," *Ethiopian Journal of Health Development*, vol. 12, pp. 231–235, 1998.

[31] R. Belay and W. Solomon, "Magnitude of Schistosoma mansoni and intestinal helminthic infections among school children in Wondo-Genet zuria, southern Ethiopia," *Ethiopian Journal of Health Development*, vol. 11, pp. 125–129, 1997.

[32] B. Erko, S. Tedla, and B. Petros, "Transmission of intestinal schistosomiasis in Bahir Dar, northwest Ethiopia," *Ethiopian Medical Journal*, vol. 29, no. 4, pp. 199–211, 1991.

[33] T. Essa, Y. Birhane, M. Endris, A. Moges, and F. Moges, "Current status of *Schistosoma mansoni* infections and associated risk factors among students in Gorgora town, Northwest Ethiopia," *ISRN Infectious Diseases*, vol. 2013, Article ID 636103, 7 pages, 2013.

[34] L. Worku, D. Damte, M. Endris, H. Tesfa, and M. Aemero, "Schistosoma mansoni infection and associated determinant factors among school children in Sanja Town, northwest Ethiopia," *Journal of Parasitology Research*, vol. 2014, Article ID 792536, 7 pages, 2014.

[35] A. Alemu, A. Atnafu, Z. Addis et al., "Soil transmitted helminths and *schistosoma mansoni* infections among school children in zarima town, Northwest Ethiopia," *BMC Infectious Diseases*, vol. 11, article 189, 2011.

[36] D. Haftu, N. Deyessa, and E. Agedew, "Prevalence and determinant factors of intestinal parasites among school children in Arba Minch town, Southern Ethiopia," *American Journal of Health Research*, vol. 2, pp. 247–254, 2014.

[37] A. E. Butterworth, "Immunological aspects of human schistosomiasis," *British Medical Bulletin*, vol. 54, no. 2, pp. 357–368, 1998.

[38] B. Gryseels, "Human resistance to Schistosoma infections: age or experience?" *Parasitology Today*, vol. 10, no. 10, pp. 380–384, 1994.

[39] J. R. Stothard, J. C. Sousa-Figueiredo, M. Betson, A. Bustinduy, and J. Reinhard-Rupp, "Schistosomiasis in African infants and preschool children: let them now be treated!," *Trends in Parasitology*, vol. 29, no. 4, pp. 197–205, 2013.

[40] F. Gashaw, M. Aemero, M. Legesse et al., "Prevalence of intestinal helminth infection among school children in Maksegnit and Enfranz Towns, northwestern Ethiopia, with emphasis on Schistosoma mansoni infection," *Parasites and Vectors*, vol. 8, no. 1, article no. 567, 2015.

[41] T. Elbaz and G. Esmat, "Hepatic and intestinal schistosomiasis: review," *Journal of Advanced Research*, vol. 4, no. 5, pp. 445–452, 2013.

[42] Z. Genming, U. K. Brinkmann, J. Qingwu, Z. Shaoji, L. Zhide, and Y. Hongchang, "The relationship between morbidity and intensity of Schistosoma japonicum infection of a community in Jiangxi Province, China," *Southeast Asian Journal of Tropical Medicine and Public Health*, vol. 28, no. 3, pp. 545–550, 1997.

[43] C. H. King, R. F. Sturrock, H. C. Kariuki, and J. Hamburger, "Transmission control for schistosomiasis—why it matters now," *Trends in Parasitology*, vol. 22, no. 12, pp. 575–582, 2006.

[44] T. Y. Sukwa, M. K. Bulsara, and F. K. Wurapa, "The relationship between morbidity and intensity of *Schistosoma mansoni* infection in a rural Zambian community," *International Journal of Epidemiology*, vol. 15, no. 2, pp. 248–251, 1986.

[45] M. G. Chipeta, B. Ngwira, and L. N. Kazembe, "Analysis of schistosomiasis haematobium infection prevalence and intensity in Chikhwawa, Malawi: an application of a two part model," *PLoS Neglected Tropical Diseases*, vol. 7, no. 3, article e2131, 2013.

[46] H. Sady, H. M. Al-Mekhlafi, M. A. K. Mahdy, Y. A. L. Lim, R. Mahmud, and J. Surin, "Prevalence and associated factors of schistosomiasis among children in Yemen: implications for an effective control programme," *PLoS Neglected Tropical Diseases*, vol. 7, Article ID e2377, 2013.

[47] A. Dabo, A. Z. Diarra, V. Machault et al., "Urban schistosomiasis and associated determinant factors among school children in Bamako, Mali, West Africa," *Infectious Diseases of Poverty*, vol. 4, no. 1, article 4, pp. 1–13, 2015.

[48] B. Matthys, A. B. Tschannen, N. T. Tian-Bi et al., "Risk factors for Schistosoma mansoni and hookworm in urban farming communities in western Côte d'Ivoire," *Tropical Medicine and International Health*, vol. 12, no. 6, pp. 709–723, 2007.

[49] A. H. A. Abou-Zeid, T. A. Abkar, and R. O. Mohamed, "Schistosomiasis and soil-transmitted helminths among an adult population in a war affected area, Southern Kordofan state, Sudan," *Parasites and Vectors*, vol. 5, no. 1, article 133, 2012.

[50] B. Alemayehu and Z. Tomass, "*Schistosoma mansoni* infection prevalence and associated risk factors among schoolchildren in Demba Girara, Damot Woide District of Wolaita Zone, Southern Ethiopia," *Asian Pacific Journal of Tropical Medicine*, vol. 8, no. 6, pp. 457–463, 2015.

4

Silent Human *Trypanosoma brucei gambiense* Infections around the Old Gboko Sleeping Sickness Focus in Nigeria

Karshima Solomon Ngutor,[1] **Lawal A. Idris,**[2] **and Okubanjo Oluseyi Oluyinka**[2]

[1]*Department of Animal Health, Federal College of Animal Health and Production Technology, PMB 001, Vom, Nigeria*
[2]*Department of Veterinary Parasitology and Entomology, Ahmadu Bello University, PMB 1045, Zaria, Nigeria*

Correspondence should be addressed to Karshima Solomon Ngutor; torkarshima@yahoo.co.uk

Academic Editor: Emmanuel Serrano Ferron

Trypanosoma brucei gambiense causes Gambian trypanosomosis, a disease ravaging affected rural parts of Sub-Saharan Africa. We screened 1200 human blood samples for *T. b. gambiense* using the card agglutination test for trypanosomosis, characterized trypanosome isolates with *Trypanosoma gambiense* serum glycoprotein-PCR (TgsGP-PCR), and analyzed our data using Chi square and odds ratio at 95% confidence interval for statistical association. Of the 1200 samples, the CATT revealed an overall infection rate of 1.8% which ranged between 0.0% and 3.5% across study sites. Age and sex based infection rates ranged between 1.2% and 2.3%. We isolated 7 (33.3%) trypanosomes from the 21 seropositive samples using immunosuppressed mice which were identified as *T. b. gambiense* group 1 by TgsGP-PCR. Based on study sites, PCR revealed an overall infection rate of 0.6% which ranged between 0.0% and 1.5%. Females and males revealed PCR based infection rates of 0.3% and 0.8%, respectively. Infection rates in adults (1.3%) and children (0.1%) varied significantly ($p < 0.05$). We observed silent *T. b. gambiense* infections among residents of this focus. Risks of disease development into the second fatal stage in these patients who may also serve as reservoirs of infection in the focus exist.

1. Introduction

Trypanosoma brucei gambiense causes the Gambian sleeping sickness, a very chronic, debilitating, complex, and fatal parasitic zoonosis ravaging affected rural parts of Sub-Saharan Africa. The disease transmitted by tsetse flies is widespread in the Sub-Saharan African region posing serious public health problems in the region and to tourists visiting tropical Africa [1, 2]. It is a prototype of a neglected zoonotic pathogen in terms of drug development and sustainable control programmes.

In natural conditions, transmission of the parasite is cyclical through bites of infected tsetse flies including *Glossina palpalis*, *G. tachinoides*, and *G. fuscipes*. These vectors are especially common at watering places like rivers or lakes where people frequently visit to collect water and do their washing and animals visit to drink water [3]. The parasite is divided into two subtypes; type 1 causes a more chronic disease and represents about 90% of all cases of Gambian trypanosomosis, while type 2 is said to be associated with an acute like disease and represents the remaining 10% of the disease [4, 5].

The pathogen is ranked 9th of the 25 human infectious diseases in Africa based on its socioeconomic impact [6] and is incriminated in over 15,000 new cases yearly with the majority of these cases ending fatally [7]. Seventy million people are continuously exposed to the risk of infection in 38 Sub-Saharan African countries where active transmission is reported and only 5–7% of the population at risk is covered by surveillance [7]. Some of these infections may be latent and may remain unnoticed until they get to the second fatal stage [8].

In Nigeria however, human infection with *T. b. gambiense* in the Gboko sleeping sickness focus was first reported in 1974, and since then no successful attempt has been made to control the disease in the affected region [9]. Considering the fact that this infection is still not yet routinely diagnosed in Nigerian hospitals even in endemic areas and the risk associated with parasites invasion of the central nervous system and producing fatal disease, we designed this study to conduct

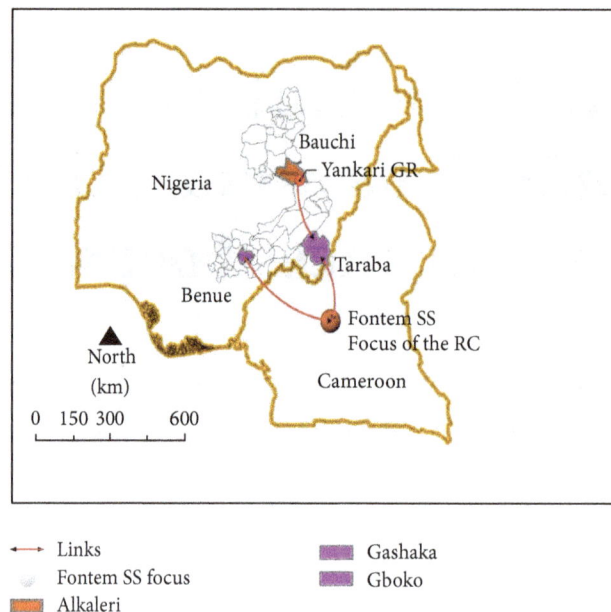

FIGURE 1: Links between the study sites and possible sources of *T. b. gambiense* infections. GR: game reserve; RC: Republic of Cameroon; SS: sleeping sickness.

an active screening of *T. b. gambiense* in humans in the old Gboko sleeping sickness focus in Nigeria and characterized isolates using TgsGP-polymerase chain reaction.

2. Materials and Methods

2.1. Study Area. This study was carried out around the old Gboko sleeping sickness focus which is located in the northern part of Nigeria between longitudes $7°47'$ and $10°00'$ east and latitudes $60°25'$ and $8°8'$ north. It shares boundaries with five other states, namely, Nasarawa (north), Taraba (east), Cross River (south), Enugu (southwest), and Kogi (west), and with the Republic of Cameroon to the southeast (Figure 1). The major occupation in this region is agriculture, particularly crop and livestock farming, as well as fishing.

2.2. Study Design. We conducted a cross-sectional study around the old Gboko sleeping sickness focus in Nigeria within six Local Government Areas (LGAs), namely, Gashaka, Gboko, Ibi, Karim Lamido, Ukum, and Vandeikya, considering their relationships with the Gboko and Fontem sleeping sickness foci and the Gashaka-Gumti and Yankari game reserves. Human subjects were sampled systematically from each household by selecting every 5th household following the order in which houses were built, and individuals were selected using balloting by names.

2.3. Sample Collection

2.3.1. Blood Sampling of Human Subjects. Two millilitres of blood was aseptically collected from each human subject via the median cubital vein using a 5 mL syringe and 21 G needle

and transferred immediately into clean labelled sample bottles containing ethylene diamine tetra-acetic acid (EDTA) at 1.5 mg/mL of blood [10] and gently shaken until the blood was properly mixed with the anticoagulant and analyzed within 1-2 hours after collection [11]. These samples were subjected to the card agglutination test for *T. b. gambiense* as described by Magnus et al. [12]. All CATT positive samples were then inoculated into mice immunosuppressed with cyclophosphamide to isolate trypanosomes.

2.4. Laboratory Analysis of Samples

2.4.1. The Card Agglutination Test for Trypanosomosis (CATT). The CATT test kits were obtained from the Institute for Tropical Medicine, Antwerp, Belgium. Blood samples were analyzed using the CATT as described by Magnus et al. [12]. This test is based on the detection of *T. b. gambiense* specific LiTat 1.3 antibodies using a purified *T. b. gambiense* variable surface antigen.

2.4.2. Mice Infection and Parasitaemia Estimation. Mice were obtained from the Small Animal Experimental Unit of the National Veterinary Research Institute, Vom, Nigeria, for the study. The mice were appropriately labelled and screened for ectoparasites and endoparasites by standard techniques [13] and acclimatized for two weeks before inoculation. The mice were immunosuppressed using intraperitoneal administration of cyclophosphamide at 200 mg per kg body weight. About 0.5 mL of each seropositive blood sample was enriched with 0.5 mL of phosphate saline glucose (PSG) buffer and 0.3 mL of the PSG buffer enriched seropositive blood from each seropositive sample was then inoculated into an intraperitoneally immunosuppressed mouse. Tail-blood samples were collected from each mouse every 24 hours to estimate parasitaemia using the rapid matching technique as described by Herbert and Lumsden [14].

2.4.3. Isolation and Purification of Trypanosomes. The bloodstream form trypanosomes isolated from humans and propagated in mice were separated from the infected mice blood using a DEAE 52 column (Whatman, Maidstone, Kent, UK) as described by Lanham and Godfrey [15]. These parasites were then stored at 4°C until needed for DNA extraction.

2.4.4. Extraction of DNA from Purified Trypanosomes. Trypanosome DNA extraction was done using GeneJET genomic DNA extraction kit (Thermo Scientific, Germany) using the method described by Oury et al. [16]. The extracted DNA was stored at −20°C until needed for PCR at the Biotechnology Laboratory of the Ahmadu Bello University, Zaria, Nigeria.

2.4.5. Detection of Trypanosoma gambiense Specific Glycoprotein (TgsGP). The TgsGP primer set with sequences forward 5′-GCTGCTGTGTTCGGAGAGC-3′ and reverse 5′-GCCATCGTGCTTGCCGCTC-3′ [17] was used to characterize the *Trypanosoma* isolates. Cycling conditions for TgsGP-PCR were denaturation at 98°C for 10 seconds to activate the Phusion Flash II DNA Polymerase, followed by 40 cycles with denaturation at 98°C for 1 second, annealing at 63°C for 30 seconds, 30-second elongation at 72°C, and a final extension at 72°C for 5 minutes as recommended by the manufacturer. All amplified products were analyzed by electrophoresis in a 2% agarose gel and UV illumination after ethidium bromide staining.

2.5. Data Analysis. All data obtained during the study were analyzed using GraphPad Prism 4.0. Infection rates of *T. b. gambiense* were calculated by dividing the number of infected individuals by the total number of individuals examined and expressed as percentages. This was done for different variables such as study sites, sex, and age. The Chi square (χ^2) test and odds ratio were used where appropriate to compare the prevalence rates based on different variables and values of $p < 0.05$ were considered significant.

3. Results

A total of 1200 human blood samples from 6 sites around the old Gboko sleeping sickness focus were analyzed using the CATT and PCR for the presence of *Trypanosoma brucei gambiense* as shown in Table 1.

The CATT revealed an overall infection rate of 1.8% of the 1200 samples studied. Based on the 6 sites studied, Gashaka, Gboko, Ibi, Karim Lamido, Ukum, and Vandeikya recorded infection rates of 3.5% (7/200), 2.5% (5/200), 0.5% (1/200), 2.5% (5/200), 0.0% (0/200), and 1.5% (3/200), respectively (Table 1). Sex based infection rates were 1.2% (7/600) and 2.3% (14/600) for females and males, respectively (Table 2).

Table 1: PCR and CATT based prevalence of *Trypanosoma brucei gambiense* in relation to study sites.

Study sites	Number examined	CATT positive	TgsGP-PCR positive
Gashaka	200	7 (3.5)	3 (1.5)
Gboko	200	5 (2.5)	2 (1.0)
Ibi	200	1 (0.5)	0 (0.0)
Karim Lamido	200	5 (2.5)	1 (0.5)
Ukum	200	0 (0.0)	0 (0.0)
Vandeikya	200	3 (1.5)	1 (0.5)
Total	**1200**	**21 (1.8)**	**7 (0.6)**
χ^2	—	**10.32**	**5.892**
p value	—	**0.066**	**0.3169**

Table 2: PCR and CATT based infection rates of *Trypanosoma brucei gambiense* in relation to sex.

Sex	Number examined	CATT positive	TgsGP-PCR positive
Female	600	7 (1.2)	2 (0.3)
Male	600	14 (2.3)	5 (0.8)
Total	**1200**	**21 (1.8)**	**7 (0.6)**
χ^2	—	**2.375**	**1.293**
p value	—	**0.1233**	**0.2554**
Odds ratio	—	**0.4941**	**0.3980**

Table 3: PCR and CATT based infection rates of *Trypanosoma brucei gambiense* in relation to age.

Age	Number examined	CATT positive	TgsGP-PCR positive
Adult (≥18 years)	450	10 (2.2)	6 (1.3)
Children (≤17)	750	11 (1.5)	1 (0.1)
Total	**1200**	**21 (1.8)**	**7 (0.6)**
χ^2	—	**0.9338**	**6.946**
p value	—	**0.3339**	**0.0082**
Odds ratio	—	**1.527**	**10.08**

Infection rates recorded by adults and children were 2.2% (7/200) and 1.5% (7/200), respectively (Table 3).

Trypanosomes were isolated from 7 (33.3%) of the 21 seropositive samples using immunosuppressed mice. These isolates were characterized using the *Trypanosoma gambiense* serum glycoprotein- (TgsGP-) PCR as group 1 of *T. b. gambiense*. PCR revealed an overall infection rate of 0.6% of the 1200 samples analyzed. PCR based infection rates in relation to study sites were 1.5% (3/200) and 1.0% (2/200) for Gashaka and Gboko, respectively (Table 1). Both Karim Lamido and Vandeikya recorded 0.5% (1/200) while 0% (0/200) infection rates were recorded by both Ibi and Ukum (Table 1). Females and males revealed infection rates of 0.3% (2/600) and 0.8% (5/600) while there was significant variation ($p < 0.05$) between the 1.3% (6/450) and 0.1% (1/750) infection rates recorded by adults and males, respectively.

4. Discussion

In our study, we inoculated human seropositive samples into mice to propagate *T. b. gambiense* considering the low parasitaemia associated with this parasite. Mice were inoculated with cyclophosphamide at 200 mg per kg body weight to suppress mice immunity due to the low parasite isolation rate reported for this parasite [18]. Isolated trypanosomes were also purified to ensure that DNA from mice blood does not contaminate trypanosome DNA which was required for polymerase chain reaction.

We observed silent *T. b. gambiense* infections in this old sleeping sickness focus. Previous report showed that Gboko was endemic for the pathogen [9]. The findings of our study 4 decades after the first report in the region still indicate that the area may still be endemic for HAT and tsetse probably due to lack of or inadequate stakeholders' efforts towards the control and eradication of the disease. This region shares border with the Fontem sleeping sickness focus of the Republic of Cameroon, thus making transboundary transmission a possibility. The riverine nature of this region makes it suitable for tsetse habitat [3] and therefore is a probable reason for the occurrence of this infection. Other factors which include availability of several wildlife species in the Gashaka-Gumti and neighbouring Yankari game reserves and the occupational hazards through farming and fishing [19] might have contributed to the occurrence of the disease. Furthermore, evidence of lack of sustained vector control measures due to lack of funds [20], the high cost and unavailability of trypanocides used against *T. b. gambiense*, and the increasing trend of treatment failure [21] might have also contributed to the occurrence of this infection in the region.

The overall seroprevalence rate of 1.8% revealed by this study is lower than the earlier report by Karshima et al. [22] in a region of Taraba State. The higher prevalence observed in Gashaka may be attributable to the presence of the Gashaka-Gumti National Park which harbours several wildlife species that can serve as reservoirs of *T. b. gambiense*. The extension of the Yankari game reserve forest into the Jevjev and Binari areas of Karim Lamido may explain also the high prevalence observed in the area. The riverine nature of this region might have contributed to the high prevalence observed as riverine areas are known to promote tsetse breeding.

The variations in the prevalence of *T. b. gambiense* in relation to sex are in agreement with the earlier report by Karshima et al. [22], who also reported higher prevalence in males than in females in a region of Taraba State. This may be due to the greater involvement of males in occupations such as farming, fishing, and hunting which put them at more risk than the females. This finding however contradicts the report of Mohammed et al. [23] who reported higher prevalence in females in Southern Sudan probably due to differences in vegetation, tsetse density, and vectorial capacities of tsetse flies in the Sudan. Although there was no significant difference in the seroprevalence in adult and children, the prevalence in children may not be unconnected with their practice of swimming around rivers and streams which may expose them to tsetse bites.

One of the major objectives of this study was to isolate and characterize trypanosome isolates from serologically positive humans. This objective was achieved by the detection of the TgsGP gene which is present in only one strain of the 3 trypanosomes pathogenic to man. This gene is specific to group 1 *T. b. gambiense* and is shown to confer resistance to human serum. We were able to isolate and characterize 7 stocks from 21 CATT seropositive humans. The high specificity of the PCR technique may be a reason for the low infection rate revealed by the technique. On the other hand it was also possible that the CATT detected inactive infections where antibodies have not completely disappeared from the blood.

The epidemiological and clinical significance of detecting only one species of human infective trypanosomes in the study area is that the risk of treatment failure associated with mixed infections of *T. b. rhodesiense* and types 1 and 2 *T. b. gambiense* which respond differently to HAT chemotherapy [24] will be overcome. It was also not surprising to have detected only type 1 *T. b. gambiense* among all human isolates since it has been reported to be associated with over 90% of cases of human African trypanosomosis [4, 5].

In conclusion, we observed silent *T. b. gambiense* infections among residents of this old sleeping sickness focus. Infections were more prevalent in children and males. The epidemiological implication of this finding is the risk of tsetse flies spreading these silent infections to uninfected population.

Consent

Prior to sampling of each human subject verbal consent was sought and individuals aged 18 years and above willing to participate were given written consent to sign. Individuals below the age of 18 years who were willing to participate in the study also signed written consents which were countersigned by at least one of their parents.

Acknowledgment

The authors acknowledged the Centre for Biotechnology, Ahmadu Bello University, Zaria, Nigeria, for allowing them to use the facilities for molecular analysis.

References

[1] J. M. Conway-Klaaseen, J. M. Wyrick-Glatzel, N. Neyrinck, and P. A. Belair, "African sleeping sickness in a young American tourist," *Laboratory Medicine*, vol. 33, pp. 783–788, 2002.

[2] T. Jelinek, Z. Bisoffi, L. Bonazzi et al., "Cluster of African trypanosomiasis in travelers to Tanzanian national parks," *Emerging Infectious Diseases*, vol. 8, no. 6, pp. 634–635, 2002.

[3] MW Service, *A Guide to Medical Entomology*, Blackwell Scientific Publishers, 3rd edition, 1980.

[4] C. Cordon-Obras, C. García-Estébanez, N. Ndong-Mabale et al., "Screening of *Trypanosoma brucei gambiense* in domestic livestock and tsetse flies from an insular endemic focus (Luba, Equatorial Guinea)," *PLoS Neglected Tropical Diseases*, vol. 4, no. 6, article e704, 2010.

[5] C. M. Wombou Toukam, P. Solano, Z. Bengaly, V. Jamonneau, and B. Bucheton, "Experimental evaluation of xenodiagnosis to detect trypanosomes at low parasitaemia levels in infected hosts," *Parasite*, vol. 18, no. 4, pp. 295–302, 2011.

[6] A. Geiger, G. Simo, P. Grébaut, J.-B. Peltier, G. Cuny, and P. Holzmuller, "Transcriptomics and proteomics in human African trypanosomiasis: current status and perspectives," *Journal of Proteomics*, vol. 74, no. 9, pp. 1625–1643, 2011.

[7] OIE, "Standardized techniques for the diagnosis of tsetse transmitted trypanosomiasis," in *OIE Terrestrial Manual*, OIE, Rome, Italy, 2008.

[8] S. L. Wastling, K. Picozzi, C. Wamboga et al., "Latent *Trypanosoma brucei gambiense* foci in Uganda: a silent epidemic in children and adults?" *Parasitology*, vol. 138, no. 12, pp. 1480–1487, 2011.

[9] B. A. Aiyedun and A. A. Amodu, "Human sleeping sickness in the Gboko endemic areas in Nigeria," *Acta Tropica*, vol. 33, no. 1, pp. 88–95, 1974.

[10] S. M. Lewis and C. T. Stoddart, "Effects of anticoagulants and containers (glass and plastic) on the blood count," *Laboratory Practice*, vol. 20, no. 10, pp. 787–792, 1971.

[11] J. B. Kennedy, K. T. Maehara, and A. M. Baker, "Cell and platelet stability in disodium and tripotassium EDTA," *The American Journal of Medical Technology*, vol. 47, no. 2, pp. 89–93, 1981.

[12] E. Magnus, T. Vervoot, and N. Van-Meirvenne, "A card-agglutination test with stained trypanosomes (C.A.T.T.) for the serological diagnosis of T. B. gambiense trypanosomiasis," *Annales de la Société Belge de Médecine Tropicale*, vol. 58, no. 3, pp. 169–176, 1978.

[13] O. W. Schalm, N. C. Jain, and E. J. Carrol, *Veterianry Haematology*, Lea and Febihger Publishers, Philadelphia, Pa, USA, 3rd edition, 1975.

[14] W. J. Herbert and W. H. R. Lumsden, "*Trypanosoma brucei*: a rapid 'matching' method for estimating the host's parasitemia," *Experimental Parasitology*, vol. 40, no. 3, pp. 427–431, 1976.

[15] S. M. Lanham and D. G. Godfrey, "Isolation of salivarian trypanosomes from man and other mammals using DEAE-cellulose," *Experimental Parasitology*, vol. 28, no. 3, pp. 521–534, 1970.

[16] B. Oury, N. Dutrait, B. Bastrenta, and M. Tibayrenc, "*Trypanosoma cruzi*: evaluation of a RAPD synapomorphic fragment as a species-specific DNA probe," *The Journal of Parasitology*, vol. 83, no. 1, pp. 52–57, 1997.

[17] M. Radwanska, F. Claes, S. Magez et al., "Novel primer sequences for polymerase chain reaction-based detection of *Trypanosoma brucei gambiense*," *American Journal of Tropical Medicine and Hygiene*, vol. 67, no. 3, pp. 289–295, 2002.

[18] P. P. Pyana, I. N. Lukusa, D. M. Ngoyi et al., "Isolation of *Trypanosoma brucei* gambiense from cured and relapsed sleeping sickness patients and adaptation to laboratory mice," *PLoS Neglected Tropical Diseases*, vol. 5, no. 4, Article ID e1025, 2011.

[19] H. M. Chapman, S. M. Olson, and D. Trumm, "An assessment of changes in the montane forests of Taraba State, Nigeria, over the past 30 years," *Oryx*, vol. 38, no. 3, pp. 282–290, 2004.

[20] S. O. Omotainse, J. O. Kalejaiye, P. Dede, and A. J. Dadah, "The current status of tsetse and animal trypanosomosis in Nigeria," *Vom Journal of Veterinary Sciences*, vol. 1, no. 1, pp. 1–7, 2004.

[21] D. M. Ngoyi, V. Lejon, P. Pyana et al., "How to shorten patient follow-up after treatment for *Trypanosoma brucei gambiense* sleeping sickness," *The Journal of Infectious Diseases*, vol. 201, no. 3, pp. 453–463, 2010.

[22] N. S. Karshima, I. Ajogi, A. I. Lawal, G. Mohammed, and O. O. Okubanjo, "Detection of *Trypanosoma brucei gambiense* specific LiTat 1.3 antibodies in humans and cattle in Taraba State, North-Eastern Nigeria," *Journal of Veterinary Advances*, vol. 2, no. 12, pp. 580–585, 2012.

[23] Y. O. Mohammed, K. H. Elmali, M. M. Mohammed-Ahmed, and I. Elrayah, "Factors influencing the sero-prevalence of *Trypanosoma brucei* gambiense sleeping sickness in Juba District, Central Equatorial State, Southern Sudan," *Journal of Public Health and Epidemiology*, vol. 2, no. 5, pp. 100–108, 2010.

[24] R. Brun, R. Schumacher, C. Schmid, C. Kunz, and C. Burri, "The phenomenon of treatment failures in Human African Trypanosomiasis," *Tropical Medicine and International Health*, vol. 6, no. 11, pp. 906–914, 2001.

5

Cysticercus fasciolaris in Brown Rats (Rattus norvegicus) in Grenada, West Indies

Ravindra Sharma, Keshaw Tiwari, Kristen Birmingham,
Elan Armstrong, Andrea Montanez, Reneka Guy, Yvette Sepulveda,
Veronica Mapp-Alexander, and Claude DeAllie

School of Veterinary Medicine, St. George's University, West Indies, Grenada

Correspondence should be addressed to Ravindra Sharma; rsharma@sgu.edu

Academic Editor: Bernard Marchand

Cat is the definitive host of *Taenia taeniaeformis (T. taeniaeformis)*. *Cysticercus fasciolaris (C. fasciolaris)*, the larval stage of *T. taeniaeformis*, develops in small rodents which act as intermediate host. The aim of this study was to estimate the prevalence of *C. fasciolaris* in brown rats *(Rattus norvegicus)* in the densely human populated parishes, St. George's and St. David's of Grenada, West Indies. One hundred and seventy rats were trapped near the residential areas from May to July, 2017 and examined for *C. fasciolaris* in their liver. Of the 170 rats 115 (67.6%, CI 95% from 60.1 to 74.6) were positive for the larval stage of *T. taeniaeformis*. One to three cysts were observed in each liver, containing a single larva in each cyst. The prevalence was 77.9% in St. George and 59.1% in St. David which is a significant difference ($p < 0.05$) between the two parishes under study. Based on gender, prevalence in males was 60.9% and females 74.7%. Significant difference was observed between young and adult rats ($p = 0.03$). Prevalence in young rats was 45.0% compared to adults (70.7%). Further study of risk assessment in the cat population in areas of the present research is strongly suggested.

1. Introduction

Taenia taeniaeformis is a cestode parasite found in the intestine of cats as final host. Wild rodents, mainly mice, various species of rats, and voles act as intermediate host for the parasite. The intermediate hosts get infected through ingestion of contaminated feed, water, and beddings from eggs of the parasite voided by cats. Eggs develop into larval form (metacestodes) in the liver of intermediate host. The larval form of *T. taeniaeformis* is called *C. fasciolaris*. *Taenia crassicollis, Hydatigera fasciolaris, Strobilocercus*, and bladder worm are synonyms of *Cysticercus fasciolaris* [1]. *C. fasciolaris* develops mainly in the liver of rodents and contains larval stages of the parasite. Occasionally cysts also develop in the abdominal wall and kidney, filled with purulent exudate without larvae [2]. A small number of fibrosarcoma cases in the liver of rats associated with cysts of *T. taeniaeformis* have been reported [3–5]. Cats get infected by ingestion of rodents infected with *C. fasciolaris*. Although rare, humans get infected with eggs of *T. taeniaeformis* from cats [6].

T. taeniaeformis has been reported in rodents and cats worldwide. The report of *C fasciolaris* particularly, in brown rats *(R. norvegicus)*, is from India [7, 8], Korea [2], Malaysia [9], Serbia [10], and USA [3]. In Grenada, during a survey conducted in 2005 for *Angiostrongylus cantonensis (A. cantonensis)* in lung/heart of *R. norvegicus* [11], lesions of *C. fasciolaris* in the liver of (29.6%) rats were also reported. As far as authors are aware, there is no published report of *C. fasciolaris* in brown rats in other Caribbean nations. The aim of this report is to estimate the prevalence of *C. fasciolaris* in brown rats from Grenada and compare with the previous report.

2. Materials and Methods

2.1. Ethical Approval. The project (Detection of zoonotic pathogens in brown rats in Grenada) was approved by the Institutional Animal Care and Use Committee (IACUC # 16009-R) of St. George's University Grenada.

TABLE 1: Prevalence of *Cysticercus fasciolaris* in brown rats of Grenada.

Parish	Number of rats examined	Number of rats infected	Percentage (%) of rats infected
St. Georges	77	60	77.9%[*]
St. David	93	55	59.1%[*]
Total	170	115	67.6%

[*] *p* value equals 0.0132.

2.2. Study Area. Grenada is the southernmost country in the Caribbean Sea with an area of 348.5 Km^2. The country with low hills, small trees and shrubs, and tropical climate is most suitable for the existence of brown rats. The country is divided into six parishes. The parishes of St. George and St. David were selected for sampling because of their dense human population compared to the other four parishes.

2.3. Species of Rat. Brown rats or Norway rats *(R. norvegicus)* belong to genus *Rattus* under the family Muridae [12]. They are also called brown rats or sewer rats. Brown rats have stocky, gray brown bodies with shorter tail than body length. Brown rats have prominent and pale ears which stick up above the head. Brown rats are larger than most other rat species [13].

2.4. Collection of Rats. One hundred and seventy rats were collected live from 1st May to 14th July, 2017, using live traps (45 cm *l* × 15 cm *w* × 15 cm *h*) with cheese and or various local fruits as bait. Attempts were made to trap the rats near the residential buildings. Trapping in both parishes was conducted near 10-meter periphery of human dwellings. Traps were placed in the evening and visited next day during the morning. Traps with rats were covered with black cloth and transported to the necropsy laboratory of the school of veterinary medicine, St. George's University, Grenada, and transferred to the anesthesia machine. Rats were anesthetized using isoflurane in oxygen via anesthesia machine (portable vet anesthesia machine isoflurane vaporizer VET CE), manufacturer DRE (Avante health Solution Company USA).

2.5. Collection of Samples. The anesthetized rats were examined physically for their health and weighed. The abdominal cavity of rats was opened using a surgical blade and a pair of forceps. Liver, lung, kidney, and abdominal cavity were examined and recorded for gross lesions of *C. fasciolaris*. Those tissues with gross lesions were fixed in 10% neutral buffered formalin, processed for paraffin embedding, sectioned at 4 μm thickness, stained with hematoxylin and eosin, and examined under the light microscope. Before fixation of tissues, the parasites were removed from the cysts and examined. Prevalence of infection was calculated as the number of infected animals divided by the number of examined animals.

2.6. Statistical Analysis. The data was analyzed by the statistical analysis: Fisher's exact test, using graphical statistical software (https://www.graphpad.com/quickcales/contingency2).

3. Results and Discussion

Trapped rats were examined physically for their body condition and signs of illness. Weak and fragile with rough hair coat were the criteria used for illness. No apparent illness was observed in any rat. Previous researchers [10, 14] also reported the healthy physical status of rats in spite of *C. fasciolaris* in their liver.

Out of 170 brown rats examined, 115 showed lesions of *C. fasciolaris* in their liver, giving 67.6% (95% CI from 0.6006 to 0.7461) positivity. The results for the prevalence are included in Table 1. The results showed 77.9% and 59.1% of positive rats in St George's and in St. David's parishes, respectively. Prevalence of *C. fasciolaris* by parish was statistically significant (*p* < 0.05). Risk factors being similar in both parishes, this difference in prevalence is not well explained. Further research involving more number of rats is suggested to answer the difference. Previous researchers reported in brown rats a prevalence of 100% in the Philippines [14], 33.3% in India [15], 33.8% in Korea [2], and 29.9% in Serbia [10]. During a study conducted by Chikweto et al. [11] in Grenada on *A. cantonensis* in brown rats, researchers found 29.6% rats also infected with *C. fasciolaris*. Variations in the prevalence of *C. fasciolaris* in different countries indicate infection risk factors, including seasonal variation in the infection pressure on the intermediate hosts [16]. Prevalence rate found in the present report in Grenada is higher compared to previous finding [11]. Since there is not much variation of the season in Grenada, the higher prevalence found in our study could be the result of the sampling areas in our study. Our samples were obtained from two densely human populated parishes, compared to previous study where samples were from all 6 parishes of the country.

On gross examination of liver of infected rats, one to three cysts were found in each liver (Figure 1). Size of cysts varied from 2.0 mm to 8.0 mm. Color of the cysts ranged from white to grayish white. Each cyst contained single larvae embedded in white turbid color fluid. Larvae were removed from the cyst to study its characteristics. The usual size of the larvae in the present study varied between 6 and 20 cm (Figure 2) but may reach up to 32 cm [17]. Jithendran and Somvanshi [1] in an experimental study showed that the size of the cyst and larvae vary with their stage of development.

Histopathology of the liver showed minimal pathological lesion in the liver parenchyma, except in and around the cysts. The cysts had a central lumen which contained *C. fasciolaris*. The wall of cysts varied in thickness from thin connective tissue capsule in mature *C. fasciolaris* and thick wall of connective tissue in juvenile *C. fasciolaris*. These findings are consistent with Lee et al. [2]. Similar to Jithendran and

TABLE 2: Prevalence of *Cysticercus fasciolaris* in brown rats of Grenada according to gender.

Parish	Male		Female	
	Number of rats examined	Number of rats infected (%)	Number of rats examined	Number of rats infected (%)
St. Georges	39	26 (66.7%)	38	34 (89.5%)
St. David	48	27 (56.3%)	45	28 (62.2%)
Total	87	53 (60.9%)*	83	62 (74.7%)*

* p value equals 0.0711.

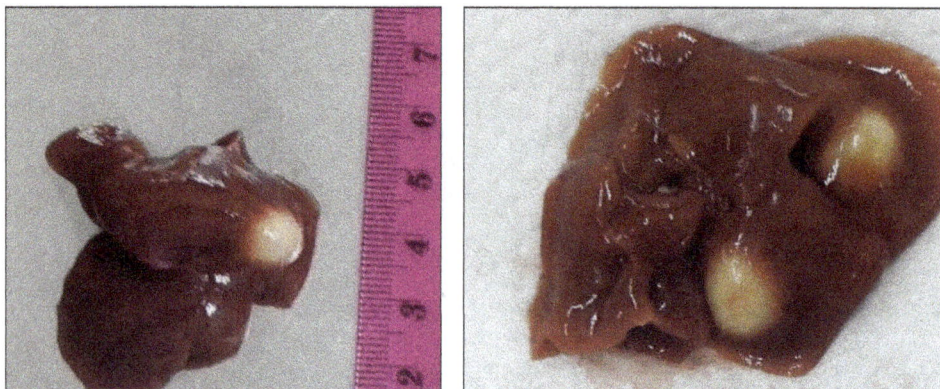

FIGURE 1: Multiple cysts of *C. fasciolaris* in liver.

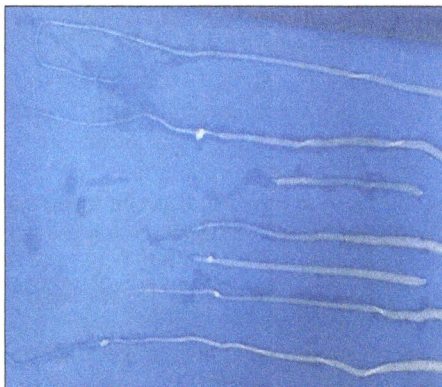

FIGURE 2: The *Cysticercus fasciolaris* taken out from the cysts.

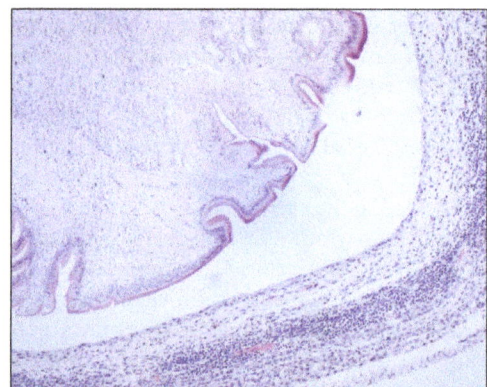

FIGURE 3: Section of liver of *Rattus norvegicus* showing thick capsule with mononuclear cell Infiltration around a *Cysticercus fasciolaris*.

Somvanshi [1] we also found lymphocytic cuffing around the cysts (Figure 3).

The prevalence of *C. fasciolaris* according to gender in the present study is included in Table 2. We report prevalence of *C. fasciolaris* in 66.7% male and 89.5% female in St George's parish compared to 56.3% male and 62.2% female in St David's parish. In our study, there was no significant statistical difference between male and female. Lee et al. [2] and Kataranovski et al. [10] also reported no difference in prevalence of *C. fasciolaris* among male and female rats. However, contrary to our findings Rodríguez-Vivas et al. [18] found higher prevalence in adult male rats. The authors did not explain the reasons for higher prevalence in adult males.

The results for prevalence of *C. fasciolaris* in young and adult rats are tabulated in Table 3. The age of the animals was determined on their weight and size. Rats below 100 G

were grouped as young and over 100 G as adult following the methodology used by previous researchers [16, 19]. The prevalence was 45% in young rats and 70.7% in adult rats. This demonstrates a higher rate of infection in adult rats. The difference in prevalence with age groups was statistically significant ($p < 0.05$). Our observation is in accordance with previous researchers [2, 16, 18]. The reason for higher prevalence in adult rats is not well explained. However, Lee et al. [2] indicated that positivity in adults may be reflecting the accumulation of infection with age. To answer this question further research is suggested.

The Grenadian community likes cats as a pet. However, these pets are not always confined inside the home resulting in roaming behavior near and around the residential areas. The population of cats in the study areas of St. George's and

TABLE 3: Prevalence of *Cysticercus fasciolaris* in brown rats of Grenada according to age.

Parish	Young		Adult	
	Number of rats examined	Number of rats infected	Number of rats examined	Number of rats infected
St. Georges	13	9 (69.2%)	64	51 (79.7%)
St. David	7	0 (0.0%)	86	55 (64.0%)
Total	*20*	*9 (45.0%)**	*150*	*106 (70.7%)**

*p value equals 0.0389.

St. David is not known. Since rats are the intermediate host and are the final host for *T. taeniaeformis*, there is a need of risk assessment of the rat as well as cat population in these two parishes. This study has found strong evidence to educate the community regarding proper maintenance of hygienic conditions in and around their dwellings to prevent the survival and proliferation of the rat population.

Acknowledgments

The authors thankfully acknowledge the funding for the project from One Health One Medicine (OHRI Grant 06-14-10) of St. George's University. Technical assistance of Ray Samuel is appreciated.

References

[1] K. P. Jithendran and R. Somvanshi, "Experimental infection of mice with Taenia taeniaformis eggs from cats - Course of infection and pathological studies," *Indian Journal of Experimental Biology (IJEB)*, vol. 36, no. 5, pp. 523–525, 1998.

[2] B.-W. Lee, B.-S. Jeon, H.-S. Kim, H.-C. Kim, and B.-I. Yoon, "Cysticercus fasciolaris infection in wild rats (Rattus norvegicus) in Korea and formation of cysts by remodeling of collagen fibers," *Journal of Veterinary Diagnostic Investigation*, vol. 28, no. 3, pp. 263–270, 2016.

[3] R. Armando, W. Alexander, and B. Matthew, "Taenia taeniaeformis- induces metastatic sarcoma in a pet rat (Rattus norvegicus)," *Journal of Exotic pet medicine*, vol. 16, no. 1, pp. 45–48, 2007.

[4] M. A. Hanes and L. J. Stribling, "Fibrosarcomas in two rats arising from hepatic cysts of *Cysticercus fasciolaris*," *Veterinary Pathology*, vol. 32, no. 4, pp. 441–444, 1995.

[5] M. Kumar, P. L. Reddy, V. Aparna et al., "Strobilocercus fasciolaris infection with hepatic sarcoma and gastroenteropathy in a Wistar colony," *Veterinary Parasitology*, vol. 141, no. 3-4, pp. 362–367, 2006.

[6] S. Ekanayake, N. D. Warnasuriya, P. S. Samarakoon, H. Abewickrama, N. D. Kuruppuarachchi, and A. S. Dissanaike, "An unusual 'infection' of a child in Sri lanka, with Taenia taeniaeformis of the cat," *Annals of Tropical Medicine and Parasitology*, vol. 93, no. 8, pp. 869–873, 1999.

[7] R. Somvanshi, G. S. C. Ganga Rao, and R. Laha, "Pathological changes associated with spontaneous Cysticercus fasciolarisinfected wild rats," *Indian Journal of Comparative Microbiology, Immunology and Infectious Diseases*, vol. 15, pp. 58–60, 1994.

[8] S. R. Ramteke, V. S. Bhaygude, and G. K. Sawale, "Occurrence of Cysticercus fasciolaris infection in stray rats (Rattus norvegicus) in Mumbai (Maharastra)," *Indian Veterinary Journal*, vol. 94, pp. 14-15, 2017.

[9] M. Elizabeth, H. Kohn, I. Camichael et al., "Larvae of Taenia Taeniaeformisin the liver of a laboratory rat (Rattus norvegicus)," *Annals of Clinical Pathology*, vol. 2, no. 3, p. 1028, 2014.

[10] M. Kataranovski, L. Zolotarevski, S. Belij et al., "First record of Calodium hepaticum and Taenia taeniaeformis liver infection in wild Norway rats (Rattus norvegicus) in Serbia," *Archives of Biological Sciences*, vol. 62, no. 2, pp. 431–440, 2010.

[11] A. Chikweto, M. I. Bhaiyat, C. N. L. Macpherson et al., "Existence of Angiostrongylus cantonensis in rats (Rattus norvegicus) in Grenada, West Indies," *Veterinary Parasitology*, vol. 162, no. 1-2, pp. 160–162, 2009.

[12] "An Age: the animal aging and longevity database," http://www.genomics.Senescence.info/species/entry.php?species-Rattus-norvegicus, 23rd October 2017.

[13] "Rattus nor ve gicus: wild about gardens," http://www.wild-aboutgardens.org.UK/wildlife/mammals/rats-brown, visited 23rd October 2017.

[14] F. G. Claveria, J. Causapin, M. A. de Guzman, M. G. Toledo, and C. Salibay, "Parasite biodiversity in Rattus spp caught in wet markets," *The Southeast Asian Journal of Tropical Medicine and Public Health*, vol. 36, pp. 146–148, 2005.

[15] L. D. Singla, N. Singla, V. R. Parshad, P. D. Juyal, and N. K. Sood, "Rodents as reservoir of parasites in India," *Integrative Zoology*, vol. 3, no. 1, pp. 21–26, 2008.

[16] P. Burlet, P. Deplazes, and D. Hegglin, "Age, season and spatiotemporal factors affecting the prevalence of echinococcus multilocularis and taenia taeniaeformis in arvicola terrestris," *Parasites & Vectors*, vol. 4, no. 1, article no. 6, 2011.

[17] T. C. Cheng, *General Parasitology*, Academic Press, Orlando, 2nd edition, 1991.

[18] R. I. Rodríguez-Vivas, J. A. Panti-May, J. Parada-López, S. F. Hernćndez-Betancourt, and H. A. Ruiz-Piña, "The occurrence of the larval cestode Cysticercus fasciolaris in rodent populations from the Cuxtal ecological reserve, Yucatan, Mexico," *Journal of Helminthology*, vol. 85, no. 4, pp. 458–461, 2011.

[19] J. A. Panti-May, S. Hernández-Betancourt, H. Ruíz-Piña, and S. Medina-Peralta, "Abundance and population parameters of commensal rodents present in rural households in Yucatan, Mexico," *International Biodeterioration & Biodegradation*, vol. 66, no. 1, pp. 77–81, 2012.

6

Molecular Evidence of High Proportion of *Plasmodium vivax* Malaria Infection in White Nile Area in Sudan

Makarim M. Adam Suliman,[1] **Bushra M. Hamad,**[1]
Musab M. Ali Albasheer,[1] **Maytha Elhadi,**[1,2] **Mutaz Amin Mustafa,**[1,2]
Maha Elobied,[3] **and Muzamil Mahdi Abdel Hamid**[1]

[1]*Department of Parasitology and Medical Entomology, Institute of Endemic Diseases, University of Khartoum, Khartoum, Sudan*
[2]*Faculty of Medicine, University of Khartoum, Khartoum, Sudan*
[3]*Faculty of Pharmacy, Al-Neelain University, Sudan*

Correspondence should be addressed to Muzamil Mahdi Abdel Hamid; mahdi@iend.org

Academic Editor: Emmanuel Serrano Ferron

Plasmodium falciparum is a predominant malaria species that infects humans in the African continent. A recent WHO report estimated 95% and 5% of *P. falciparum* and *P. vivax* malaria cases, respectively, in Sudan. However many laboratory reports from different areas in Sudan indicated otherwise. In order to verify, we selected four hundred suspected malaria cases from Aljabalain area located in the White Nile state, central Sudan, and diagnosed them with quality insured microscopy and species-specific nested PCR. Our results indicated that the proportion of *P. vivax* infections among suspected malaria cases was high. We found that on average 20% and 36.5% of malaria infections in both study areas were caused by *P. vivax* using both microscopy and PCR, respectively. This change in pattern is likely due to the recent demographic changes and high rate of immigration from neighbouring countries in the recent years. This is the first extensive clinical study of its kind that shows rising trend in *P. vivax* malaria cases in White Nile area, Sudan.

1. Background

Malaria is an infectious disease of humans and other animals caused by *Plasmodium* parasites [1]. According to the latest WHO estimates, there were about 198 million cases of malaria in 2013 (with an uncertainty range of 124 million to 283 million) and an estimated 584 000 deaths (with an uncertainty range of 367 000 to 755 000) in the world [2]. Malaria mortality rates have fallen by 47% globally since 2000 and by 54% in the WHO African Region [2].

Five species of plasmodia can infect humans. The vast majority of deaths in Sub-Saharan regions are caused by *P. falciparum*, while *P. vivax*, *P. ovale*, *P. malariae*, and *P. knowlesi* cause generally milder form of malaria which other than *P. knowlesi* is rarely fatal [1, 3].

P. vivax is responsible for most malaria cases in Asia and Latin America but it is almost absent from most of central Africa due to the absence of Duffy antigen, the receptor which *P. vivax* uses to invade human erythrocytes [4]. In eastern and southern Africa, *P. vivax* represents around 10% to 40% of malaria cases but <1% of cases in western and central Africa [4, 5].

In Sudan until recently the majority of malaria cases were caused by *P. falciparum*. *P. vivax* is relatively rare; 95% of cases are caused by *P. falciparum* and the other 5% are caused by *P. vivax* [6]. However, in recent years many clinicians observed recurrent relapses of malaria infections in different areas in Sudan suggesting perhaps a higher than expected transmission of non-falciparum malaria parasites (most likely *P. vivax* since it is the second most important malaria parasite species in Sudan). The objective of this study was to document the suggested rise in the proportion of *P. vivax* infections among suspected malaria cases in White Nile state in Sudan.

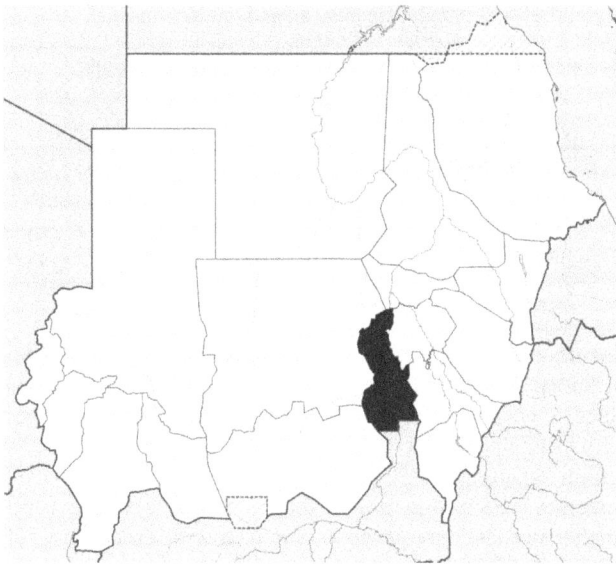

FIGURE 1: Study site: White Nile area in Sudan.

TABLE 1: The mean age, gender and haemoglobin level in Aljabalain hospital and military hospital.

	Aljabalain hospital	Military hospital
Number of patients	200	200
Mean age ± SD, range (years)	30 ± 15 (2–80)	31 ± 13 (1–70)
Less than 15 (%)	19 (9.5%)	17 (8.5%)
15–30 (%)	93 (46.5%)	92 (46.3%)
More than 30 (%)	88 (44%)	90 (45.2%)
Gender		
Males (%)	70 (34.7%)	128 (64%)
Females (%)	130 (64.4%)	72 (36%)
Mean Hb level ± SD, range (g/dL)	11.8 ± 2.2 (4–18)	12 ± 2.1 (5–19)
Anaemia* (%)	135 (67.5%)	146 (73%)
Parasitaemia**		
Low (%)	180 (90%)	185 (92.5%)
High (%)	20 (10%)	15 (7.5%)

* Anaemia was defined as Hb level less than 13.5 g/dL for males and 12.5 g/dL for females.
** Low parasitaemia was defined as number of asexual parasites ≤500/μL of blood. High parasitaemia: >500 asexual parasites/μL of blood.

2. Materials and Methods

2.1. Ethical Considerations. The study was approved by the ethical committee of the Institute of endemic diseases, University of Khartoum. Informed consent was obtained from each patient before participation in the study.

2.2. Study Area and Sample Collection. This study was a cross-sectional study carried out in the White Nile area which is one of the central Sudan states, Figure 1. It lies between latitudes 33 and 30-31 north and longitude 13 and 30–12 east, occupying an area of around 675000 km², with population of 1.675 million. The study was conducted in Aljabalain area, 80 km south of Rabak town, the capital of White Nile state. The area is considered mesoendemic for malaria; transmission follows mainly the rainy season (July to October). Four hundred suspected malaria cases based on patient's symptoms (fever, headache, sweating, nausea, and vomiting) and signs (pyrexia, pallor) were chosen randomly from Aljabalain hospital and Aljabalain military hospital aged >1 year old regardless of gender. Samples were collected during the rainy season (July–September) in 2012. All suspected malaria cases were treated with a standard regimen (artemisinin based combination therapy) by the local malaria control program according to the Sudanese Malaria Treatment Guidelines in Sudan National Malaria Control Program (NMCP), and the National Protocol for Treatment of Malaria (Federal Ministry of Health, Khartoum, Sudan, 2013, unpublished).

2.3. Diagnosis of Malaria. Two and a half mL of venous blood was obtained from each patient. Malaria was diagnosed using blood film microscopy and confirmed with PCR. Both thick and thin blood films were used, fields were read at least twice, and the procedure was followed according to quality control guidelines of WHO. PCR was done for both *P. falciparum* and *P. vivax*. The PCR was performed at the

Institute of Endemic Diseases, University of Khartoum, with quality control in place (both positive and negative control were used). Parasite genomic DNA was extracted from whole blood samples using Chelex method. A fragment of the plasmodial 18S rRNA gene was amplified by PCR and species identification was performed with species-specific oligoprobes using the following primers: for *P. falciparum*, rPLU5: 5′ CTTGTTGTTGCCTTAAACTTC-3′, rPLU6: 5′-TTAAAATTGTTGCATTAAAACG-3′; for *P. vivax*, rVIV1: 5′-CGCTTCTAGCTTAACCACATAACTGATAC-3′, rVIV2: 5′-ACTTCCAAGCCGAAG CAAAGA AAG TCC TTA-3′, as described previously [7].

2.4. Statistical Analysis. Data were analyzed using SPSS (statistical package for the social sciences) version twentieth software.

3. Results

In our study, both males and females are affected by malaria; however more females were represented in Aljabalain hospital (64.4%) and more males were represented from Aljabalain military hospital (64%). The average haemoglobin level in all patients from the two study sites was 11.9 g/dL. The prevalence of anaemia among malaria patients was high, more than two-thirds in both study sites. The majority of malaria cases (more than 90%) had low parasite level (≤500 parasite/μL of blood), Table 1.

In both study sites, the proportion of *P. vivax* infections among suspected malaria cases was high. Microscopy results showed that 30 (15%) and 50 (25%) of malaria infections in Aljabalain hospital and Aljabalain military hospital were

TABLE 2: Positive results of blood films and PCR for *P. falciparum* and *P. vivax* in Aljabalain hospital and military hospital.

	Aljabalain hospital (200)		Military hospital (200)	
	Microscopy	PCR	Microscopy	PCR
	N (%)	N (%)	N (%)	N (%)
P. falciparum	107 (53.5%)	115 (57.5%)	93 (46.5%)	112 (56%)
P. vivax	30 (15%)	66 (33%)	50 (25%)	80 (40%)
Mixed (*P. falciparum* + *P. vivax*)	0 (0%)	5 (2.5%)	0 (0%)	4 (2%)
Negative	63 (31.5%)	14 (7%)	57 (28.5%)	4 (2%)

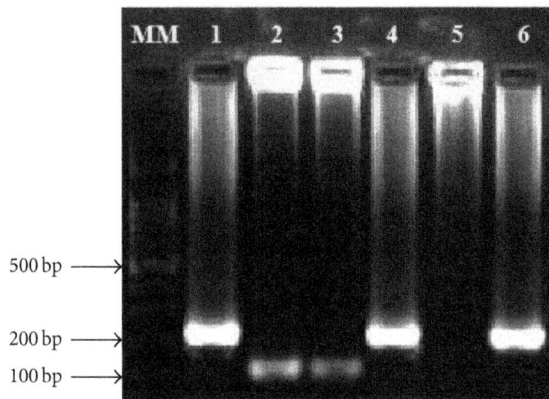

FIGURE 2: Detection of *Plasmodium* 18S rRNA gene using nested PCR from Sudanese malaria patients; MM: 100 bp ladder (iNtRON Biotechnology, South Korea), 1, 4, 6: *P.* falciparum, 2, 3: *P. vivax*, and 5: negative control.

caused by *P. vivax*. The results were even higher with PCR (Figure 2); 66 (33%) and 80 (40%) of samples from Aljabalain hospital and Aljabalain military hospital were positive for *P. vivax*, respectively. Mixed infections (*P. falciparum* + *P. vivax*) were detected in 2.25% of samples in average from both study areas, Table 2.

4. Discussion

This study was carried out in an area characterized by seasonal and unstable malaria transmission. The most remarkable result in this study was the unexpected high proportion (about 40% by PCR) of *P. vivax* infections among suspected malaria cases, eight times more than that previously reported in Sudan [4]. This change in pattern is most likely due to the recent varied composition of the community resulting from several migrations of people from several Asian and African countries to work at petroleum and new sugar companies in White Nile area especially from Ethiopia where high prevalence of *P. vivax* infection (31%) among malaria cases was found [6].

Another suggested explanation for the emergence of *P. vivax* is parasite's development of alternative mechanisms to invade human erythrocytes other than the Duffy antigen. This is a plausible explanation since *P. vivax* infection of Duffy negative genotype was reported previously in many African countries [8–16]. This is the first study of its kind to

document the significant rise in malaria *P. vivax* transmission in Sudan. And it has important health policy implications since *P. vivax* infection requires eradication of liver stages with primaquine due to presence of dormant hypnozoites within hepatocytes [17, 18]. Recent data showed that *P. vivax* infection is becoming more severe especially in children, and further studies are required to understand the exact causes of this pattern [19].

The sampled population (four hundred cases) of this study was selected based on their highly suspected symptoms of malaria and positive blood film microscopy done in hospitals' laboratories. Blood films were later proved only 50% positive in more quality insured settings. This obvious result of malaria overdiagnosis by a factor of two is probably due to lack of training in general hospitals, low quality control microscopy, and high work load. The same result was previously found in Tanzania where malaria was overdiagnosed by a factor of five [20, 21].

PCR diagnosis for malaria is accurate especially for differentiating between plasmodia species, but it is expensive and needs well-trained personnel. In this study quality control microscopy allowed for 20% of *Plasmodium vivax* diagnosis, while by PCR it was 38.7%. This may be due to difficulty in differentiation between species based on a single ring form especially for untrained personnel. In this study more than 90% of slides had low parasite density (≤ 500 parasite/μL); this also makes differentiation between plasmodia species very difficult.

5. Conclusion

Our study confirmed the observed high percentage of *P. vivax* infections in White Nile area, central Sudan. This result has important implications for the malaria control and necessitates modification of current guidelines for the treatment of malaria in Sudan.

Abbreviations

BFFM: Blood film for malaria
PCR: Polymerase chain reaction
Pf: *Plasmodium falciparum*
Pv: *Plasmodium vivax*.

Competing Interests

All authors declare no competing interests.

Authors' Contributions

Muzamil Mahdi Abdel Hamid, Makarim M. Adam Suliman, and Bushra M. Hamad made substantial contributions to the conception and the design of the study. Musab' M. Ali Albasheer did the lab work. Analysis and interpretation of data was done by Muzamil Mahdi Abdel Hamid, Makarim M. Adam Suliman, and Bushra M. Hamad. Muzamil Mahdi Abdel Hamid, Maha Elobied, and Mutaz Amin Mustafa revised the manuscript and Mutaz Amin Mustafa wrote it. Muzamil Mahdi Abdel Hamid gave final approval of the version to be published; all authors read and approved the final manuscript.

Acknowledgments

The authors send their gratitude to the patients and families who participated in this study. This study was funded by TWAS research Grant agreement no. 13-145 RG/BIO/AF/AC_G.

References

[1] Centers for Disease Control and Prevention, "Malaria Worldwide," September 2015 http://www.cdc.gov/malaria/malaria_worldwide/index.html.

[2] WHO, "Malaria," 2015, http://www.who.int/mediacentre/factsheets/fs094/en/.

[3] J. K. Baird, "Evidence and implications of mortality associated with acute plasmodium vivax malaria," *Clinical Microbiology Reviews*, vol. 26, no. 1, pp. 36–57, 2013.

[4] K. Mendis, B. J. Sina, P. Marchesini, and R. Carter, "The neglected burden of *Plasmodium vivax* malaria," *The American Journal of Tropical Medicine and Hygiene*, vol. 64, no. 1-2, supplement, pp. 97–106, 2001.

[5] L. Golassa, F. N. Baliraine, N. Enweji, B. Erko, G. Swedberg, and A. Aseffa, "Microscopic and molecular evidence of the presence of asymptomatic *Plasmodium falciparum* and *Plasmodium vivax* infections in an area with low, seasonal and unstable malaria transmission in Ethiopia," *BMC Infectious Diseases*, vol. 15, no. 1, article 310, 2015.

[6] E. Lo, D. Yewhalaw, D. Zhong et al., "Molecular epidemiology of Plasmodium vivax and Plasmodium falciparum malaria among duffy-positive and duffy-negative populations in Ethiopia," *Malaria Journal*, vol. 14, article 84, 2015.

[7] G. Snounou and B. Singh, "Nested PCR analysis of Plasmodium parasites," *Methods in molecular medicine*, vol. 72, pp. 189–203, 2002.

[8] K. Sondén, E. Castro, L. Trönnberg, C. Stenström, A. Tegnell, and A. Färnert, "High incidence of Plasmodium vivax malaria in newly arrived Eritrean refugees in Sweden since may 2014," *Eurosurveillance*, vol. 19, no. 35, pp. 1–4, 2014.

[9] H. M. Mathews and J. C. Armstrong, "Duffy blood types and vivax malaria in Ethiopia," *American Journal of Tropical Medicine and Hygiene*, vol. 30, no. 2, pp. 299–303, 1981.

[10] D. Ménard, C. Barnadas, C. Bouchier et al., "Plasmodium vivax clinical malaria is commonly observed in Duffy-negative Malagasy people," *Proceedings of the National Academy of Sciences of the United States of America*, vol. 107, no. 13, pp. 5967–5971, 2010.

[11] C. Mendes, F. Dias, J. Figueiredo et al., "Duffy negative antigen is no longer a barrier to Plasmodium vivax—molecular evidences from the African West Coast (Angola and Equatorial Guinea)," *PLoS Neglected Tropical Diseases*, vol. 5, no. 6, article e1192, 2011.

[12] O. Mercereau-Puijalon and D. Ménard, "Plasmodium vivax and the Duffy antigen: a paradigm revisited," *Transfusion Clinique et Biologique*, vol. 17, no. 3, pp. 176–183, 2010.

[13] H. G. Ngassa Mbenda and A. Das, "Molecular evidence of *Plasmodium vivax* mono and mixed malaria parasite infections in Duffy-negative native Cameroonians," *PloS one*, vol. 9, no. 8, Article ID e103262, 2014.

[14] J. R. Ryan, J. A. Stoute, J. Amon et al., "Evidence for transmission of Plasmodium vivax among a Duffy antigen negative population in Western Kenya," *American Journal of Tropical Medicine and Hygiene*, vol. 75, no. 4, pp. 575–581, 2006.

[15] T. G. Woldearegai, P. G. Kremsner, J. R. F. J. Kun, and B. Mordmüller, "Plasmodium vivax malaria in duffy-negative individuals from Ethiopia," *Transactions of the Royal Society of Tropical Medicine and Hygiene*, vol. 107, no. 5, pp. 328–331, 2013.

[16] N. Wurtz, K. Mint Lekweiry, H. Bogreau et al., "Vivax malaria in Mauritania includes infection of a Duffy-negative individual," *Malaria Journal*, vol. 10, article 336, 2011.

[17] D. Fernando, C. Rodrigo, and S. Rajapakse, "Primaquine in vivax malaria: an update and review on management issues," *Malaria Journal*, vol. 10, article 351, 2011.

[18] L. Hulden and L. Hulden, "Activation of the hypnozoite: a part of *Plasmodium vivax* life cycle and survival," *Malaria Journal*, vol. 10, article 90, 2011.

[19] H. Mahgoub, G. I. Gasim, I. R. Musa, and I. Adam, "Severe *Plasmodium vivax* malaria among Sudanese children at New Halfa Hospital, Eastern Sudan," *Parasites & Vectors*, vol. 5, article 154, 2012.

[20] K. Harchut, C. Standley, A. Dobson et al., "Over-diagnosis of malaria by microscopy in the Kilombero Valley, Southern Tanzania: an evaluation of the utility and cost-effectiveness of rapid diagnostic tests," *Malaria Journal*, vol. 12, no. 1, article 159, 2013.

[21] H. Reyburn, R. Mbatia, C. Drakeley et al., "Overdiagnosis of malaria in patients with severe febrile illness in Tanzania: A Prospective Study," *British Medical Journal*, vol. 329, no. 7476, pp. 1212–1215, 2004.

7

Neobenedenia melleni Parasite of Red Snapper, *Lutjanus erythropterus*, with Regression Statistical Analysis between Fish Length, Temperature, and Parasitic Intensity in Infected Fish, Cultured at Jerejak Island, Penang, Malaysia

Rajiv Ravi and Zary Shariman Yahaya

School of Biological Sciences, Universiti Sains Malaysia, Minden, 11800 Penang, Malaysia

Correspondence should be addressed to Rajiv Ravi; rajiv_ravi86@yahoo.com

Academic Editor: Bernard Marchand

The fish parasites collected from *Lutjanus erythropterus* fish species showed a correlation with parasitic intensity, fish size, and temperature, and statistical model summary was produced using SPSS version 20, statistical software. Statistical model summary concluded that among the variables which significantly predict the prevalence of *Neobenedenia melleni* parasites are fish length and water temperature, both significant at 1% and 5%. Furthermore, the increase in one unit of fish length, holding other variables constant, increases the prevalence of parasite by approximately 1 ($0.7 \approx 1$) unit. Also, increasing the temperature from 32°C to 33°C will positively increase the number of parasites by approximately 0.32 units, holding other variables constant. The model can be summarized as estimated number of *Neobenedenia melleni* parasites = 8.2 + 0.7 ∗ (fish length) + 0.32 ∗ (water temperature). Next, this study has also shown the DNA sequence and parasitic morphology of *Neobenedenia melleni*. Nucleotide sequence for 18s ribosomal gene RNA in this study showed 99% similarity with *N. melleni* EU707804.1 from GenBank. Finally, all the sequence of *Neobenedenia melleni* in this study was deposited in GenBank with accession numbers of KU843501, KU843502, KU843503, and KU843504.

1. Introduction

Information and quantitative data on cultured fishes are limited in Southeast Asia. However, the existing data explains closely that similar species of fishes are cultured throughout the Southeast Asian region and the dominant parasites found infecting each species of these cultured marine fishes are similar [1–5]. Numerous studies on the parasitic fauna of marine fishes have indicated that the dominant parasites in each fish species are the same regardless of the wild or cultured [3, 6]. The main difference between the wild and cultured, diseased marine fishes is that the number and variety of parasites in both groups of cultured fishes greatly exceed those found in the wild fishes [7].

Monogenean parasites have been recognized as serious pathogens of fish in sea cage aquaculture [8–10]. Monogenea parasites have no intermediate host, predominantly parasitise the external surfaces of fish, and display two distinctive diets that traditionally divide them into two subclasses, the blood feeding polyopisthocotylea and the epithelial feeding monopisthocotylea [11]. These are sometimes named Heteronchoinea and Polyonchoinea, respectively [12]. These subclasses are united by various morphological synapomorphic larvae with three ciliated zones, adults, and larvae with two pairs of pigmented eyes, one pair of ventral anchors (hamuli), and one egg filament [13]. Inference about the Monogenea parasite is monophyletic, which has been ubiquitous for decades [13–18]. *Neobenedenia melleni* (MacCallum, 1927) Yamaguti, 1963, a capsalid monogenean of the subfamily *Benedenia* sp., is disreputable as a widespread pathogen of many teleost species in aquaculture [19]. This parasite feeds on epithelial cells mucus of host fish, which gives increased effects towards irritation and mucus hyperproduction of their hosts [20]. Like most of other monogenean groups, benedenids have traditionally been identified to species on the basis of morphological characters such as the shape of posterior hamuli, the type of anterior attachment organ, and the length of uterus, vitelline reservoir, and the type and relative size of testes [21]. Though it has been argued for a long time that morphological characters based identification of parasite can be affected, to a large extent, by extrinsic factors such as the age of parasite, environmental temperature, and even artifacts caused by various dealings for specimen processing, as discussed by Li et al. [22], most monogeneans could be appropriately distinguished because of their high level of host specification. However, *Neobenedenia melleni* does not obey the rule because it has been reported from more than 100 teleost fish species belonging to more than 30 families with worldwide distributions [23].

To date, there is no reference yet that has been done on the correlations of *Neobenedenia melleni* parasite infestations to the fish size, temperature, and salinity factors in Malaysia. This is an important aspect of research as it will benefit fish farmers for aquaculture industry to predict any fish parasite infestation in their farm and to take initiatives to prevent parasitic infections. Thus, the objective of this study is to show the prevalence and statistical analysis of *Neobenedenia melleni* parasite to the fish size and water temperature in *Lutjanus erythropterus* fish species sampled from cage culture Jerejak Island, Penang, Peninsular Malaysia. Furthermore, we have successfully identified the parasite species using morphology and molecular approach.

2. Materials and Methods

2.1. Sampling Locality and Parasite Collection. The experiment was carried out with 400 fish specimens of cultured *Lutjanus erythropterus* fish species from Jerejak Island, Penang, Peninsular Malaysia (5.320097 longitude, 100.3189185 latitude). The length (cm) of each fish was measured prior to parasite examination. Fresh water medium was used as anesthetics to reduce the stress as well as for easy handling. After the fish has been anaesthetized, presence of ectoparasite was examined via external fish body examination and direct observation under light microscope [24]. The site specificity of parasite was obtained from head, body, and both sides of inner operculum.

First morphological identification of parasite was done by first staining the parasite with a few drops of lactophenol solutions (200 mL lactic acid, 200 g/L phenol, 400 mL glycerol, and 200 mL deionized water). Upon staining, slides were observed under the compound microscope (Leica, USA). Parasite found was taken out carefully from the infected area, and then the number of parasites obtained from each fish was recorded, preserved with 70% ethanol solution in universal bottle for further examination. After the pictures of parasites had been taken, identification of parasites collected was done by morphological observation using identification keys as suggested by Kua et al. [25, 26].

2.2. Morphological Method Using Scanning Electron Microscope. Second morphological identification was done using the Supra 50vp ultra high resolution LEO analytical Fesem, scanning electron microscope. Electron microscopic sample preparation was done as suggested by protocol of Supra 50vp ultra high resolution LEO analytical Fesem, scanning electron microscope guide manual. Firstly, suspended samples in ethanol were put into serial dilution of 90%, 80%, and 70% ethanol. Then, a droplet of the suspension was placed on a carbon film coated 400-mesh copper grid for 1–3 minutes. The droplet is then dried using pieces of filter paper. The grid was then placed in a filter paper lined Petri dish for preservation in desiccator. Finally, imaging would be carried out after 3 days of preservation.

2.3. Molecular Method Using DNA Identification. The genomic DNA extraction and purification of the parasite was performed using the procedures provided by Qiagen DNeasy Blood and Tissue Kit (Qiagen, Inc., Valencia, CA, USA). Purified genomic DNA was eluted by adding $100 \mu L$ of buffer AE to the same spin column in a new Eppendorf tube and centrifuged at $5200 g$ for 1 min. The centrifuge step was repeated again for a total of $200 \mu L$ sample volume. DNA sample was stored at $-20°C$ and concentration measured with ACT-Gene NanoDrop spectrophotometer (ASP 2680, Taiwan).

Ribosomal RNA 18s partial sequences were amplified from purified genomic DNA using the specific primers 18sF (5′-GCG CGA GAG GTG AAA TTC AT-3′) as forward primer and 18sR (5′-AGT TTA CCC AGC CCT TTC GA-3′) as reverse primer, as discussed by Dang et al. [27] synthesized by MyTACG Bioscience (Malaysia). Polymerase chain reaction (PCR) was carried out using a total volume of $25 \mu L$ master mix solutions ($14 \mu L$ of ddH_2O, $2.5 \mu L$ of Promega PCR buffer, $3 \mu L$ of Promega $MgCl_2$ solutions, $1 \mu L$ of Promega dNTP, $1 \mu L$ of each forward primer and reverse primer, $2 \mu L$ of DNA template, and $0.5 \mu L$ of Promega Go Taq DNA polymerase). Standard cycle conditions for PCR were set accordingly by initial denaturation for 10 min at $95°C$, followed by 35 cycles of 30 s at $95°C$, 30 s at $50°C$, 60 s at $72°C$, and final elongation of 7 minutes at $72°C$. The whole PCR was carried out in MyCycler thermal cycler Bio-Rad PCR systems (USA). Purification of PCR product was performed using the procedure and materials provided in a QIAquick PCR Purification Kit (Qiagen, Inc.). Amplification products were sequenced in both directions by MyTACG Bioscience Company (Malaysia).

2.4. DNA Alignment and Phylogenetic Analysis. Alignment analysis of nucleic acid sequences was performed using ClustalW2 MEGA 5. Distance-based tree approach to species identification was conducted using MEGA 5 software. A BLAST search was conducted with DNA sequence that was amplified. Using MEGA 5, as discussed by Tamura et al. [28], a distance-based tree approach to species identification was carried out by neighbour-joining the 18s sequences of recorded species from the BLAST search and those analyzed in this study. Pairwise distance calculation is done using MEGA 5 analysis tools and the Kimura 2-parameter [29]; method serves as the substitution model. In addition, the

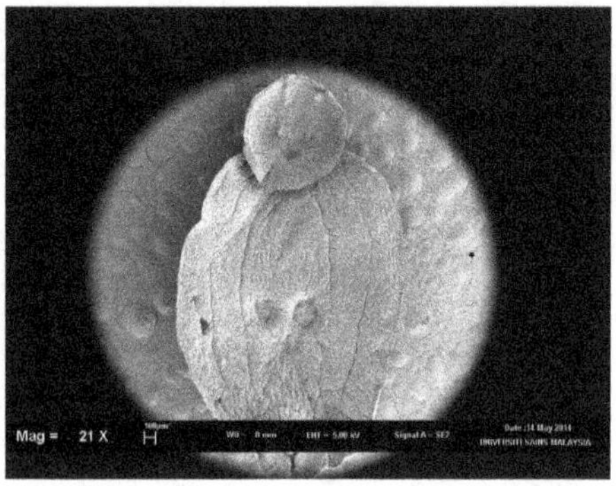

FIGURE 1: Ventral view of whole in SEM, *Neobenedenia melleni*.

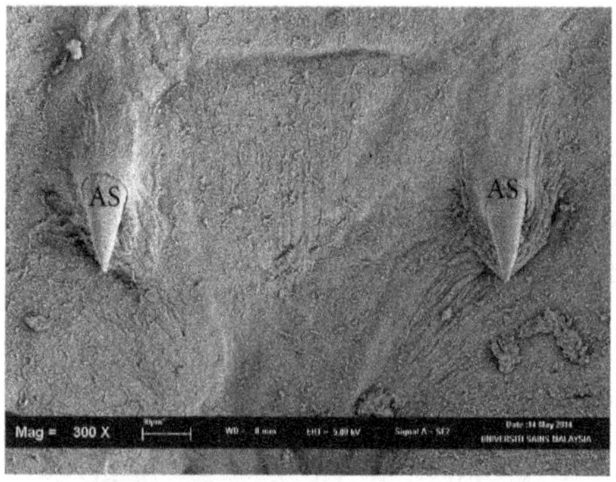

FIGURE 2: Ventral view of *Neobenedenia melleni* in SEM; AS: accessory sclerite; 40 μm.

bootstrap method was deployed as test of phylogeny using 1000 bootstrap replications. Finally, all the sequences were submitted in GenBank according to submission protocols.

2.5. Statistical Analysis and Water Parameters Records. Statistical analysis in this study was performed using the Statistical Package for Social Sciences software, SPSS version 20. The multiple regression analysis was employed and in all cases, the significance level is set at 5% as discussed by Field [30]. The water parameters were measured in sea cage using Saltwater Master Test Kit, Aquarium Pharmaceuticals Index (API), USA. The procedure for each test was done according to manufacturer's instructions.

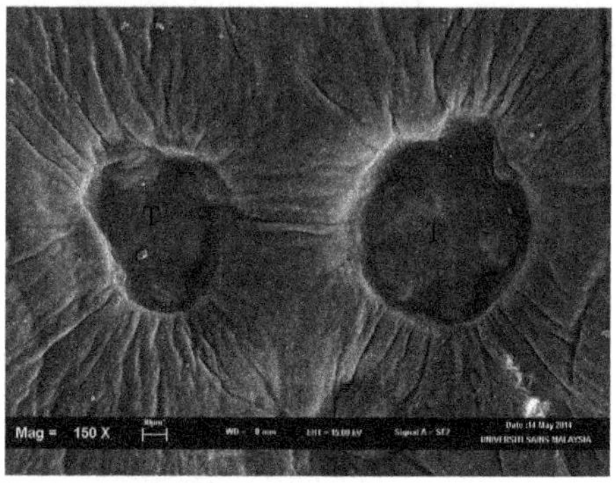

FIGURE 3: Ventral view of *Neobenedenia melleni* in SEM; T = testis organs.

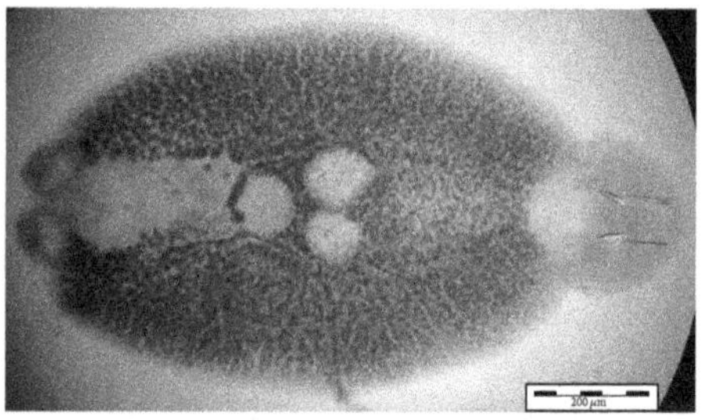

3. Results and Discussion

3.1. Morphological Analysis of Neobenedenia melleni. Using morphological key as described by Lawler [31] and Bullard et al. [32, 33]. In revising the generic diagnosis for *Neobenedenia*, Whittington and Horton [23] noted a variety of forms, which were more than 80 specimens attributed to *Neobenedenia melleni* from various host species. We are able to identify the parasite collected as *Neobenedenia melleni* according to Figures 1–6, AS: accessory sclerite, 40 μm; T: testis organs; AO: anterior attachment organ; P: pigmented eye; MA: male accessory gland reservoir; G: gland of Goto; V: vitelline reservoir; A: anterior hamulus, 150 μm; P: posterior hamulus, 40 μm. Total length of a sample specimen, *Neobenedenia melleni*, in this study, was recorded as 1050 μm. The width length is recorded as 700 μm.

3.2. DNA and Phylogenetic Analysis. Based on the results obtained upon gel electrophoresis analysis of the DNA template and PCR product, clearly visible bands were detected around 700 bp sequence, whereby this analysis was referred to Lucigen 1 kb DNA marker (USA) (Figure 7). Besides that, the optical density (OD) ratio of DNA was 2.0 with concentration of 135 (ng/μL) for genomic DNA. The DNA sequence was further analyzed using Clustal W, Bioedit Software. The 18s sequence was successfully analyzed, recovered from all *Neobenedenia melleni* individuals. Nucleotide BLAST sequence for 18s ribosomal RNA gene from this study has shown 99% similarity with *N. melleni* EU707804.1 from

FIGURE 4: Ventral view of whole *Neobenedenia melleni* in compound microscope.

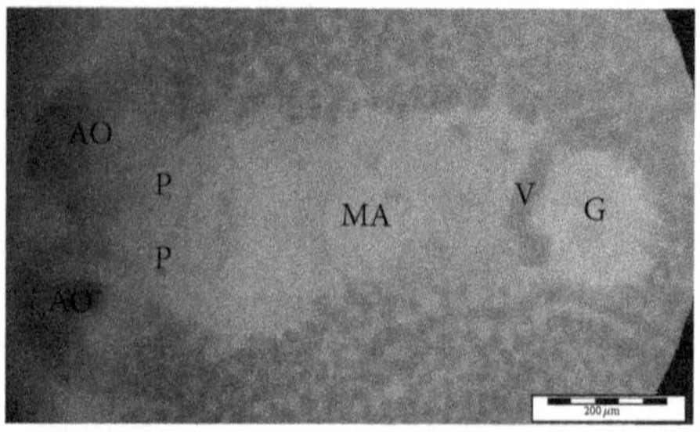

FIGURE 5: Ventral view of *Neobenedenia melleni* in compound microscope; AO: anterior attachment organ; P: pigmented eye; MA; male accessory gland reservoir; G: gland of Goto; V: vitelline reservoir.

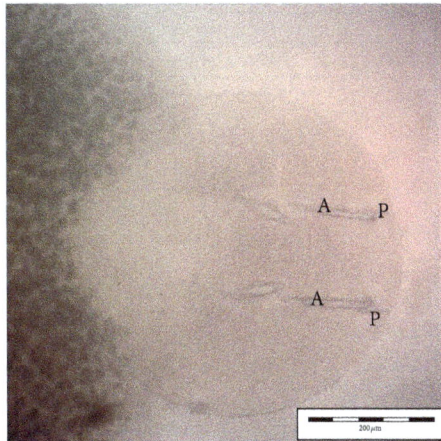

FIGURE 6: Ventral view of *Neobenedenia melleni* in compound microscope; A: anterior hamulus 150 μm; P: posterior hamulus 40 μm.

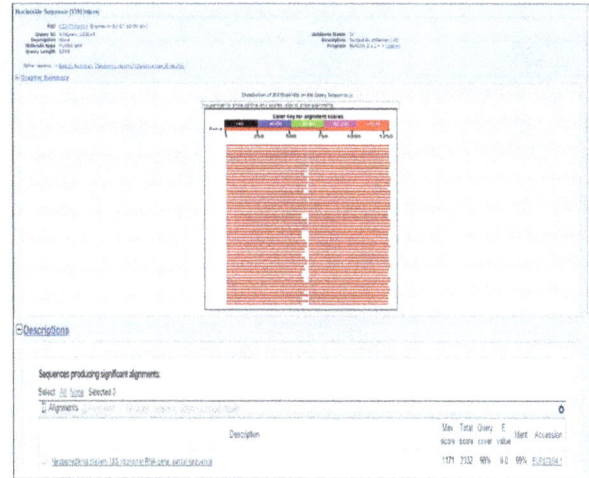

FIGURE 8: Nucleotide BLAST sequence for PCR product.

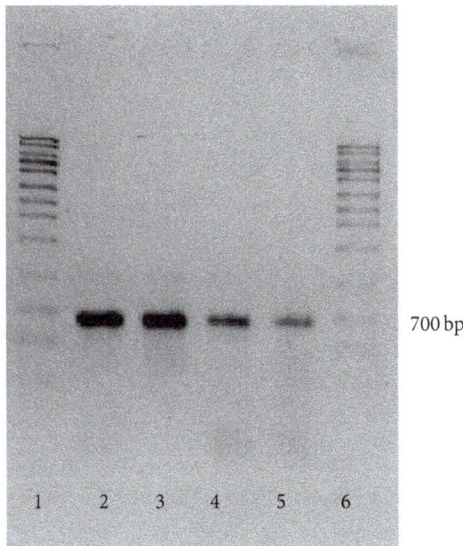

700 bp

Lanes 1, 6: DNA ladder 1 kb
Lane 2-5: *Neobenedenia melleni*

FIGURE 7: Gel electrophoresis of PCR 18s ribosomal RNA gene.

TABLE 1: The table shows the descriptive statistics of dependent and predictor variables.

Characteristics	Frequency, n	Prevalence (%)	Mean	SD
Dependent variables				
Neobenedenia melleni	379	94.8[a]	25.24	2.8
Predictor variables				
Fish length	375	93.8[a]	24.3	3.7
Categorical variables				
Water temperature				
32.0	154	54.4		
33.0	129	45.6		
Water salinity				
32.0	168	59.4		
33.0	115	40.6		

Note. [a]Total percentage is not 100% due to missing values; SD: standard deviation.

GenBank dataset, as shown in Figure 8. Meanwhile, Figure 9 shows the constructed phylogenetic tree which shows two closely related clades between species.

All the individuals of *Neobenedenia melleni* recorded in this study showed a close relationship between species that was recorded from NCBI, *Neobenedenia melleni* EU707804.1, as 96% bootstrap value, followed by 76% of similarity between *Allobenedenia epinepheli* EU707800.1. The least similarity was recorded with *Encotyllabe chironemi* AJ228774.1, *Benedenia epinepheli* EU707802.1, and *Neobenedenia girellae* AY551326. Finally, all the sequence of *Neobenedenia melleni* in this study was deposited in GenBank with accession numbers KU843501, KU843502, KU843503, and KU843504. All this sequence is available as public database in GenBank.

3.3. Statistical Modeling for Neobenedenia melleni. A total of all 379 fishes were infected by *Neobenedenia melleni* parasite out of 400 examined fishes in natural sea culture cage environment. Table 1 shows the descriptive statistics of dependent variable and predictor variables involved in this study. The average number of the *Neobenedenia melleni* parasite found in examination of fishes is approximately 25 with a standard deviation of 2.8. The mean value of fish length is 24.3 cm with a standard deviation of 3.7. Meanwhile, the binary coded variables, water temperature (0 = 32°C, 1 = 33°C) and salinity (0 = 32 ppt, 1 = 33 ppt), both have a higher percentage of low temperature (54.4%) and salinity (59.4%) compared to its counterpart.

Table 2 shows that the variables are positively correlated with one another and are significant at 1%. A large correlation of 0.92 is observed between fish length and prevalence of *Neobenedenia melleni* (increase in fish length will increase the prevalence of *Neobenedenia melleni* parasite) and 0.437

TABLE 2: Correlations between variables using bivariate analysis.

Variables	Prevalence of *Neobenedenia melleni*	Fish length	Water temperature	Water salinity
Prevalence of *Neobenedenia melleni*	1			
Fish length	0.917***	1		
Water temperature	0.298***	0.270***	1	
Water salinity	0.437***	0.428***	0.384***	1

Note: *** significant at 1%.

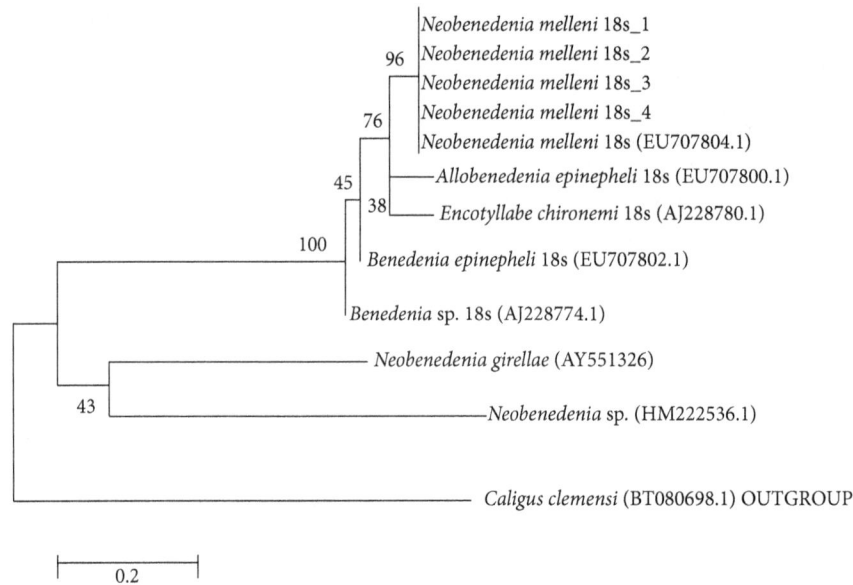

FIGURE 9: Constructed phylogenetic tree for *Neobenedenia melleni*.

TABLE 3: Variance inflation factor values of predictor variables.

Variables	Tolerance statistics	Variance inflation factor (VIF)
Fish length	0.751	1.332
Temperature	0.778	1.286
Salinity	0.659	1.518

between water salinity and prevalence of *Neobenedenia melleni* parasite (increase in water salinity increases the prevalence of *Neobenedenia melleni* parasite). Finally, a moderate correlation of 0.3 is observed between water temperature and prevalence of *Neobenedenia melleni* parasite [30].

According to Table 3, the variance inflation factor (VIF) values are less than 10.0 or 2.0 and the tolerance statistics are above 0.2 [30]. The tolerance statistics is the reciprocal of VIF or 1/VIF. Multicollinearity issues are negated because the values met more than the requirement of VIF, variance inflation factor, and tolerance statistics.

Table 4 shows the multiple correlation coefficients R, the correlation among all the independent variables (temperature, fish length, and salinity), and the dependent variable which is at value 0.912. The R-Square value shows that all the predictors account for 84.5% of variation in the prevalence of

parasite. The adjusted R-Square value (0.843) is similar to that of R-Square indicating that if these data were collected from the population rather than a sample it would have a similar result. Therefore, the result from this sample is generalized to the entire population of *Neobenedenia melleni* parasite infesting in fishes, as discussed in Field [30].

Table 5 shows that the model is a significant fit to the data, at less than 5%. Thus, the model is significantly improved to the ability to predict the dependent variable, prevalence of *Neobenedenia melleni* parasite infesting in fishes [34].

Table 6 shows the parameter estimates of multiple regression modeling. Among the variables that significantly predict the prevalence of *Neobenedenia melleni* parasite are fish length and water temperature, both significant at 1% and 5%; however salinity is not a significant predictor of *Neobenedenia melleni* in this analysis. Furthermore, the increase in one unit of fish length, holding other variables constant, increases the prevalence of parasite by approximately 1 ($0.7{\approx}1$) unit. Also, increasing the temperature from 32 to 33 degrees Celsius increases the number of parasite by approximately 0.32 units, holding other variables constant. The 95% confidence interval (CI) conforms to the results obtained by observing the parameter estimate for fish length and temperature within the confidence interval and within a positive confidence interval bound.

TABLE 4: The model summary.

Multiple correlation coefficient, R	R-Square	Adjusted R-Square	Standard error of the estimate
0.912	0.845	0.843	1.124

TABLE 5: The model fit values.

	Sum of squares	df	Mean square	F-statistics	P value
Regression	1829.444	4	457.361	361.868	0.000
Residual	334.930	265	1.264		
Total	2164.374	269			

TABLE 6: Fitted values of the predictor variables via multiple regression analysis.

Variables	Parameter estimates, β	Standard error, (SE)	95% Confidence interval (CI)	
			Lower bound	Upper bound
Constant	8.244	0.465	7.329	9.159
Fish length	0.669***	0.021	0.628	0.710
Temperature	0.319**	0.156	0.013	0.625
Salinity	0.124	0.172	−0.214	0.462

Note. Significant at ***1% and **5% significance level.

The model can be rewritten as

Estimated number of *Neobenedenia melleni* parasites

$$= 8.2 + 0.7 * (\text{Fish Length}) + 0.32 \quad (1)$$

$$* (\text{Water Temperature}).$$

Figure 10 shows the histogram of the residuals data which has a bell shaped curve indicating that the residuals are normally distributed. This is further verified by visualizing the normal P-P plot in Figure 11 which also shows that the points lie along a diagonal line indicating that the residuals are normally distributed. Figure 12 shows that the points are randomly and evenly dispersed throughout the plot and concur with the assumptions of linearity and homoscedasticity of the residuals has been met as discussed in Field [30].

In this study, we have deployed multiple regression analysis method to observe biotic and abiotic factors that have influenced the miscellany of parasites in hosts, like fish length, water temperature, parasites count, and salinity [9]. These multiple variables are predicted to influence cultured fish and to come across rates with parasites and with the number of parasites that can endure in populations. A positive relationship is predicted among fish length, temperature, salinity, and parasite diversity because larger fish represent larger infection surface area for parasitic colonization [34, 35]. Besides that, temperature is mainly important as an environmental factor which merely controls the development period of parasitic copepods. Parasites, growth rates, egg production, survival rate, and conscription are reported to be high at higher water temperatures [21, 23]. The multiple regression analysis is integrated with the objective to produce a model that would best predict the optimal number of parasitic infestation based on observed values of three independent variables which were the length of fish, mean temperature, and salinity.

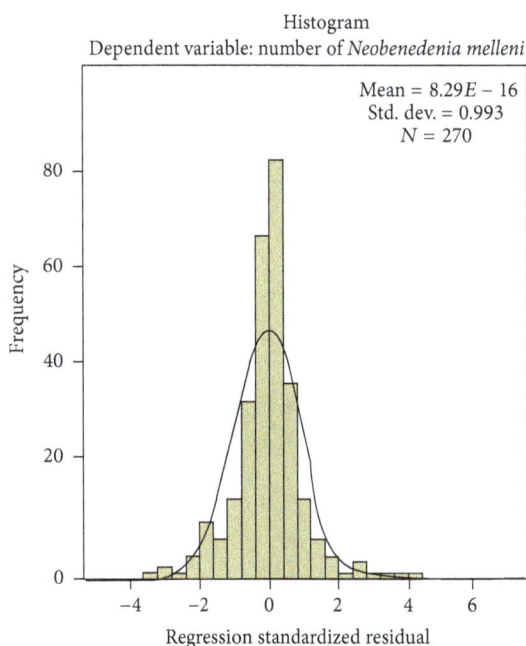

FIGURE 10: Histogram of the residuals data shows that the histogram has a bell shaped curve indicating that the residuals are normally distributed.

Several monogenean species exhibit short life cycles in warm temperatures [9, 23]. Accelerated parasitic life cycles will increase the metabolic and development rate associated with warm conditions [8, 14, 15]. Presently, the reason for the unpredictable and irregular nature of *Neobenedenia melleni* infection is unknown. Steps that can be implemented in reducing this rapid parasitic infestations are to have more attentive, frequent fish stock monitoring during warm, high temperature water conditions [36]. The major role of this

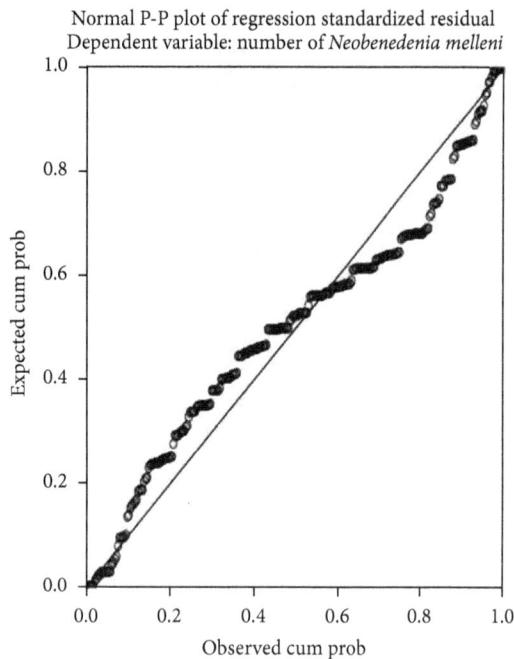

FIGURE 11: P-P plot shows that the points which represent the residuals lie along the diagonal line showing that the residuals are normally distributed.

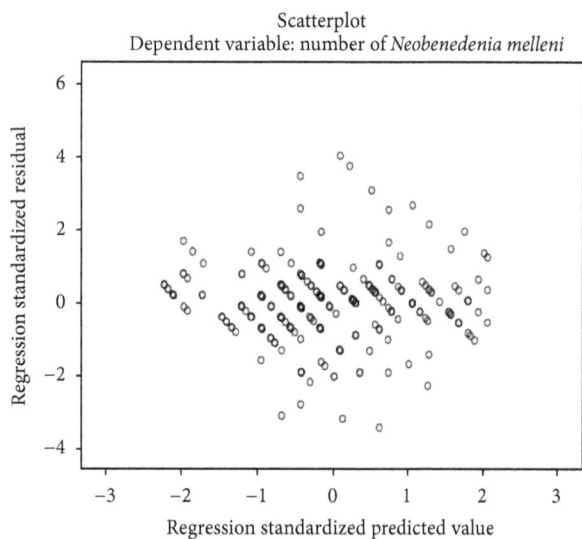

FIGURE 12: The figure shows that the points are randomly and evenly dispersed and do not have any specific pattern indicating that there is homoscedasticity or homogeneity of variance and linearity assumption is met.

temperature factors has been previously described by studies of life cycle of *Neobenedenia melleni*. Ogawa and Yokoyama [10] explained that *Neobenedenia melleni* took only 10 days to complete at 30°C as opposed to 20 days at 20°C in seawater.

Accordingly to Grau et al. [8], the hatching survival rate of *Neobenedenia melleni* eggs was less than 12% when incubated at salinity less than 18 ppt for 4 days.

4. Conclusion

In summary, this study has established an overview with statistical analysis for correlations of fish length and temperature that influences the number of fish parasites present in *Lutjanus erythropterus* fish species. Furthermore, morphology and DNA sequence identification were shown for *Neobenedenia melleni* parasite found in this cultured fish from Jerejak Island, Penang, Malaysia.

Competing Interests

All the authors declare that there are no competing interests in this paper.

Acknowledgments

Funding for the study was provided by Universiti Sains Malaysia Research University Grant (1001/PBIOLOGI/811259). The authors acknowledged GST group of companies for allowing them to perform these experiments at their fish farm. Furthermore, Malaysian MyBrain, MyPhD Scholarship Program is acknowledged. Mr. Johari at the SEM Unit, School of Biological Sciences, Universiti Sains Malaysia, is acknowledged for assisting with the photography sessions and Ms. Azirah Akbhar Ali, Ms. Yanie Zain, and Ms. Fatynn Amirah Firuz are acknowledged for assisting in fish sampling works.

References

[1] Y. C. Chong and T. M. Chao, "Quarantine treatment of imported *Epinephelus tauvina* fry," *Singapore Veterinary Journal*, vol. 8, pp. 2G–26G, 1984.

[2] Y. C. Chong and T. M. Chao, *Common Diseases of Marine Food Fish*, Fish Handbook, Primary Production Department of the Ministry of National Development, Singapore, 2011.

[3] T. S. Leong, *Parasites and Diseases of Cultured Marine Finfishes in Southern Asia*, Universiti Sains Malaysia, 1994.

[4] T. S. Leong and S. Y. Wong, "Parasites of wild and cultured golden snapper, *Lutjanus johni* (Bloch) in Malaysia," *Tropical Biomedicine*, vol. 6, pp. 73–76, 1989.

[5] T. S. Leong and S. Y. Wong, "Parasites of marine finfishes cultured in ponds and cages in Indonesia," in *Tropical Fish Health Management in Aquaculture*, J. S. Langdon, G. L. Enriquez, and S. Sukimin, Eds., vol. 48 of *Special Publication*, Biotrop, Bogor, Indonesia, 1992.

[6] T.-S. Leong and S.-Y. Wong, "A comparative study of the parasite fauna of wild and cultured grouper (Epinephelus malabaricus Bloch & Schneider) in Malaysia," *Aquaculture*, vol. 68, no. 3, pp. 203–207, 1988.

[7] T. S. Leong and S. Y. Wong, "Parasites of healthy and diseased juvenile grouper (*Epinephelus malabaricus* Bloch & *Schneider*) and seabass (*Lates calcarifer* Bloch) in floating cages in Penang, Malaysia," *Asian Fisheries Science*, vol. 3, no. 3, pp. 319–327, 1990.

[8] A. Grau, S. Crespo, E. Pastor, P. González, and E. Carbonell, "High infection by *Zeuxapta seriolae* (Monogenea: Heteraxinidae) associated with mass mortalities of amberjack Seriola dumerili Risso reared in sea cages in the Balearic Islands

(Western Mediterranean)," *Bulletin of the European Association of Fish Pathologists*, vol. 23, no. 3, pp. 139–142, 2003.

[9] L. A. Tubbs, C. W. Poortenaar, M. A. Sewell, and B. K. Diggles, "Effects of temperature on fecundity in vitro, egg hatching and reproductive development of *Benedenia seriolae* and *Zeuxapta seriolae* (Monogenea) parasitic on yellowtail kingfish *Seriola lalandi*," *International Journal for Parasitology*, vol. 35, no. 3, pp. 315–327, 2005.

[10] K. Ogawa and H. Yokoyama, "Parasitic diseases of cultured marine fish in Japan," *Fish Pathology*, vol. 33, no. 4, pp. 303–309, 1998.

[11] D. T. J. Littlewood, K. Rohde, R. A. Bray, and E. A. Herniou, "Phylogeny of the Platyhelminthes and the evolution of parasitism," *Biological Journal of the Linnean Society*, vol. 68, no. 1-2, pp. 257–287, 1999.

[12] W. A. Boeger and D. C. Kritsky, "Phylogenetic relationships of the Monogenoidea," in *Interrelationships of the Platyhelminthes*, D. T. J. Littlewood and R. A. Bray, Eds., pp. 92–102, Taylor & Francis, London, UK, 2001.

[13] A. E. Lockyer, P. D. Olson, and D. T. J. Littlewood, "Utility of complete large and small subunit rRNA genes in resolving the phylogeny of the Neodermata (Platyhelminthes): implications and a review of the cercomer theory," *Biological Journal of the Linnean Society*, vol. 78, no. 2, pp. 155–171, 2003.

[14] B. E. Bychowsky, *Monogenetic Trematodes, Their Systematics and Phylogeny*, Academy of Sciences, USSR, Moscow, Russia, English translation by W. J. Hargis Jr., and P. C. Oustinoff, American Institute of Biological Sciences, Washington, DC, USA, 1957 (Russian).

[15] J.-L. Justine, "Cladistic study in the Monogenea (Platyhelminthes), based upon a parsimony analysis of spermiogenetic and spermatozoal ultrastructural characters," *International Journal for Parasitology*, vol. 21, no. 7, pp. 821–838, 1991.

[16] J.-L. Justine, "Non-monophyly of the monogeneans?" *International Journal for Parasitology*, vol. 28, no. 10, pp. 1653–1657, 1998.

[17] J. Llewellyn, "Taxonomy, genetics and evolution of parasites," *The Journal of Parasitology*, vol. 56, no. 4, pp. 287–504, 1970.

[18] I. Mollaret, B. G. M. Jamieson, and J.-L. Justine, "Phylogeny of the Monopisthocotylea and Polyopisthocotylea (Platyhelminthes) inferred from 28S rDNA sequences," *International Journal for Parasitology*, vol. 30, no. 2, pp. 171–185, 2000.

[19] D. T. J. Littlewood, R. A. Bray, and K. A. Clough, "A phylogeny of the Platyhelminthes: towards a total-evidence solution," *Hydrobiologia*, vol. 383, pp. 155–160, 1998.

[20] T. S. Leong and S. Y. Wong, "Parasites of grouper, *Epinephelus suillus* from Pulau Langkawi and Kelantan, Malaysia," *Journal Bioscience*, vol. 6, article 5457, 1995.

[21] S. Yamaguti, *Systema Helminthum, Volume IV: Monogenea and Aspidocotylea*, Interscience Publishers, New York, NY, USA, 1963.

[22] A.-X. Li, X.-Y. Wu, X.-J. Ding et al., "PCR-SSCP as a molecular tool for the identification of *Benedeniinae* (Monogenea: Capsalidae) from marine fish," *Molecular and Cellular Probes*, vol. 19, no. 1, pp. 35–39, 2005.

[23] I. D. Whittington and M. A. Horton, "A revision of *Neobenedenia* Yamaguti, 1963 (Monogenea: Capsalidae) including a redescription of *N. melleni* (MacCallum, 1927) Yamaguti," *Journal of Natural History*, vol. 30, no. 8, pp. 1113–1156, 1963.

[24] Z. Kabata, *Parasitic Copepoda of British Fishes*, The Ray Society, London, UK, 1979.

[25] B. C. Kua, S. Z. Abdullah, M. F. Abtholuddin, N. F. Mohd, and N. N. Mansor, "Marine leech isolated from cage-cultured sea bass (Lates calcarifer) fingerlings: a parasite or Vector," in *Abstract in Programme Book of the 15th Scientific Conference Proceeding*, Terengganu, Malaysia, December 2006.

[26] B. C. Kua and H. Faizul, "Scanning electron microscopy of three species of *Caligus* (Copepoda:Caligidae) parasitized on cultured marine fish at Bukit Tambun, Penang," *Malaysian Journal of Microscopy*, vol. 6, pp. 9–13, 2010.

[27] B. T. Dang, A. Levsen, C. Schander, and G. A. Bristow, "Some *Haliotrema* (Monogenea: Dactylogyridae) from cultured grouper (*Epinephelus* Spp.) with emphasis on the phylogenetic position of *Haliotrema cromileptis*," *Journal of Parasitology*, vol. 96, no. 1, pp. 30–39, 2010.

[28] K. Tamura, J. Dudley, M. Nei, and S. Kumar, "MEGA4: Molecular Evolutionary Genetics Analysis (MEGA) software version 4.0," *Molecular Biology and Evolution*, vol. 24, no. 8, pp. 1596–1599, 2007.

[29] Kimura, "Estimation of evolutionary distance between nucleotide sequences," *Molecular Biology and Evolution*, vol. 3, pp. 269–285, 1984.

[30] A. Field, *Discovering Statistics Using SPSS*, 2nd edition, 2005.

[31] A. R. Lawler, *Zoogeography and Host-Specificity of the Superfamily Capsaloidea Price, 1936 (Monogenea: Monopisthocotylea). An Evaluation of the Host-Parasite Locality Records of the Superfamily Capsaloidea Price, 1936, and Their Utility in Determinations of Host-Specificity and Zoogeography*, Special Papers in Marine Science no. 6, 1981.

[32] S. A. Bullard, G. W. Benz, and J. S. Braswell, "*Dionchus postoncomiracidia* (Monogenea: Dionchidae) from the skin of blacktip sharks, *Carcharhinus limbatus* (Carcharhinidae)," *Journal of Parasitology*, vol. 86, no. 2, pp. 245–250, 2000.

[33] S. A. Bullard, G. W. Benz, R. M. Overstreet, E. H. Williams Jr., and J. Hemdal, "Six new host records and an updated list of wild hosts for *Neobenedenia melleni* (MacCallum) (Monogenea: Capsalidae)," *Comparative Parasitology*, vol. 67, no. 2, pp. 190–196, 2000.

[34] R. Ravi and Z. S. Yahaya, "Relationship between size of fish, temperature and parasitic intensity in snakehead fish species from Kepala Batas, Penang, Peninsular Malaysia," *Pertanika Journal of Tropical Agricultural Science*, vol. 38, no. 2, pp. 295–307, 2015.

[35] R. Ravi and Z. S. Yahaya, "DNA barcoding of marine leeches (*Zeylanicobdella argumensis*) from crimson red snapper (*Lutjanus eryhtopterus*), peninsular Malaysia and their phylogenetic analysis," in *Proceedings of the 9th Regional IMT-GT UNINET Conference*, Penang, Malaysia, November 2014.

[36] K. L. Main and C. Rosenfeld, Eds., *Culture of Highvalue Marine Fishes in Asia and the United States*, The Oceanic Institute, Honolulu, Hawaii, USA, 1995.

8

Genetic Polymorphism of *msp*1 and *msp*2 in *Plasmodium falciparum* Isolates from Côte d'Ivoire versus Gabon

William Yavo,[1,2] **Abibatou Konaté,**[1,2] **Denise Patricia Mawili-Mboumba,**[3]
Fulgence Kondo Kassi,[2,4] **Marie L. Tshibola Mbuyi,**[3] **Etienne Kpongbo Angora,**[2]
Eby I. Hervé Menan,[2,4] **and Marielle K. Bouyou-Akotet**[3]

[1] *Malaria Research and Control Centre, National Institute of Public Health, BPV 47, Abidjan, Côte d'Ivoire*
[2] *Faculty of Pharmacy, Department of Parasitology and Mycology, Félix Houphouët-Boigny University, BPV 34, Abidjan, Côte d'Ivoire*
[3] *Faculty of Medicine, Department of Parasitology and Mycology, University des Sciences de la Santé, BP 4009, Libreville, Gabon*
[4] *Parasitology and Mycology Laboratory of Diagnosis and Research Centre on AIDS and Other Infectious Diseases,*
 01 BPV 03, Abidjan, Côte d'Ivoire

Correspondence should be addressed to William Yavo; yavowilliam@yahoo.fr

Academic Editor: José F. Silveira

Introduction. The characterization of genetic profile of *Plasmodium* isolates from different areas could help in better strategies for malaria elimination. This study aimed to compare *P. falciparum* diversity in two African countries. *Methods.* Isolates collected from 100 and 73 *falciparum* malaria infections in sites of Côte d'Ivoire (West Africa) and Gabon (Central Africa), respectively, were analyzed by a nested PCR amplification of *msp*1 and *msp*2 genes. *Results.* The K1 allelic family was widespread in Côte d'Ivoire (64.6%) and in Gabon (56.6%). For *msp*2, the 3D7 alleles were more prevalent (>70% in both countries) compared to FC27 alleles. In Côte d'Ivoire, the frequencies of multiple infections with *msp*1 (45.1%) and *msp*2 (40.3%) were higher than those found for isolates from Gabon, that is, 30.2% with *msp*1 and 31.4% with *msp*2. The overall complexity of infection was 1.66 (SD = 0.79) in Côte d'Ivoire and 1.58 (SD = 0.83) in Gabon. It decreased with age in Côte d'Ivoire in contrast to Gabon. *Conclusion.* Differences observed in some allelic families and in complexity profile may suggest an impact of epidemiological facies as well as immunological response on genetic variability of *P. falciparum*.

1. Background

The World Health Organization (WHO) reported recently a significant decline in the global malaria burden over the last 15 years, achieving the 2000 Millennium Development Goals. Fifty-seven countries have reduced their malaria cases by 75%, in line with the World Health Assembly's target for 2015. Malaria mortality rates in Africa have fallen by 66% among all age groups and by 77% among children under 5. The progress made in reduction of malaria burden is attributed to massive rollout of effective prevention and treatment tools, including insecticide-treated nets and implementation of Artemisinin-based Combination Therapies (ACTs) as the first-line treatment [1].

Regardless of the progress made, Africa continues to shoulder the majority of malaria burden. Côte d'Ivoire and Gabon are two sub-Saharan Africa (SSA) countries with different malaria endemicity. In Côte d'Ivoire, located in West Africa, malaria accounts for 43% of all causes of outpatient visits. The disease is responsible for one-third of reported deaths in health facilities in the country [2]. In Gabon, a Central African country, malaria prevalence varied between 2005 and 2011. After a trend of a reduction between 2005 and 2008, a rebound of malaria cases prevalence was observed in the rural and urban areas [3]. Further, malaria risk shifted towards older children and adults who are thought to have acquired premunition [3, 4]. Some of the factors that hinder progress in malaria control and slow down the elimination agenda include the emergence and spread of drug-resistant parasite strains and development of vectors resistant to insecticide [5–7]. Development of a malaria vaccine would be an ideal addition to the existing tools being deployed towards malaria control and elimination. However, one of the limitations of the development of malaria vaccine is the extensive genetic diversity in parasite populations limiting the efficacy of acquired protective immunity to malaria [8]. Individuals are often simultaneously infected by multiple parasite clones that are related to the transmission intensity and are described as a factor determining the host immune status [9–11]. This situation may have an impact on the efficacy of malaria vaccine and clinical issue of the disease. It is therefore important to characterize the parasite populations in order to adapt malaria control and elimination strategies.

To be successful, malaria elimination will require knowledge of parasite genome variation in different geographical locations and a better understanding of the factors that determine gene flow between locations [12]. Merozoite surface protein-1 (msp-1) and merozoite surface protein-2 (msp-2) which are asexual blood stage antigens are considered as prime candidates for the development of a malaria vaccine and are also suitable markers for the identification of genetically distinct *Plasmodium falciparum* parasite populations [8, 11, 13].

There are a limited number of studies that have compared genetic profiles of malaria parasites from different African countries particularly by genotyping *msp*1 and *msp*2 genes. This study aimed to compare the genetic diversity of *P. falciparum* in infected patients living in two different areas, Côte d'Ivoire and Gabon, by using these two highly polymorphic genes [13–15].

2. Methods

2.1. Study Areas. This study was carried out in Côte d'Ivoire and in Gabon.

In Côte d'Ivoire, isolates have been collected in four of the six sentinel sites of malaria surveillance of the National Malaria Control Program (NMCP), namely, Abidjan, the economic capital city located in the south-east, Abengourou in the east, San Pedro in the southwest, and Yamoussoukro in the center of the country. Abidjan and San Pedro are coastal and forest areas with a hot and humid climate. Abengourou is a forest area with a subequatorial climate. Yamoussoukro is a forest transition zone with a tropical climate. In Côte d'Ivoire, malaria is endemic and predominantly caused by *Plasmodium falciparum* [2, 16]. The transmission is perennial with peaks during rainy seasons (from April to August and from October to November).

In Gabon, parasites were collected at the Centre Hospitalier Universitaire de Libreville (CHL) and the Regional Hospital of Melen (RHM), two sentinel sites for malaria surveys of the NMCP. The CHL is located in Libreville, the capital city, and the RHM is in a suburban area located 11 km north of Libreville. Malaria transmission, predominantly caused by *P. falciparum*, is perennial without significant fluctuation throughout the year in Gabon [3]. The climate is equatorial in the whole country.

2.2. Study Design and Patients Enrolment. Patients from Côte d'Ivoire were enrolled during a prospective randomized control survey assessing Artemisinin-based Combination Therapies (ACTs) according to the standard WHO 2003 efficacy assessment protocol conducted in 2012 at Abengourou, San Pedro, and Yamoussoukro [17] and in 2013-2014 at Abidjan (Yavo et al., nonpublished data).

In Gabon, prospective cross-sectional surveys were conducted at Libreville and Melen in 2011. In these sentinel sites, the study team collected data during at least one rainy and one dry season. In each site, febrile outpatients and inpatients are routinely screened for *P. falciparum* infection. Body temperature, history of fever, age, sex, bed net use, home treatment with antimalarial drug, and location were collected. Data on self-medication have been collected through a detailed CRF. Patients were asked about previous history of fever and drug intake the month before the consultation, the type of molecule taken, and the duration of the treatment during each survey throughout the study period [4].

Before inclusion, written informed consent was obtained from the patient or the patient's legal guardian (for children). Approvals were obtained from the national ethic committee in Côte d'Ivoire and the Ministry of Health in Gabon.

2.3. Parasite DNA Extraction. After malaria diagnostic tests, thick and thin blood smears and parasitized blood samples were collected on filter paper in both countries. DNA from *P. falciparum* isolates from Côte d'Ivoire was extracted using Chelex method as previously described [18]. Nucleic acids extraction from Gabon samples was performed using the QIAamp kit (QIAGEN®) according to the manufacturer's instructions.

2.4. P. falciparum Genotyping. For all *P. falciparum* isolates genotyping, the two polymorphic loci *msp*1 and *msp*2 were analyzed together by using nested PCR. Primary amplifications followed by secondary PCR reactions using specific primers for K1, Mad20, and RO33 (for *msp*1) and 3D7 and FC27 (for *msp*2) allelic families were performed. The PCR primers sequences and cycles were previously described [13]. All isolates have been amplified at the Diagnosis and Research Centre on AIDS and Other Infectious Diseases of Abidjan, Côte d'Ivoire, with the same reagents and bench. Each genotype was identified based on the size of the PCR products using a 1.5% agarose gel. A 0.5 µg/mL of ethidium bromide was used for the UV detection. Electrophoresis conditions were 100 mV for 30 min.

Bands were visualized under an UV transilluminator (BIOCOM™) and fragment sizes estimated by comparison to the 1 kb plus DNA ladder (SmartLadder™ small fragment, EUROGENTEC).

2.5. Definitions. The detection of a single PCR fragment for each locus was classified as an infection with one parasite genotype. The detection of more than one PCR fragment for either *msp*1 or *msp*2 loci (i.e., an infection with more than one parasite genotype) was defined as a multiple *P. falciparum* infection. The number of patients with more than one parasite genotype out of the number of infected population was

TABLE 1: Distribution of the different allelic families of *msp*1 and *msp*2 genes.

	Côte d'Ivoire	Gabon	p^*
*msp*1	($N1_{CI}$ = 82)	($N1_{Gab}$ = 53)	
K1, *n* (%)	53 (64.6)	30 (56.6)	*0.349*
Mad20, *n* (%)	33 (40.2)	23 (43.4)	*0.716*
RO33, *n* (%)	29 (35.4)	17 (32.1)	*0.694*
*msp*2	($N2_{CI}$ = 72)	($N2_{Gab}$ = 51)	
3D7 *n* (%)	53 (73.6)	37 (72.5)	*0.896*
FC27 *n* (%)	37 (51.4)	24 (47.1)	*0.636*

*Chi-square test.

defined as the frequency of multiple infections (MI). The complexity was defined as the mean number of parasite genotypes per infected patient [19].

2.6. Statistical Analysis. All data were recorded using Epi data version 3.1 and analyzed with SPSS for windows (version 16.0). The frequency of each allele was estimated over all the alleles detected. Allelic family distribution and the number of genotypes detected in each infected patient were calculated according to the site. The complexity of infection was determined according to the site and the age. The complexity of infection and the frequency of multiple infections were calculated by combining the *msp*1 and *msp*2 PCR genotyping results. The highest number of bands detected, whatever the locus, was used to calculate the value for the overall complexity of infection.

Differences between groups were assessed using Chi-square or Mann-Whitney tests. The level of significance for statistical tests was set at 0.05.

3. Results

A total of 173 parasite DNA samples were analyzed: 100 from Côte d'Ivoire and 73 from Gabon.

3.1. Characteristics of Both Study-Sites Patients. The median age of patients from Côte d'Ivoire was 11 years (min. = 1 year; max. = 74 years). It was 16 years for the patients from Gabon (min. = 1 year and max. = 80 years).

3.2. msp1 and msp2 Genes Genotyping

*3.2.1. Allelic Families Frequency. msp*1 and *msp*2 genes have been successfully amplified in 82% (82/100) and 72% (72/100) samples from Côte d'Ivoire, respectively. Likewise, amplification rate was of 72.6% (53/73) for *msp*1 gene and 69.9% (51/73) for *msp*2 gene in isolates from Gabon.

The distribution of the different allelic families of *msp*1 and *msp*2 genes is shown in Table 1. In both areas, K1-type alleles and 3D7 type alleles were the most frequent. Allelic families' frequencies were not statistically different between areas from both countries.

*3.2.2. Allelic Diversity. msp*1 gene analysis showed that 12 K1 type alleles with a size ranging from 100 to 500 bp, 8 Mad20

type alleles (100–300 bp), and 8 RO33 type alleles (120–300 bp) were identified in *P. falciparum* isolates from Côte d'Ivoire. In Gabon, 15 K1 type alleles with a size ranging from 160 to 900 bp, 14 Mad20 type alleles (170–820 bp), and 10 RO33 type alleles (140–920 bp) were found. K1 200 bp, Mad 200 bp, and RO33 150 bp alleles were predominant in Côte d'Ivoire. In Gabon, K1 850 bp, Mad 790 bp, and RO33 (150 bp and 900 bp) alleles were the most frequent (Figures 1(a), 1(b), and 1(c)).

Based on *msp*2 gene analysis, 12 3D7 type alleles with a size ranging from 200 to 700 bp and 9 FC27 type alleles (200–600 bp) were detected in Côte d'Ivoire. In Gabon, 16 3D7 type alleles (290–900 bp) and 11 FC27 type alleles (300–880 bp) were identified. 3D7 300 bp and FC27 500 bp alleles were the most frequent in Côte d'Ivoire while in Gabon, 3D7 800 bp and FC27 600 bp alleles were predominant (Figures 2(a) and 2(b)).

The prevalences of common alleles in the two countries were 35.7% (10/28) for *msp*1 and 38.7% (10/26) for *msp*2.

3.2.3. Multiple Infections. A total of 236 (129 for *msp*1 and 107 for *msp*2) and 151 (78 for *msp*1 and 73 for *msp*2) individual *msp* fragments were, respectively, found in Côte d'Ivoire and Gabon.

The numbers of genotypes per isolate were 1 to 3 and 1 to 4 for *msp*1 gene in Côte d'Ivoire and Gabon, respectively. For *msp*2 gene, there were 1 to 5 genotypes per isolate from Côte d'Ivoire versus 1 to 4 per isolate from Gabon.

For *msp*1 gene, the overall proportion of multiple infections (MI) was 45.1% in Côte d'Ivoire versus 30.2% in Gabon (p = 0.083). There was the same trend with *msp*2 gene: 40.3% versus 31.4% of multiple infections in Côte d'Ivoire and Gabon, respectively (p = 0.312). The distributions of MI and complexity of infection (COI) according to the age of patients were shown in Table 2. Overall, the COI was 1.66 (SD = 0.79) in Côte d'Ivoire and 1.58 (SD = 0.83) in Gabon (p = 0.293).

4. Discussion

This study enabled a comparison of the genetic diversity of *P. falciparum* from infected patients living in Côte d'Ivoire versus Gabon using the most polymorphic regions of *msp*1 and *msp*2 genes. Thus, a high genetic diversity of the population of *P. falciparum* isolates from both areas was found.

Within *msp*1 gene, the high diversity is compatible with the high level of malaria transmission in both areas. Data from Gabon underline the heterogeneity of *P. falciparum* strains in this country based on genetic diversity. Indeed, a high diversity was reported in isolates from Libreville (30 alleles) during the year 2011-2012 [20] while in Franceville in the southeast only 9 alleles have been identified 12 years before [11]. It could be more accurate to conduct studies including several samples collection at different time point within the same region to assess and compare the genetic profile of parasites circulating in endemic areas [21] in an attempt to avoid intra- and interindividual variation in the number of parasite genotypes detected in the different episodes of malaria [22]. Moreover, a single blood sample may not be enough to show the whole diversity of parasites carried by

TABLE 2: MI and COI according to the age of patients.

Age (years)	Côte d'Ivoire			Gabon		
	<5	≥5		<5	≥5	
msp1						
			p			*p*
MI, %	70.6	38.5	0.018*	30.8	30	0.958*
COI (SD)	2 (0.73)	1.46 (0.72)	0.010**	1.46 (0.82)	1.47 (0.80)	0.980**
msp2						
			p			*p*
MI, %	43.7	39.3	0.748*	7.7	39.5	0.031*
COI (SD)	1.5 (0.71)	1.48 (0.71)	0.754**	1.08 (0.71)	1.55 (0.73)	0.032**
msp1 and *msp2*						
			p			*p*
MI, %	84	61.7	0.063*	58.8	57.1	0.902*
COI (SD)	2 (0.8)	1.58 (0.8)	0.036**	1.41 (0.84)	1.63 (0.83)	0.371**

*Chi-square test. **Mann-Whitney test.

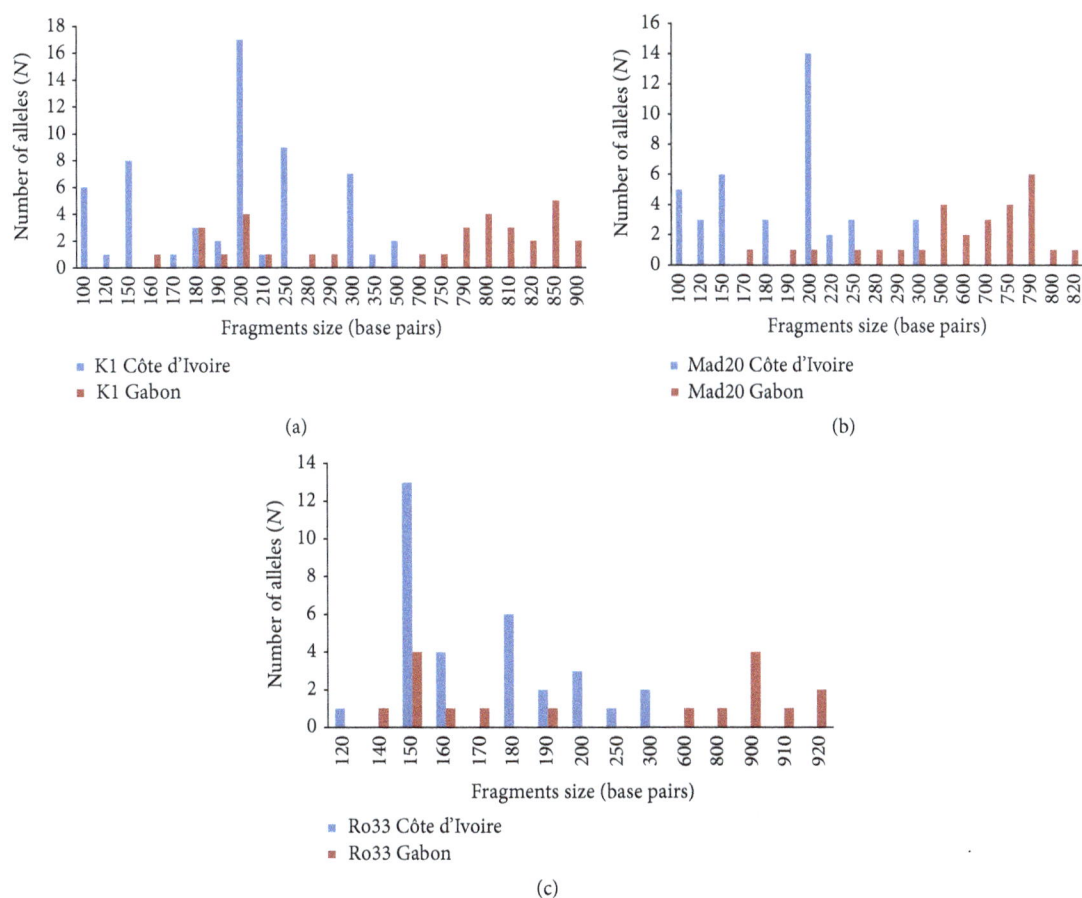

FIGURE 1: (a) *P. falciparum msp1* K1 type alleles classified according to the length (in base pairs) in Côte d'Ivoire and Gabon. (b) *P. falciparum msp1* Mad20 type alleles classified according to the length (in base pairs) in Côte d'Ivoire and Gabon. (c) *P. falciparum msp1* RO33 type alleles classified according to the length (in base pairs) in Côte d'Ivoire and Gabon.

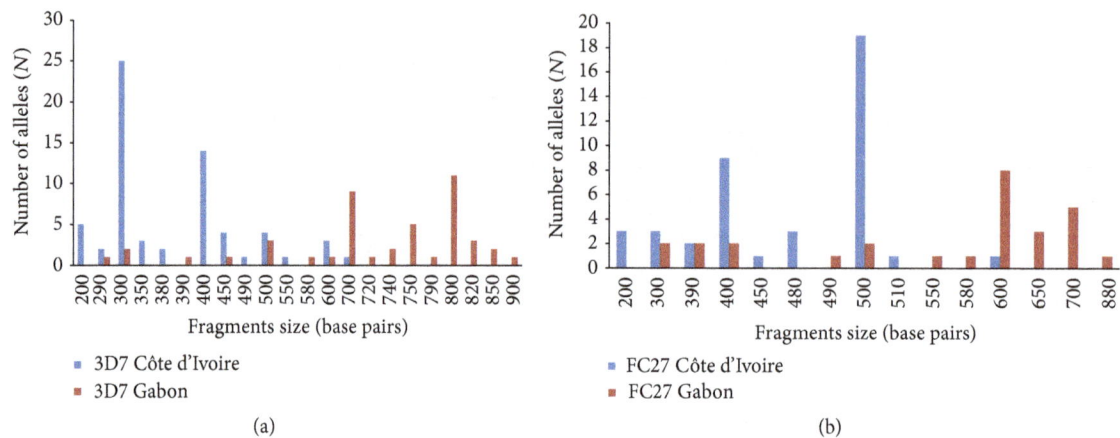

FIGURE 2: (a) *P. falciparum msp*2 3D7 type alleles classified according to the length (in base pairs) in Côte d'Ivoire and Gabon. (b) *P. falciparum msp*2 FC27 type alleles classified according to the length (in base pairs) in Côte d'Ivoire and Gabon.

an individual, since genotypes can appear and disappear in a very short time [23, 24]. The predominance of K1 allelic family in malaria endemic countries is frequently reported as found here [13, 20, 21, 25]. It has been previously shown that the RO33 family was poorly polymorphic in Gabon [20, 26], in contrast to our findings with 10 different alleles.

Considering *msp*2 allelic diversity, a predominance of 3D7 allelic family was found in both areas. This has also been reported in previous studies conducted in Benin [27], Burkina Faso [13], and Congo-Brazzaville [25]. However, the present data differ from those reported in Nigeria [28] and recently in Benin [21].

In our study, the size of *msp*1 and *msp*2 PCR products obtained from isolates collected in Gabon was longer compared to the one of the alleles detected in Côte d'Ivoire. Such difference between both areas may suggest a specific immune response against these alleles or a random event due to genetic drift [29]. However, an underestimation of such difference can be expected. One limit of the use of any marker based on DNA fragment size is convergence. In fact, similar sized fragments (but not identical) can be scored as identical leading to a false impression of similarity. Within allele families, alleles of the same size may have different amino acids motifs [10, 30]. Nevertheless, markers such as *msp*1 and *msp*2 genes are enough robust markers of polymorphism and can be used successfully to characterize genetic *P. falciparum* strains populations [13, 14].

The allelic variations as well as the presence of common alleles should be better explored by sequencing the isolates from both countries. This could be useful for designing a malaria vaccine based on study sites specificity.

For each gene, the MI was higher in Côte d'Ivoire than in Gabon but the genotypes numbers per isolate were almost equal. In Côte d'Ivoire, based on *msp*2 gene genotyping, Mara et al. [31] have found a number of clonal infections per individual, ranging from 1 to 8, with a mean of 2.88. In Gabon, 1 to 6 genotypes per isolate were reported from the *msp*1 gene analysis [20]. The number of clones coinfecting a single host

can be used as an indicator of the level of malaria transmission or the level of host acquired immunity that is related to the endemicity [32–34]. During *P. falciparum* infection with several genotypes, a selection for effective transmission of sexual gametocyte stages to mosquitoes may occur and can probably be considered as a result of the presence of specific alleles which mediate the survival of the parasite inside the mosquito. Studies reported that the genetic diversity of natural populations of parasites could vary in remote sites far from each other by a few kilometers [13, 35]. Likewise, multiclonal infections may vary depending on the clinical status of the individual (asymptomatic versus symptomatic), the severity of the disease (simple cases versus severe), the malaria endemicity, the age of the patient, and the acquired immunity [13, 20, 36–38]. Although from our data further analyses are limited regarding COI, it seems that the different genotypes circulating in both areas may undergo different selective pressure leading to a variation of COI in each context. Thus, the significant decrease of the overall COI with age in Côte d'Ivoire could result in the development of premunition in highly endemic areas and the maintenance of the population's immunity through perennial exposure to mosquito bites [20]. This trend was already reported for *msp*2 gene in a rural setting of Côte d'Ivoire [39].

Considering *msp*1 gene only, it was shown that the relation between the complexity and age could change according to the site with a trend of increase in rural area and decrease in urban area [40]. In our study, we observed a significant decrease of COI with age for *msp*1 gene in Côte d'Ivoire.

In contrast, the overall COI tended to increase with age in Gabon as reported in Burkina Faso and Benin [13, 21] suggesting that younger children might still have protection from maternal antibodies or could simply mean a lower risk of multiple infections due to a lower exposure time at risk of infection [13]. This trend was significant with *msp*2 gene in our study in Gabon.

However, some authors did not find any association between the COI and age [41, 42].

5. Conclusion

The polymorphism in *P. falciparum* clinical isolates from both malaria endemic areas was high. The differences observed in some allelic families and in the complexity profile may suggest an impact of epidemiological facies as well as immunological response on *P. falciparum* genetic diversity.

Abbreviations

MI: Multiple infections
bp: Base pairs
COI: Complexity of infections.

Competing Interests

The authors declare that they have no competing interests.

Authors' Contributions

William Yavo and Marielle K. Bouyou-Akotet are the principal investigators of the study. Abibatou Konaté, Denise Patricia Mawili-Mboumba, Fulgence Kondo Kassi, Marie L. Tshibola Mbuyi, and Eby I. Hervé Menan supervised the study. Etienne Kpongbo Angora and Abibatou Konaté conducted the statistical analysis. All authors contributed to the drafting of the paper.

Acknowledgments

The authors hereby thank the study team and their administrative staff in each site. The authors are also grateful to the patients who took part in this study, to the administrative staff of the different health facilities, and to Dr Edwin Kamau for his great help correcting this paper.

References

[1] World Health Organization, *World Malaria Report*, WHO, Geneva, Switzerland, 2015.

[2] Ministère de la Santé et de la Lutte Contre le SIDA, *PNLP, Revue des Performances du Programme National de Lutte Contre le Paludisme*, PNLP, Abidjan, Côte d'Ivoire, 2013.

[3] M. K. Bouyou-Akotet, C. L. Offouga, D. P. Mawili-Mboumba, L. Essola, B. Madoungou, and M. Kombila, "*Falciparum* malaria as an emerging cause of fever in adults living in Gabon, Central Africa," *BioMed Research International*, vol. 2014, Article ID 351281, 7 pages, 2014.

[4] D. P. Mawili-Mboumba, M. K. Bouyou-Akotet, E. Kendjo et al., "Increase in malaria prevalence and age of at risk population in different areas of Gabon," *Malaria Journal*, vol. 12, article 3, 2013.

[5] World Health Organization, *Global Plan for Insecticide Resistance Management in Malaria Vectors*, World Health Organization, Geneva, Switzerland, 2012.

[6] C. V. Plowe, "The evolution of drug-resistant malaria," *Transactions of the Royal Society of Tropical Medicine and Hygiene*, vol. 103, supplement 1, pp. S11–S14, 2009.

[7] E. A. Ashley, M. Dhorda, R. M. Fairhurst et al., "Tracking resistance to artemisinin collaboration (TRAC)," *The New England Journal of Medicine*, vol. 371, no. 5, pp. 411–423, 2014, Erratum in: *The New England Journal of Medicine*, vol. 371, no. 8, p. 786, 2014.

[8] B. Genton, I. Betuela, I. Felger et al., "A recombinant blood-stage malaria vaccine reduces *Plasmodium falciparum* density and exerts selective pressure on parasite populations in a phase 1-2b trial in Papua New Guinea," *Journal of Infectious Diseases*, vol. 185, no. 6, pp. 820–827, 2002.

[9] D. Arnot, "Unstable malaria in Sudan: the influence of the dry season. Clone multiplicity of *Plasmodium falciparum* infections in individuals exposed to variable levels of disease transmission," *Transactions of the Royal Society of Tropical Medicine and Hygiene*, vol. 92, no. 6, pp. 580–585, 1998.

[10] S. L. Takala, A. A. Escalante, O. H. Branch et al., "Genetic diversity in the Block 2 region of the merozoite surface protein 1 (MSP-1) of *Plasmodium falciparum*: additional complexity and selection and convergence in fragment size polymorphism," *Infection, Genetics and Evolution*, vol. 6, no. 5, pp. 417–424, 2006.

[11] M.-T. Ekala, H. Jouin, F. Lekoulou, S. Issifou, O. Mercereau-Puijalon, and F. Ntoumi, "*Plasmodium falciparum* merozoite surface protein 1 (MSP1): genotyping and humoral responses to allele-specific variants," *Acta Tropica*, vol. 81, no. 1, pp. 33–46, 2002.

[12] A. Ghansah, L. Amenga-Etego, A. Amambua-Ngwa et al., "Monitoring parasite diversity for malaria elimination in sub-Saharan Africa," *Science*, vol. 345, no. 6202, pp. 1297–1298, 2014.

[13] I. Soulama, I. Nébié, A. Ouédraogo et al., "*Plasmodium falciparum* genotypes diversity in symptomatic malaria of children living in an urban and a rural setting in Burkina Faso," *Malaria Journal*, vol. 8, article 135, 2009.

[14] M. U. Ferreira and D. L. Hartl, "*Plasmodium falciparum*: worldwide sequence diversity and evolution of the malaria vaccine candidate merozoite surface protein-2 (MSP-2)," *Experimental Parasitology*, vol. 115, no. 1, pp. 32–40, 2007.

[15] M. A. Pacheco, M. Cranfield, K. Cameron, and A. A. Escalante, "Malarial parasite diversity in chimpanzees: the value of comparative approaches to ascertain the evolution of *Plasmodium falciparum* antigens," *Malaria Journal*, vol. 12, article 328, 2013.

[16] W. Yavo, K. N. Ackra, E. I. H. Menan et al., "Etude comparative de quatre techniques de diagnostic biologique du paludisme utilisées en Côte d'Ivoire," *Bulletin de la Société de Pathologie Exotique*, vol. 95, no. 4, pp. 238–240, 2002.

[17] W. Yavo, A. Konaté, F. K. Kassi et al., "Efficacy and safety of Artesunate-Amodiaquine versus Artemether-Lumefantrine in the treatment of uncomplicated *Plasmodium falciparum* malaria in sentinel sites across Côte d'Ivoire," *Malaria Research and Treatment*, vol. 2015, Article ID 878132, 8 pages, 2015.

[18] R. E. L. Paul, I. Hackford, A. Brockman et al., "Transmission intensity and *Plasmodium falciparum* diversity on the northwestern border of Thailand," *The American Journal of Tropical Medicine and Hygiene*, vol. 58, no. 2, pp. 195–203, 1998.

[19] G. Peyerl-Hoffmann, T. Jelinek, A. Kilian, G. Kabagambe, W. G. Metzger, and F. Von Sonnenburg, "Genetic diversity of *Plasmodium falciparum* and its relationship to parasite density in an area with different malaria endemicities in West Uganda," *Tropical Medicine and International Health*, vol. 6, no. 8, pp. 607–613, 2001.

[20] M. K. Bouyou-Akotet, N. P. M'Bondoukwé, and D. P. Mawili-Mboumba, "Genetic polymorphism of merozoite surface protein-1 in *Plasmodium falciparum* isolates from patients with mild to severe malaria in Libreville, Gabon," *Parasite*, vol. 22, article 12, 2015.

[21] A. Ogouyèmi-Hounto, D. K. Gazard, N. Ndam et al., "Genetic polymorphism of merozoite surface protein-1 and merozoite

surface protein-2 in *Plasmodium falciparum* isolates from children in South of Benin," *Parasite*, vol. 20, article 37, 2013.

[22] P. I. Mayengue, A. J. Luty, C. Rogier, M. Baragatti, P. G. Kremsner, and F. Ntoumi, "The multiplicity of *Plasmodium falciparum* infections is associated with acquired immunity to asexual blood stage antigens," *Microbes and Infection*, vol. 11, no. 1, pp. 108–114, 2009.

[23] P. Druilhe, P. Daubersies, J. Patarapotikul et al., "A primary malarial infection is composed of a very wide range of genetically diverse but related parasites," *Journal of Clinical Investigation*, vol. 101, no. 9, pp. 2008–2016, 1998.

[24] S. Jafari-Guemouri, C. Boudin, N. Fievet, P. Ndiaye, and P. Deloron, "*Plasmodium falciparum* genotype population dynamics in asymptomatic children from Senegal," *Microbes and Infection*, vol. 8, no. 7, pp. 1663–1670, 2006.

[25] P. I. Mayengue, M. Ndounga, F. V. Malonga, M. Bitemo, and F. Ntoumi, "Genetic polymorphism of merozoite surface protein-1 and merozoite surface protein-2 in *Plasmodium falciparum* isolates from Brazzaville, Republic of Congo," *Malaria Journal*, vol. 10, article 276, 2011.

[26] A. Aubouy, F. Migot-Nabias, and P. Deloron, "Polymorphism in two merozoite surface proteins of *Plasmodium falciparum* isolates from Gabon," *Malaria Journal*, vol. 2, article 12, 2003.

[27] S. Issifou, S. Djikou, A. Sanni, F. Lekoulou, and F. Ntoumi, "Pas d'influence de la saison de transmission ni de l'âge des patients sur la complexité et la diversité génétique des infections dues à *Plasmodium falciparum* à Cotonou (Bénin)," *Bulletin de la Société de Pathologie Exotique*, vol. 94, no. 2, pp. 195–198, 2001.

[28] M. K. Oyebola, E. T. Idowu, Y. A. Olukosi et al., "Genetic diversity and complexity of *Plasmodium falciparum* infections in Lagos, Nigeria," *Asian Pacific Journal of Tropical Biomedicine*, vol. 4, supplement 1, pp. S87–S91, 2014.

[29] H. Bogreau, F. Renaud, H. Bouchiba et al., "Genetic diversity and structure of African *Plasmodium falciparum* populations in urban and rural areas," *The American Journal of Tropical Medicine and Hygiene*, vol. 74, no. 6, pp. 953–959, 2006.

[30] S. Takala, O. Branch, A. A. Escalante, S. Kariuki, J. Wootton, and A. A. Lal, "Evidence for intragenic recombination in *Plasmodium falciparum*: identification of a novel allele family in block 2 of merozoite surface protein-1: asembo Bay Area Cohort Project XIV," *Molecular and Biochemical Parasitology*, vol. 125, no. 1-2, pp. 163–171, 2002.

[31] S. E. Mara, K. D. Silué, G. Raso et al., "Genetic diversity of *Plasmodium falciparum* among school-aged children from the Man region, western Côte d'Ivoire," *Malaria Journal*, vol. 12, article 419, 2013.

[32] T. Smith, I. Felger, M. Tanner, and H. P. Beck, "Premunition in *Plasmodium falciparum* infection: insights from the epidemiology of multiple infections," *Transactions of the Royal Society of Tropical Medicine and Hygiene*, vol. 93, supplement 1, pp. 59–64, 1999.

[33] T. J. C. Anderson, B. Haubold, J. T. Williams et al., "Microsatellite markers reveal a spectrum of population structures in the malaria parasite *Plasmodium falciparum*," *Molecular Biology and Evolution*, vol. 17, no. 10, pp. 1467–1482, 2000.

[34] D. K. Raj, B. R. Das, A. P. Dash, and P. C. Supakar, "Genetic diversity in the merozoite surface protein 1 gene of *Plasmodium falciparum* in different malaria-endemic localities," *American Journal of Tropical Medicine and Hygiene*, vol. 71, no. 3, pp. 285–289, 2004.

[35] L. Konaté, J. Zwetyenga, C. Rogier et al., "5. Variation of *Plasmodium falciparum* msp1 block 2 and msp2 allele prevalence and of infection complexity in two neighbouring Senegalese villages with different transmission conditions," *Transactions of the Royal Society of Tropical Medicine and Hygiene*, vol. 93, supplement 1, pp. 21–28, 1999.

[36] A. Mayor, F. Saute, J. J. Aponte et al., "*Plasmodium falciparum* multiple infections in Mozambique, its relation to other malariological indices and to prospective risk of malaria morbidity," *Tropical Medicine and International Health*, vol. 8, no. 1, pp. 3–11, 2003.

[37] K. Tanabe, G. Zollner, J. A. Vaughan et al., "*Plasmodium falciparum*: genetic diversity and complexity of infections in an isolated village in western Thailand," *Parasitology International*, vol. 64, no. 3, pp. 260–266, 2015.

[38] M. A. Pacheco, M. Lopez-Perez, A. F. Vallejo et al., "Multiplicity of infection and disease severity in *Plasmodium vivax*," *PLoS Neglected Tropical Diseases*, vol. 10, no. 1, Article ID e0004355, 2016.

[39] K. D. Silué, I. Felger, J. Utzinger et al., "Prévalence, diversité antigénique et multiplicité d'infections de *Plasmodium falciparum* en milieu scolaire au centre de la Côte d'Ivoire," *Médecine Tropicale*, vol. 66, pp. 149–156, 2006.

[40] D. P. Mawili-Mboumba, N. Mbondoukwe, E. Adande, and M. K. Bouyou-Akotet, "Allelic diversity of MSP1 gene in *Plasmodium falciparum* from rural and urban areas of Gabon," *Korean Journal of Parasitology*, vol. 53, no. 4, pp. 413–419, 2015.

[41] J. Zwetyenga, C. Rogier, A. Tall et al., "No influence of age on infection complexity and allelic distribution in *Plasmodium falciparum* infections in Ndiop, a Senegalese village with seasonal, mesoendemic malaria," *The American Journal of Tropical Medicine and Hygiene*, vol. 59, no. 5, pp. 726–735, 1998.

[42] D. A. Müller, J. D. Charlwood, I. Felger, C. Ferreira, V. Do Rosario, and T. Smith, "Prospective risk of morbidity in relation to multiplicity of infection with *Plasmodium falciparum* in São Tomé," *Acta Tropica*, vol. 78, no. 2, pp. 155–162, 2001.

9

Horizontal Gene Transfers from Bacteria to *Entamoeba* Complex: A Strategy for Dating Events along Species Divergence

Miguel Romero, R. Cerritos, and Cecilia Ximenez

Faculty of Experimental Medicine, Experimental Immunology Laboratory, National Autonomous University of Mexico, Dr. Balmis 148, Colonia Doctores, 06720 Mexico City, Mexico

Correspondence should be addressed to R. Cerritos; renecerritos@gmail.com

Academic Editor: Swapnil Sinha

Horizontal gene transfer has proved to be relevant in eukaryotic evolution, as it has been found more often than expected and related to adaptation to certain niches. A relatively large list of laterally transferred genes has been proposed and evaluated for the parasite *Entamoeba histolytica*. The goals of this work were to elucidate the importance of lateral gene transfer along the evolutionary history of some members of the genus *Entamoeba*, through identifying donor groups and estimating the divergence time of some of these events. In order to estimate the divergence time of some of the horizontal gene transfer events, the dating of some *Entamoeba* species was necessary, following an indirect dating strategy based on the fossil record of plausible hosts. The divergence between *E. histolytica* and *E. nuttallii* probably occurred 5.93 million years ago (Mya); this lineage diverged from *E. dispar* 9.97 Mya, while the ancestor of the latter separated from *E. invadens* 68.18 Mya. We estimated times for 22 transferences; the most recent occurred 31.45 Mya and the oldest 253.59 Mya. Indeed, the acquisition of genes through lateral transfer may have triggered a period of adaptive radiation, thus playing a major role in the evolution of the *Entamoeba* genus.

1. Introduction

Entamoeba genus is formed by morphologically similar amoebas; most of them are intestinal parasites that can infect several hosts [1]. *Entamoeba histolytica* is one of the most important intestinal protozoan parasites in humans causing amoebic colitis; they can also invade the liver causing amoebic liver abscess. It is estimated that this parasite causes 70,000 deaths worldwide each year [2]. Furthermore, the *E. dispar* species is morphologically almost identical to *E. histolytica*. However, until today, this has been considered as a commensal organism of the human gut. Nevertheless, *E. dispar* has been detected in patients with symptomatic amoebic colitis and also in the material of amoebic liver abscesses [3]. Very recently, a novel lineage of the *Entamoeba* genus has been detected in the intestine of rhesus macaques *Macaca mulatta*. Moreover, it has been proposed as a candidate to revive the name *E. nuttallii* for this lineage, particularly due to its genetic characteristics.

E. nuttallii infects captive and wild macaques and is capable of causing abscesses in hamster's livers [4]. The species of *Entamoeba invadens* infects reptiles and causes colitis, liver abscesses, and, sometimes, acute death. It has been used as the main encystation model for *Entamoeba* species, since the in vitro culture of *E. dispar* can excyst producing the trophozoites and, thereafter, these trophozoites can undergo encystation in vitro. Phylogenetic reconstructions performed by Stensvold et al., in 2011, based on sequences of the gene for the small subunit of rRNA, clustered together *E. histolytica* and *E. nuttallii* and, basal to the latter node, branched those from *E. dispar*. SSURNA sequences from *E. invadens* branched together with those of *E. ranarum*. Both sequences formed the sister group of a node consisting of more than two-thirds of the *Entamoeba* species included in the analysis [1].

It is well known that horizontal gene transfer (HGT, or lateral gene transfer, LGT), of genetic material between unrelated individuals, has played a significant role in prokaryotic gene acquisition and genome evolution [5, 6]. Over the past few years, its importance in eukaryotic evolution has been reevaluated as it has been found in a higher frequency than expected and related with adaptation to certain niches [7]. Despite its presence in multicellular organisms such as

Bdelloid rotifers [8], it is more likely to occur in unicellular eukaryotes [9]. Alsmark et al., in 2013, analyzing several genomes of protozoa found that *Leishmania mayor, Entamoeba histolytica*, and *Trypanosoma brucei* have the major percentage of genes acquired by lateral transfer with 0.96, 0.68, and 0.47, respectively [10]. Although the phylogenetic discrepancy has been the most reliable method to identify horizontally transferred genes, this latter procedure has been criticized, due to the following arguments: it is known that it might be modified due to methodological artifacts such as substitution saturation or long-branch attraction. Because there are only four bases that constitute nucleic acids, there is a relatively high probability that two nucleotide sequences might share the same bases in a random site by mere chance. This phenomenon is caused regularly by the high molecular substitution rate present in the locus, and its particular unwanted results are the loss of phylogenetic information and the possible high similarity between unrelated sequences. As a whole, this phenomenon is known as substitution saturation and is one of the main problems when analyzing molecular data [11]. HGT may be inferred amiss due to substitution saturation and it must be taken into account on every phylogenetic analysis.

The divergence time estimation for protozoan species is commonly a challenging endeavor, especially at the node calibration step. Even though some protozoan taxa might have fossil record, the most common strategy to calibrate date estimates is the indirect calibration based on animal or plant fossils with a specific underlying biological hypothesis [12]. In fact, the calibration of time estimates performed with protozoan fossil record has proven to be unpractical for extant taxa [13]. Although it has been suggested from gene comparisons that the divergence time between *E. histolytica* and *E. dispar* may have occurred some tenths of millions of years ago [14–16], until now no exhaustive research has been performed on the subject.

In the first annotation of the genome of *E. histolytica* HM1:IMSS reported by Loftus et al., a list of 96 HGT candidates was included, many from bacterial donors [17]. Later, in 2007, Clark et al. updated the analyses and sorted these 96 candidates into different categories according to their consistency in Bayesian and maximum likelihood distance bootstrap trees [18]. From the 96 original candidates, 41 remained strongly supported; 27 turned to be more weakly supported than before; the lateral gene transfer hypothesis of 14 candidates was weakened by increased taxonomic sampling; 9 candidates were found in other microbial eukaryotes; and, in the remaining 5 cases, vertical gene transfer is now the simplest explanation for the observed topology. Horizontal gene transfer remains the strongest hypothesis to explain 68 of the 96 original topologies [18]. But the number of LGT candidates can change according to phylogenetic methodology. For example, recently, in a study by Grant and Katz, in 2014, it is concluded that there are 116 genes of *Entamoeba* having a bacterial or archaeal origin [19]. In laboratory, Field et al. showed that the acetyl-CoA synthetase and the adh1 genes of *E. histolytica* share a common evolutionary history, more related to prokaryotes than other eukaryotes, and suggested that these genes were transferred early [20]. Hand in hand,

Nixon et al. tried to demonstrate that genes for the anaerobic metabolism in *Giardia* and *Entamoeba* genera were obtained laterally; while there was no enough data available to achieve that goal, the authors did reject the amitochondriate fossil and the hydrogen hypotheses to explain the resemblance of these genes to prokaryotic sequences [21].

The objective of this study is, primarily, to estimate the divergence time between *E. histolytica, E. dispar*, and *E. invadens* and then date HGT events of a representative genes, thereof, through the evolution of these species of *Entamoeba*. Representative gene was taken from the list of 68 candidates mentioned above and an additional analysis carried out to distinguish different levels of saturation rates in the DNA sequence, hypothesizing convergence or ancient HGT.

2. Methods

2.1. Gene Selection and Sequence Alignments. Accession numbers of the 68 well supported candidates were obtained from the list in Clark et al., 2007, and then searched in the Amoeba DB database, http://amoebadb.org/amoeba/ [22], in order to get the amino acid and coding sequences.

The following genes were selected to carry out the *Entamoeba* divergence time estimations: DNA-directed RNA polymerase I subunit RPA2 (EHI_186020), elongation factor 2 (EHI_189490), actin (EHI_131230), tubulin gamma chain (EHI_008240), and clathrin heavy chain (EHI_201510) since they are single-copy, housekeeping genes that were not obtained by HGT.

Each one of the HGT candidate amino acid sequences was used as a query against the NCBI Protein Reference Sequence Database (RefSeq) [23] with the BLASTp algorithm [24], using default parameters. The top 50 blast hits were collected for further analyses.

The homologous amino acid sequence from *E. nuttallii* was included in the analyses to calibrate the HGT divergence time estimates. Each *E. histolytica* HGT candidate was used as a query against the *E. nuttallii P19* open reading frame translation database with the BLASTp algorithm. Only the top hit was collected and discarded, if the query coverage and/or identity were less than 60%.

In order to estimate the divergence time of housekeeping genes, sequences from *E. dispar* were downloaded and then corresponding orthologs were searched against the open reading frame translation database of *E. histolytica, E. nuttallii*, and *E. invadens* (available at http://amoebadb.org/common/downloads/) using the BLASTp algorithm, with default parameters. In addition, homologs of the former were looked for in the amino acid sequence database from *Dictyostelium discoideum* (available at http://dictybase.org/) [25] also using a BLASTp search. Only the top hit was collected and discarded if the query coverage or identity was less than 60%.

Each amoeba sequence was aligned with its amoebic ortholog (when a report existed in the Amoeba DB database) and with the prokaryotic sequences found by the BLAST search. In every study, amino acid sequences were aligned using the program Clustal W [26] and then their codon sequences according to the amino acid alignment with Biopython scripting [27]. Sequence alignments were inspected manually and edited to remove synapomorphies and codons with sequencing errors.

2.2. Substitution Saturation Test. Distance matrices were built for the nucleotide alignments. The nucleotide substitution model used was the Maximum Composite Likelihood, with a gamma distribution for the rate variation among sites. All codon positions were included and ambiguous positions were removed for each sequence pair. This analysis was conducted in MEGA software [28].

For each evolutionary distance matrix, sequences with low distance values (equal to or less than 0.1 standard deviations) according to the *E. histolytica* sequences were selected to make a shorter sequence alignment, including only closely related sequences, according to the distance matrices.

Substitution saturation indexes Iss. and Iss.c [12, 29] were calculated for each alignment, considering the three positions of each codon. Whenever a sequence alignment showed substantial saturation after the first analysis, the indexes were calculated again, though we remove the third position of each codon to avoid the possible substitution saturation due to the degeneracy of the genetic code. In these assays, statistical significance value was set at $P < 0.05$. These tests were executed using the package DAMBE [30].

Tree topology examination was necessary to decide whether the alignment was phylogenetically informative for those alignments that showed substantial saturation when excluding the third position of each codon.

2.3. Phylogenetic Analyses. In order to evaluate the phylogenetic relevance of the shorter alignments that presented substantial saturation when removing the third position of each codon, the substitution saturation tests were introduced to the program MrBayes 3.2 [31]. The number of run MCMC generations was 500,000, excluding the third position of each codon; every 125 generations a tree was sampled. Whenever a tree from the latter resulted in asymmetrical topology, the HGT candidate was discarded from the analysis.

In all cases the GTR+I+G nucleotide substitution model was employed, and 25% of tree samples were discarded as burn-in. A consensus tree was constructed from the remaining samples, and then it was inspected manually and edited using Dendroscope [32].

Constructions of consensus trees for donor group designation were made using two different approaches: maximum likelihood and Bayesian phylogenetics. 61 sequence alignments were introduced to the program jModeltest2 [33], in order to find the sequence substitution model that best fitted the observed alignment. Eleven substitution schemes were used, along with relative frequencies per base, proportion of invariable sites, and the variation of substitution rates along the alignment. The base tree to perform each analysis was built with the BioNJ algorithm and the NNI search algorithm. Finally the AICc criterion was used to select the best model for each alignment.

Maximum likelihood trees were built for each alignment by the program PhyML 3 [34]. In each case, the base tree was built with BioNJ and the best tree whether from NNI or SPR search algorithms was selected. One hundred bootstrap tests

were executed per alignment. The same strategy was included in the input for the PhyML software for the candidates that passed the saturation tests, when ignoring the third position of each codon.

Similarly, the alignments that presented no substitution saturation in the first saturation analysis were used as input to construct trees by the program MrBayes [31]. 1,000,000 MCMC generations were run sampling a tree every 200 generations. Also, 1,000,000 MCMC generations were run excluding the third position of each codon for candidates that presented no substitution saturation in the second saturation analysis; a tree was sampled every 200 generations. In both cases, the GTR+I+G substitution model was used, and a consensus tree was built after discarding 25% of the resulting topologies as burn-in. For each of the 61 candidates, the bootstrap values of the coincident nodes in the maximum likelihood trees were added manually to the resulting topology of the Bayesian phylogenetic analyses using the program Dendroscope (Supplementary Information, in Supplementary Material available online at http://dx.doi.org/10.1155/2016/3241027).

2.4. Bayesian Divergence Time Estimates. Two sets of estimations were performed: first, *Entamoeba* divergence time was calculated using a set of five housekeeping genes: DNA-directed RNA polymerase I subunit RPA2, elongation factor 2, actin, tubulin gamma chain, and clathrin heavy chain. Orthologs from *E. histolytica*, *E. nuttallii*, *E. dispar*, *E. invadens*, and *D. discoideum* were included in the dataset. The input tree used for the analysis was the following: ((((*E. nuttallii*, *E. histolytica*), *E. dispar*), *E. invadens*), *D. discoideum*).

Then, the HGT event dates were evaluated only with selected candidates, after considering three criteria: well supported branching in the Bayesian phylogenies, assigned donor group at least at the phylum level, and the presence of an ortholog in *E. nuttallii*. The dataset for each estimation included sequences contained in the alignments and used for the phylogenetic reconstructions from (i) *E. histolytica*, (ii) *E. dispar* and/or *E. invadens*, (iii) up to four randomly chosen sequences from the resulting sister group of *Entamoeba* cluster in the Bayesian phylogenies referred to as "a," "b," "c," and "d," (iv) up to three randomly chosen sequences from the resulting out-group in the Bayesian phylogenies referred to as "x," "y," and "z." Moreover, the homologous sequence from *E. nuttallii* was aligned by-eye with the rest of the dataset. A common user-input tree would look like this: (((((*E. nuttallii*, *E. histolytica*), *E. dispar*), *E. invadens*), ((a, b),(c, d))), ((x, y), z)), even though the relationships within the sister group (a, b, c, and d) might vary.

The estimations were carried out using the programs Estbranches and Multidivtime [35, 36], following the step by step manual by Rutschmann [37]. The node between *E. nuttallii* and *E. histolytica* was used to calibrate the divergence estimations. Since *E. nuttallii* has only been isolated in rhesus macaques and *E. histolytica* has been found in feces from wild baboons (*Papio* sp.) [38, 39], we assumed that the *E. nuttallii* lineage diverged from *E. histolytica* at the same time that the primate lineages *Macaca* and *Papio* did. Paleontological evidence suggests that this divergence occurred after 8 Mya, but before 4 Mya [40, 41]. Consequently the node was calibrated between 5.5 and 6.5 Mya. For each alignment, the program Baseml [42] with the F84+G model was used to estimate nucleotide frequencies, transition/transversion rate ratio (parameter κ), and rate heterogeneity among sites (shape parameter α). Then, the maximum likelihood of the branch lengths of the tree and the variance-covariance matrix were estimated by the Estbranches program. Finally, a Bayes MCMC analysis was performed with the program Multidivtime, to approximate the posterior distributions of substitution rates and divergence times. A total of 5,100,000 generations were run, 100,000 were discarded as burn-in, and then a sample was taken every 100 generations.

For the *Entamoeba* divergence time, the five housekeeping genes were analyzed simultaneously, 5,100,000 generations were run, 100,000 were discarded as burn-in, and then a sample was taken every 100 generations. Time units were set to million years and referred to as "million years ago" (Mya). For the prior parameters, we selected 100 time units between the tip and the root of the tree, with a standard deviation of 50 time units, and an oldest time value of 300. For each candidate, the mean and standard deviation of prior distribution for the rate of molecular evolution at the in-group root node were set as the median of the evolution rates provided by Estbranches. The divergence time estimates were carried out in triplicate to confirm similar results of the analysis between repetitions. Results are showed as Mya ± the standard deviation provided by Multidivtime.

3. Results

3.1. Substitution Saturation Tests. Substitution saturation indexes Iss. and Iss.c [12, 28] were calculated for each alignment considering the first two or the three positions of each codon. The index Iss. is a measure of entropy of a given nucleotide sequence alignment. The index Iss.c is the measure of entropy of a simulated sequence alignment that shares the number of sequences and number of sites with the former but has a random distribution of nucleotide bases. Hence, if the Iss. value approaches that of Iss.c, it is a signal that the sequence alignment holds high substitution saturation. Both indexes were calculated for each shorter alignment. When using the three sites of each codon, 38 alignments displayed lower Iss. values than their respective Iss.c values; these differences were statistically significant ($P < 0.05$), implying little saturation. In 14 cases, the differences between Iss. and Iss.c were not statistically significant. Other 10 candidates display the same behavior, and in the remaining 6 cases the value of Iss. index was higher than Iss.c and, also, differences were statistically significant. In the second essay, in which the first 38 sequences were not included, every third position of each codon was ignored, in order to avoid the possible substitution saturation observed due to the degeneracy of the genetic code. Altogether, 15 alignments resulted in a significantly lower Iss. index, and other 4 alignments had a significantly higher Iss. than their respective Iss.c. The remaining 11 alignments showed nonsignificant differences; therefore tree topology was needed to evaluate their phylogenetic usefulness. To this end, 11 trees were built:

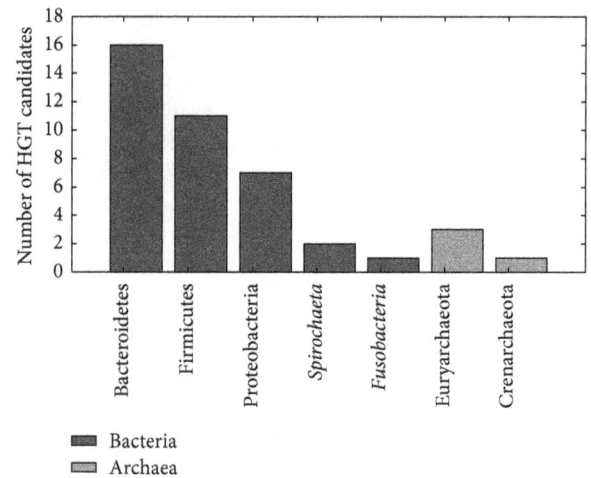

FIGURE 1: Number of genes obtained by each donor phylum identified in phylogenetic trees.

3 of them showed asymmetrical topology and 8 presented symmetrical topology. The respective HGT candidates from the 3 asymmetric trees, alongside those candidates from the 4 alignments whose Iss.c values were significantly higher in the second test, were permanently discarded from the research, since our results strongly denote that these alignments lack phylogenetic information.

3.2. Bayesian Phylogenetics and Putative Donor Groups. The assays were carried out with the lingering 61 candidates, using the complete coding sequence alignments. Phylogenetic analyses were made with the program MrBayes [31], 1,000,000 MCMC generations were run sampling a tree every 200 generations, using the first two or the three positions of each codon. When evaluating donor groups in the 38 trees constructed with complete codons, it was possible to locate a donor group at least at phylum level (Figure 1), as well as in 15 trees built, excluding the third position.

On the other hand, in three different cases, it was only possible to assign a donor group at the level of domain because the sister group of the *Entamoeba* cluster was formed by sequences that belonged to different phyla but from the same domain. A total of 61 consensus trees were built (Supplementary Information). Donor taxa could not be identified in the remaining five topologies, due to the different domains included and no apparent association with the amoebic genes. The domain Archaea was assigned to only four out of the 61 analyzed candidates, 3 of which belonged to the Euryarchaeota phylum and only one branched exclusively with sequences from Methanococcales. The bulk of the genes branched with bacterial sequences, from which 12 had no clear association with any phylum. Bacteroidetes was the most prevalent donor group with sixteen donated genes; moreover, ten of them were probably transferred from the order Bacteroidales. In 6 trees the *Entamoeba* genes branched inside a larger cluster with high posterior probability values. Alternatively, in 9 other cases amoebic genes were separated from their basal group by a large evolutionary distance, but being branched with their sister group always showed high supporting values. The second most abundant donor group was the phylum Firmicutes, even though only in 4—out of the eleven candidates—a donor order could be designated. One gene branched strongly with sequences from Bacillales and the other 3 branched with Clostridiales. In most cases, the posterior probability of every node between the *Entamoeba* genes and their sister group was close to 1.0, with the exception of the type A flavoprotein (EHI_09671), which grouped with several bacterial clusters through a polytomy. The phylum Proteobacteria was designated as the donor group for a total of seven genes; this was the phylum which presented the highest diversity of orders: Campylobacterales, Pseudomonadales, Burkholderiales, and Enterobacteriales. Despite the fact that just one gene belonged to each donor order, each one of them grouped with high posterior probability, with the exception of Fe-S cluster assembly protein (EHI_049620) whose posterior probability was 0.7 in the node between the amoebic cluster and sequences from Campylobacterales. Although one candidate branched poorly (posterior probability: 0.64) with sequences from Fusobacteria, it was not possible to determine a donor order. In up to 5 trees, the sequence of *E. invadens* did not branch with its *Entamoeba* ortholog but with prokaryotic sequences or as basal group instead.

3.3. Divergence Time Estimates. *Entamoeba* divergence time was calculated with the following set of five housekeeping genes: DNA-directed RNA polymerase I subunit RPA2, elongation factor 2, actin, tubulin gamma chain, and clathrin heavy chain. Orthologs from *E. histolytica*, *E. nuttallii*, *E. dispar*, *E. invadens*, and *D. discoideum* made up the dataset.

The median rate of molecular evolution among the five amoebic genes provided by Estbranches was 0.02364 substitutions per site per million years, which was then used as the mean and standard deviation of prior distribution for the rate of molecular evolution at the in-group root node for the Multidivtime program. Finally, the estimates for the split ages between these lineages were the following: split date between *E. nuttallii* and *E. histolytica*, 5.93 ± 0.28 Mya, between *E. histolytica* and *E. dispar*, 9.97 ± 1.37 Mya, and between *E. histolytica* and *E. invadens*, 68.18 ± 16.04 Mya (Figure 3).

Twenty-two *Entamoeba* candidates were selected after considering three criteria: well supported branching in the Bayesian phylogenies, assigned donor group at least at the phylum level, and the presence of an ortholog in *E. nuttallii*. The most recent transference was that of gene endo-1,4-beta-xylanase (EHI_096280) from Bacteroidetes dated 31.45 ± 15.69 Mya. The oldest transferences occurred 253.59 ± 28.91 Mya when the gene tartrate dehydrogenase (EHI_143560) was donated by Proteobacteria (Figure 2). The median of molecular evolution for this gene was 0.00089 substitutions per site per million years; interestingly, this rate is smaller than the final rate of substitutions for the five amoebic housekeeping genes, which was 0.0014 substitutions per site per million years. This slow rate explains why such an old HGT event is still detectable and also why a donor group could still be determined for this gene. This is interesting because it is possible that some other HGT events may have been masked because of higher nucleotide substitution rates

and the homogenization of the xenolog gene to the recipient genome.

Several overlapping transference dates were found, some of them from the same donor group: alpha-1,2-mannosidase (EHI_009520), mannose-1-phosphate guanylyltransferase (EHI_052810), and fructokinase (EHI_054510) from Bacteroidetes ranging from 55.53 Mya to 77.8715 Mya, nicotinate-phosphoribosyltransferase (EHI_023260) and hypothetical protein (EHI_072640) from Bacteroidales ranging from 94.98 Mya to 164.16 Mya, and Fe-S cluster assembly protein NifU (EHI_049620) and metallo-beta-lactamase family protein (EHI_068560) from Proteobacteria ranging from 119.89 Mya to 176.9304 Mya. Gene synteny in the genome of *E. histolytica* and functional group information were still necessary to define simultaneous horizontal gene transfer events.

4. Discussion

In this study the substitution saturation of the well supported HGT gene candidates from the genome of *E. histolytica* was verified and assigned a putative donor group for each candidate through phylogenetic reconstruction. In addition, a first approach into the divergence time estimation of some species of *Entamoeba* through indirect node calibration was presented, using the fossil record of their feasible hosts. Finally the gene transfer events of some HGT candidates were dated, revealing gene losses, postdivergence transfers, and a simultaneous transfer of two genes. The BLAST search results were able to provide a glimpse of the analysis outcome, since the top hits that resulted in the highest e-values (e-13 and e-16 for EHI_085050 and EHI_156240, resp.) belonged to candidates discarded because of substitution saturation. Moreover, for the 3 candidates whose top hits were sequences from Archaea, the latter domain was the putative donor group after inspecting the Bayesian phylogenies. It is interesting that most of the *Entamoeba* HGT candidate genes have no or few paralogs in the genome of the different species of *Entamoeba* included in the analyses, while some others had at least 3 paralogs in each genome. The former diversification may be result of neutral evolution in the case of the gene that encodes the metallo-beta-lactamase superfamily protein (EHI_115720), considering that beta-lactam antibiotics induce bacterial cell wall degradation and therefore are innocuous to *Entamoeba* species [43]. On the other hand, two of the candidates with the largest gene families, hypothetical protein and aldehyde-alcohol dehydrogenase 2 (EHI_104900 and EHI_160940, resp.), were discarded after the substitution saturation tests. It is likely that these gene families are result of an ancient HGT and have lost phylogenetic information, due to mutational saturation; or they have been acquired through vertical descent and they are similar to bacterial sequences, as a result of the same mechanism.

The procedures here presented managed to find donor groups at the order level, including the following: Bacteroidales, Clostridiales, Spirochaetales, Campylobacterales, Burkholderiales, Bacillales, Flavobacteriales, Methanococcales, and Enterobacteriales. Most of the donor taxa can be found in the gut of vertebrates, it is well known that the three bacterial phyla are major part of the gut microbiota,

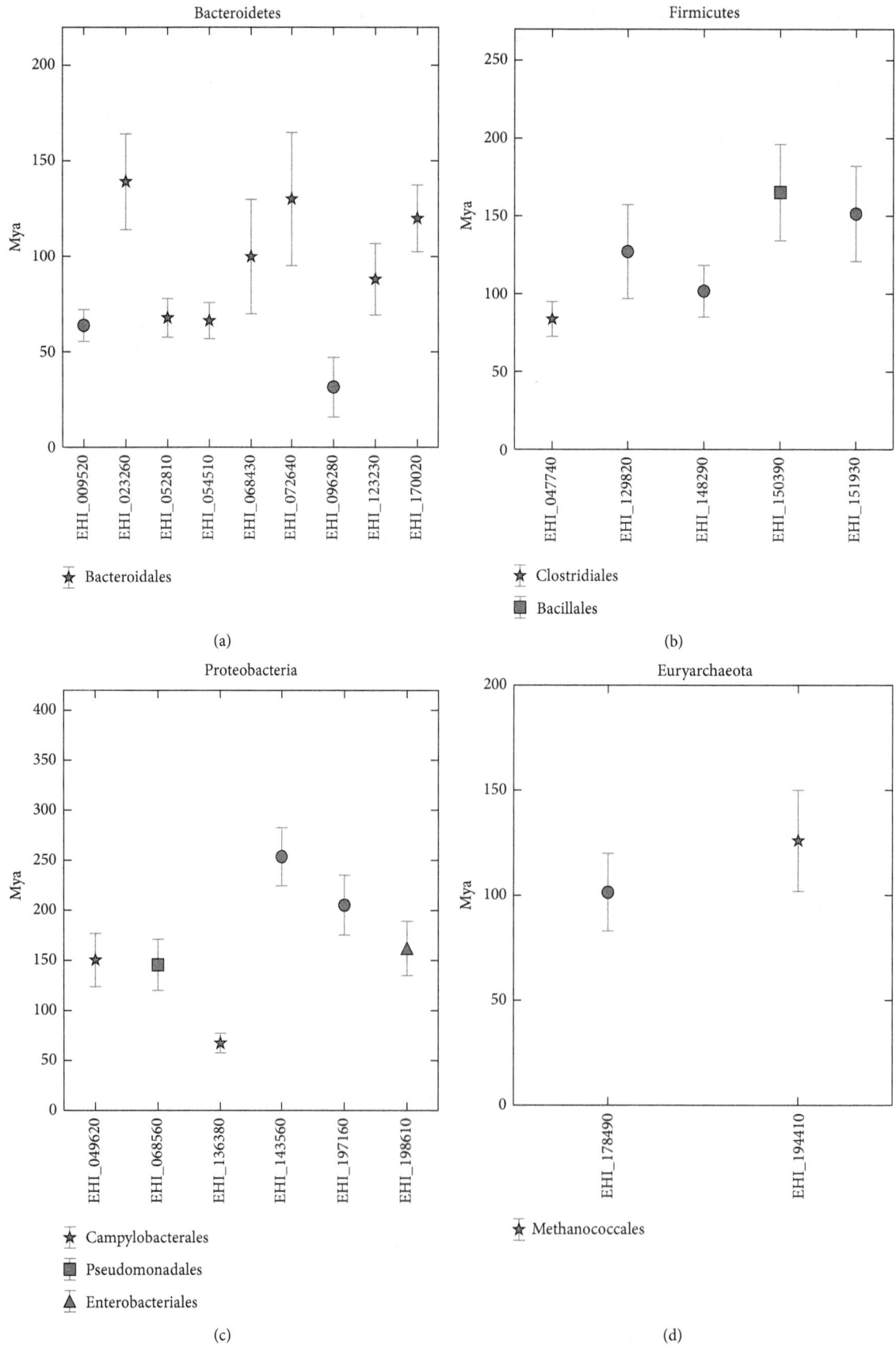

FIGURE 2: Times of horizontal gene transfer by phylogenetic groups. Measures of dispersion by standard deviation. Some simultaneous events specifically in group of Bacteroidales can be proposed.

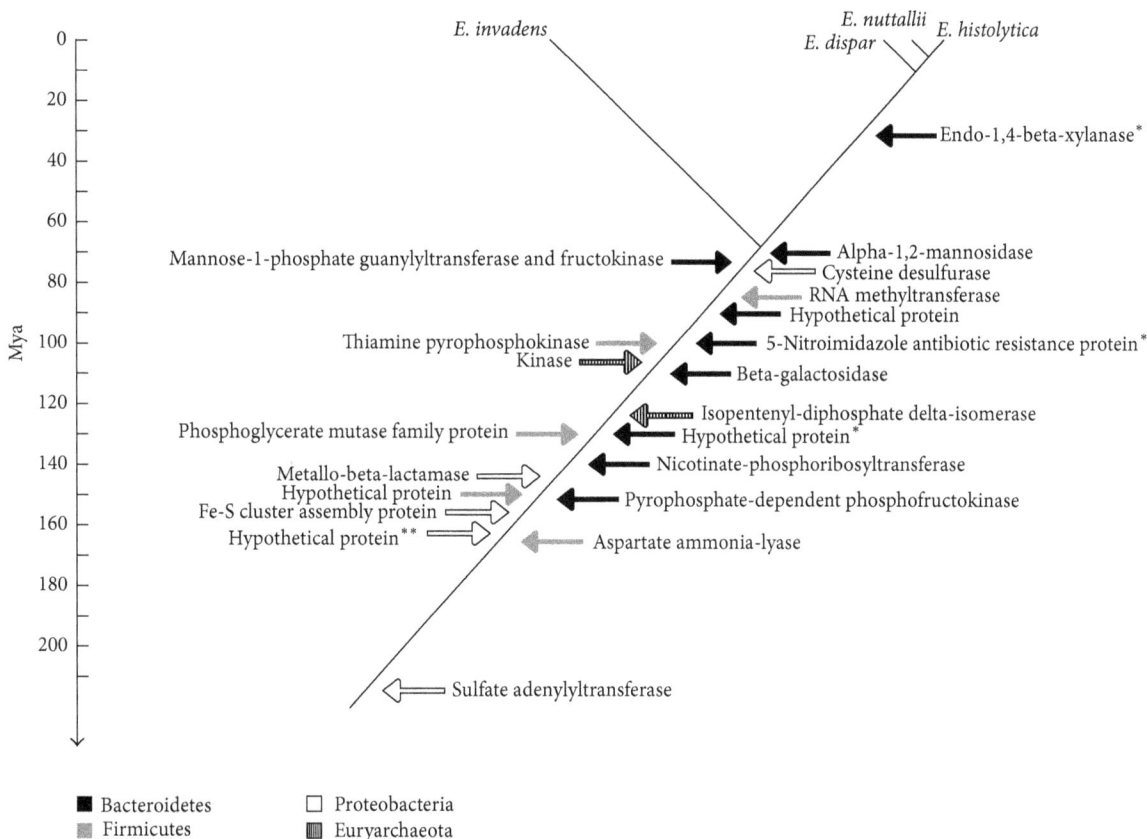

FIGURE 3: Divergence times of *Entamoeba* species, as well as horizontal transfers events. The units are Map (million years ago). *Genes without counterparts in the genome of *E. invadens*. **No homologous genes in the genome of *E. dispar*.

but other less abundant groups have donated genetic material to *Entamoeba* species such as anaerobic Archaea. Although most members of the *Entamoeba* genus are parasitic or commensal organisms, lineages of free living *Entamoeba* like *E. ecuadoriensis* have been found [1], and some of these xenolog genes have been acquired from free living donor groups.

The Bacteroidetes phylum monopolizes lateral gene donations to the *Entamoeba* species included in this analysis, and the second most abundant group is the phylum Firmicutes and then Proteobacteria (Figure 1). These results are in contrast with those reported previously [44], in which they found 16 candidates closely related to Proteobacteria; nevertheless we found only 8. Likewise, we found no traces of HGT from Actinobacteria. These differences may result of the increased sampling due to the growth of biological databases and the substitution saturation tests. Consistent results are the frequency of LGT from Bacteroidetes, Spirochaetes, and Fusobacteria. In fact, the phylum Bacteroidetes has been found as potential donor to other horizontally transferred genes in different organisms, such as Ciliates [45] and Dinoflagellates [46], thus confirming the promiscuity of this taxon.

Several studies have highlighted the ecological relationship between *E. histolytica* and bacteria, specifically during pathogenesis [47, 48]. This study supports the importance of these associations as they can provide evolutionary innovations to the genus, and although no virulence factors have

been transferred, antibiotic resistance genes are among the 61 candidates. In fact, the gene 5-nitroimidazole antibiotic resistance protein (EHI_068430) has been transferred from Bacteroidales. As most HGT events were millions of years ago, it is unlikely that these genes have functioned as acquired adaptations against antibiotics. These genes might have had other functions such as secondary metabolite degradation or no function at all, before the antibiotics were a selective pressure in the human gut.

Although the alignment for protein serine acetyltransferase (EHI_202040) resulted in little saturation while excluding the third position of each codon and its phylogeny showed a symmetric topology, it is very unlikely that an ancestor of the *Entamoeba* genus had obtained this from halophilic archaea and probably other bacteria group can be the donor.

Since estimating the age of the gene transfer events was one of the main objectives of this study, it was necessary to approximate the divergence times of the species of *Entamoeba*. To accomplish this aim, the identification of *E. nuttallii* as a separate lineage provided crucial information. The fact that *E. nuttallii* has only been isolated from rhesus macaques and *E. histolytica* has been found in feces from wild baboons [38, 39] led to the assumption that the *E. nuttallii* and *E. histolytica* lineages were separated simultaneously with their hosts at some time between 4 and 8 Mya according to the fossil record, although it has been assumed that this interval is narrower [49]. The results from the amoebic species split date calculations might be underestimated by the fact that only one node was calibrated, as there is no direct fossil record of species of *Entamoeba*. Although other authors have made studies regarding the age of the Amoebozoa phylum as a whole using animal and plant fossils, it was not possible to use their results because of differences in time scale and classification [13]. Some divergence time estimates of HGT candidates have overlapping standard deviations; this is particularly interesting when the donor groups coincide, because these genes could have been transferred at the same time. To determine if these genes were transferred simultaneously, some characteristics were taken into account: gene transfer age, donor group, metabolic context, and location in the genome. It has been suggested that functionally related genes might be located closely especially in prokaryotic genomes [50], and it should be expected that a simultaneous horizontal gene transfer would result in one or more xenologs positioned near one to another and functionally related. Two genes shared most of these attributes: the genes for the mannose-1-phosphate guanylyltransferase (EHI_052810) and the fructokinase (EHI_054510) were donated by Bacteroidales, probably between 60 and 70 million years in the past. Both genes are involved in fructose, mannose, amino sugar, and nucleotide sugar metabolism.

The dating of some transfers explained why certain genes are absent in some of the genomes of the amoebic species, included in this analysis (Figure 3). The transfer of the endo-1,4-beta-xylanase (EHI_096280) occurring 31.45 ± 15.69 Mya is in fact more recent than the divergence between the lineage of *E. invadens* and the ancestor of the other species of *Entamoeba* (68.18 ± 16.04 Mya); thus this gene was never present in the genome of the immediate ancestor of *E. invadens*. Conversely, the coding sequence for the 5-nitroimidazole-resistance protein (EHI_068430), which was transferred earlier (99.82 ± 29.94 Mya), was lost afterwards by *E. invadens*. The same conclusion could be applied to the gene coding for the hypothetic protein (EHI_198610), which was obtained 162.08 ± 27.07 Mya from Proteobacteria, which is now absent in the genome of *E. dispar*.

E. histolytica remains as one of the protists with the highest number of laterally transferred genes from bacterial origin in their genomes along with *Trichomonas vaginalis* and *Giardia lamblia* [9] or along with *Leishmania mayor* and *Trypanosoma brucei* [19]. This large uptake of bacterial genes, which in general took place relatively early in the evolutionary history of the *Entamoeba* genus, may have functioned as a trigger for adaptive evolution. The latter assertion may be palpable in the case of the genes coding for the acetyl-CoA synthetase and the adh1; but other genes gained through HGT, whose functions are unknown or obscured by biased annotations, may have been also important in the evolution of these organisms. The ancestor of the genus *Entamoeba*, which in our point of view, might as well be the ancestor of *Endolimax*, equipped with this newly acquired genes, might have tried exploring new ecosystems and forms of life and eventually settled in the gut of vertebrates.

Competing Interests

The authors declare that they have no competing interests.

Acknowledgments

The authors thank the partial support by Grant no. 79220 from the National Council of Science and Technology (CONACyT). Thanks are due to Valeria Zermeño, Lilia González-Ceron, and Rocío Incera for proofreading and translation review.

References

[1] C. R. Stensvold, M. Lebbad, E. L. Victory et al., "Increased sampling reveals novel lineages of entamoeba: consequences of genetic diversity and host specificity for taxonomy and molecular detection," *Protist*, vol. 162, no. 3, pp. 525–541, 2011.

[2] World Health Organization, *The World Health Report 1998—Life in the 21st Century: A Vision for All*, World Health Organization, Geneva, Switzerland, 1998.

[3] C. Ximénez, R. Cerritos, L. Rojas et al., "Human amebiasis: breaking the paradigm?" *International Journal of Environmental Research and Public Health*, vol. 7, no. 3, pp. 1105–1120, 2010.

[4] H. Tachibana, T. Yanagi, C. Lama et al., "Prevalence of *Entamoeba nuttalli* infection in wild rhesus macaques in Nepal and characterization of the parasite isolates," *Parasitology International*, vol. 62, no. 2, pp. 230–235, 2013.

[5] E. V. Koonin, K. S. Makarova, and L. Aravind, "Horizontal gene transfer in prokaryotes: quantification and classification," *Annual Review of Microbiology*, vol. 55, no. 1, pp. 709–742, 2001.

[6] P. R. Marri, W. Hao, and G. B. Golding, "The role of laterally transferred genes in adaptive evolution," *BMC Evolutionary Biology*, vol. 7, no. supplement 1, article S8, 2007.

[7] P. J. Keeling and J. D. Palmer, "Horizontal gene transfer in eukaryotic evolution," *Nature Reviews Genetics*, vol. 9, no. 8, pp. 605–618, 2008.

[8] E. A. Gladyshev, M. Meselson, and I. R. Arkhipova, "Massive horizontal gene transfer in bdelloid rotifers," *Science*, vol. 320, no. 5880, pp. 1210–1213, 2008.

[9] J. W. Whitaker, G. A. McConkey, and D. R. Westhead, "The transferome of metabolic genes explored: analysis of the horizontal transfer of enzyme encoding genes in unicellular eukaryotes," *Genome Biology*, vol. 10, no. 4, article R36, 2009.

[10] C. Alsmark, P. G. Foster, T. Sicheritz-Ponten, S. Nakjang, T. M. Embley, and R. P. Hirt, "Patterns of prokaryotic lateral gene transfers affecting parasitic microbial eukaryotes," *Genome Biology*, vol. 14, article R19, 2013.

[11] X. Xia, Z. Xie, M. Salemi, L. Chen, and Y. Wang, "An index of substitution saturation and its application," *Molecular Phylogenetics and Evolution*, vol. 26, no. 1, pp. 1–7, 2003.

[12] O. Fiz-Palacios, M. Romeralo, A. Ahmadzadeh, S. Weststrand, P. E. Ahlberg, and S. Baldauf, "Did terrestrial diversification of amoebas (amoebozoa) occur in synchrony with land plants?" *PLoS ONE*, vol. 8, no. 9, Article ID e74374, 2013.

[13] C. Berney and J. Pawlowski, "A molecular time-scale for eukaryote evolution recalibrated with the continuous microfossil record," *Proceedings of the Royal Society B: Biological Sciences*, vol. 273, no. 1596, pp. 1867–1872, 2006.

[14] C. G. Clark and L. S. Diamond, "Ribosomal RNA genes of 'pathogenic' and 'nonpathogenic' *Entamoeba histolytica* are

distinct," *Molecular and Biochemical Parasitology*, vol. 49, no. 2, pp. 297–302, 1991.

[15] S. Novati, M. Sironi, S. Granata et al., "Direct sequencing of the PCR amplified SSU rRNA gene of *Entamoeba dispar* and the design of primers for rapid differentiation from *Entamoeba histolytica*," *Parasitology*, vol. 112, no. 4, pp. 363–369, 1996.

[16] D. Sehgal, V. Mittal, S. Ramachandran, S. K. Dhar, A. Bhattacharya, and S. Bhattacharya, "Nucleotide sequence organisation and analysis of the nuclear ribosomal DNA circle of the protozoan parasite *Entamoeba histolytica*," *Molecular and Biochemical Parasitology*, vol. 67, no. 2, pp. 205–214, 1994.

[17] B. Loftus, I. Anderson, R. Davies et al., "The genome of the protist parasite *Entamoeba histolytica*," *Nature*, vol. 433, no. 7028, pp. 865–868, 2005.

[18] C. G. Clark, U. C. M. Alsmark, M. Tazreiter et al., "Structure and content of the entamoeba histolytica genome," *Advances in Parasitology*, vol. 65, pp. 51–190, 2007.

[19] J. R. Grant and L. A. Katz, "Phylogenomic study indicates widespread lateral gene transfer in *Entamoeba* and suggests a past intimate relationship with parabasalids," *Genome Biology and Evolution*, vol. 6, no. 9, pp. 2350–2360, 2014.

[20] J. Field, B. Rosenthal, and J. Samuelson, "Early lateral transfer of genes encoding malic enzyme, acetyl-CoA synthetase and alcohol dehydrogenases from anaerobic prokaryotes to *Entamoeba histolytica*," *Molecular Microbiology*, vol. 38, no. 3, pp. 446–455, 2000.

[21] J. E. J. Nixon, A. Wang, J. Field et al., "Evidence for lateral transfer of genes encoding ferredoxins, nitroreductases, NADH oxidase, and alcohol dehydrogenase 3 from anaerobic prokaryotes to *Giardia lamblia* and *Entamoeba histolytica*," *Eukaryotic Cell*, vol. 1, no. 2, pp. 181–190, 2002.

[22] C. Aurrecoechea, A. Barreto, J. Brestelli et al., "AmoebaDB and MicrosporidiaDB: functional genomic resources for Amoebozoa and Microsporidia species," *Nucleic Acids Research*, vol. 39, supplement 1, pp. D612–D619, 2011.

[23] K. D. Pruitt, T. Tatusova, G. R. Brown, and D. R. Maglott, "NCBI Reference Sequences (RefSeq): current status, new features and genome annotation policy," *Nucleic Acids Research*, vol. 40, no. 1, pp. D130–D135, 2012.

[24] M. Johnson, I. Zaretskaya, Y. Raytselis, Y. Merezhuk, S. McGinnis, and T. L. Madden, "NCBI BLAST: a better web interface," *Nucleic Acids Research*, vol. 36, pp. W5–W9, 2008.

[25] P. Fey, R. J. Dodson, S. Basu, and R. L. Chisholm, "One stop shop for everything *Dictyostelium*: dicty base and the Dicty Stock Center in 2012," in *Dictyosteliumdiscoideum Protocols*, L. Eichinger and F. Rivero, Eds., pp. 59–92, Humana Press, New York, NY, USA, 2013.

[26] M. A. Larkin, G. Blackshields, N. P. Brown et al., "Clustal W and clustal X version 2.0," *Bioinformatics*, vol. 23, no. 21, pp. 2947–2948, 2007.

[27] P. J. A. Cock, T. Antao, J. T. Chang et al., "Biopython: freely available python tools for computational molecular biology and bioinformatics," *Bioinformatics*, vol. 25, no. 11, pp. 1422–1423, 2009.

[28] K. Tamura, D. Peterson, N. Peterson, G. Stecher, M. Nei, and S. Kumar, "MEGA5: molecular evolutionary genetics analysis using maximum likelihood, evolutionary distance, and maximum parsimony methods," *Molecular Biology and Evolution*, vol. 28, no. 10, pp. 2731–2739, 2011.

[29] X. Xia and P. Lemey, "Assessing substitution saturation with DAMBE," in *The Phylogenetic Handbook: A Practical Approach to DNA and Protein Phylogeny*, P. Lemy, M. Salemi, and A. M. Vandamme, Eds., pp. 615–630, Cambridge University Press, Cambridge, UK, 2009.

[30] X. Xia and Z. Xie, "DAMBE: software package for data analysis in molecular biology and evolution," *Journal of Heredity*, vol. 92, no. 4, pp. 371–373, 2001.

[31] F. Ronquist, M. Teslenko, P. van der Mark et al., "Mrbayes 3.2: efficient bayesian phylogenetic inference and model choice across a large model space," *Systematic Biology*, vol. 61, no. 3, pp. 539–542, 2012.

[32] D. H. Huson, D. C. Richter, C. Rausch, T. Dezulian, M. Franz, and R. Rupp, "Dendroscope: an interactive viewer for large phylogenetic trees," *BMC Bioinformatics*, vol. 8, no. 1, article 460, 2007.

[33] D. Darriba, G. L. Taboada, R. Doallo, and D. Posada, "JModelTest 2: more models, new heuristics and parallel computing," *Nature Methods*, vol. 9, no. 8, p. 772, 2012.

[34] S. Guindon and O. Gascuel, "A simple, fast, and accurate algorithm to estimate large phylogenies by maximum likelihood," *Systematic Biology*, vol. 52, no. 5, pp. 696–704, 2003.

[35] H. Kishino, J. L. Thorne, and W. J. Bruno, "Performance of a divergence time estimation method under a probabilistic model of rate evolution," *Molecular Biology and Evolution*, vol. 18, no. 3, pp. 352–361, 2001.

[36] J. L. Thorne, H. Kishino, and I. S. Painter, "Estimating the rate of evolution of the rate of molecular evolution," *Molecular Biology and Evolution*, vol. 15, no. 12, pp. 1647–1657, 1998.

[37] F. Rutschmann, *Bayesian Molecular Dating Using PAML/Multidiv Time. A Step-by-Step Manual*, University of Zurich, Zurich, Switzerland, 2005, http://www.plant.ch.

[38] R. E. Kuntz and B. J. Myers, "Parasites of baboons (*Papio doguera* (Pucheran, 1856)) captured in Kenya and Tanzania, East Africa," *Primates*, vol. 7, no. 1, pp. 27–32, 1966.

[39] T. F. Jackson, P. G. Sargeaunt, P. S. Visser, V. Gathiram, S. Suparsad, and C. B. Anderson, "Entamoeba histolytica: naturally occurring infections in baboons," *Archivos de Investigacion Medica*, vol. 21, supplement 1, pp. 153–156, 1990.

[40] E. Delson, I. Tattersall, J. A. Van Couvering, and A. S. Brooks, *Encyclopedia of Human Evolution and Prehistory*, Garland Pub & Grill, New York, NY, USA, 2000.

[41] N. G. Jablonski, "Fossil old world monkeys: the late neogene radiation," in *The Primate Fossil Record*, Cambridge Studies in Biological and Evolutionary Anthropology, pp. 255–299, Cambridge University Press, Cambridge, UK, 2002.

[42] Z. Yang, "PAML 4: phylogenetic analysis by maximum likelihood," *Molecular Biology and Evolution*, vol. 24, no. 8, pp. 1586–1591, 2007.

[43] K. Kitano and A. Tomasz, "Triggering of autolytic cell wall degradation in *Escherichia coli* by beta-lactam antibiotics," *Antimicrobial Agents and Chemotherapy*, vol. 16, no. 6, pp. 838–848, 1979.

[44] U. C. Alsmark, T. Sicheritz-Ponten, P. G. Foster, R. P. Hirt, and T. M. Embley, "Horizontal gene transfer in eukaryotic parasites: a case study of *Entamoeba histolytica* and *Trichomonas vaginalis*," *Methods in Molecular Biology*, vol. 532, pp. 489–500, 2009.

[45] G. Ricard, N. R. McEwan, B. E. Dutilh et al., "Horizontal gene transfer from bacteria to rumen ciliates indicates adaptation to their anaerobic, carbohydrates-rich environment," *BMC Genomics*, vol. 7, no. 1, article 22, 2006.

[46] K. Moszczyński, P. MacKiewicz, and A. Bodyl, "Evidence for horizontal gene transfer from bacteroidetes bacteria to

dinoflagellate minicircles," *Molecular Biology and Evolution*, vol. 29, no. 3, pp. 887–892, 2012.

[47] H.-I. Cheun, S.-H. Cho, J.-H. Lee et al., "Infection status of hospitalized diarrheal patients with gastrointestinal protozoa, bacteria, and viruses in the Republic of Korea," *The Korean Journal of Parasitology*, vol. 48, no. 2, pp. 113–120, 2010.

[48] J. M. Galván-Moroyoqui, M. del Carmen Domínguez-Robles, E. Franco, and I. Meza, "The interplay between *Entamoeba* and enteropathogenic bacteria modulates epithelial cell damage," *PLoS Neglected Tropical Diseases*, vol. 2, no. 7, article e266, 2008.

[49] M. E. Steiper and N. M. Young, "Primate molecular divergence dates," *Molecular Phylogenetics and Evolution*, vol. 41, no. 2, pp. 384–394, 2006.

[50] T. Dandekar, B. Snel, M. Huynen, and P. Bork, "Conservation of gene order: a fingerprint of proteins that physically interact," *Trends in Biochemical Sciences*, vol. 23, no. 9, pp. 324–328, 1998.

10

A Brief View of the Surface Membrane Proteins from *Trypanosoma cruzi*

Ángel de la Cruz Pech-Canul,[1] Victor Monteón,[2] and Rosa-Lidia Solís-Oviedo[1,2]

[1]*Centre for Biomolecular Sciences, The University of Nottingham, University Park, University Blvd, Nottingham NG7 2RD, UK*
[2]*Investigaciones Biomédicas, Universidad Autónoma de Campeche, Av. Patricio Trueba s/n, Col. Lindavista, 24039 Campeche, CAM, Mexico*

Correspondence should be addressed to Rosa-Lidia Solís-Oviedo; solisoviedo@gmail.com

Academic Editor: José F. Silveira

Trypanosoma cruzi is the causal agent of Chagas' disease which affects millions of people around the world mostly in Central and South America. *T. cruzi* expresses a wide variety of proteins on its surface membrane which has an important role in the biology of these parasites. Surface molecules of the parasites are the result of the environment to which the parasites are exposed during their life cycle. Hence, *T. cruzi* displays several modifications when they move from one host to another. Due to the complexity of this parasite's cell surface, this review presents some membrane proteins organized as large families, as they are the most abundant and/or relevant throughout the *T. cruzi* membrane.

1. Introduction

Trypanosoma cruzi is a protozoan causative of Chagas' disease, a pathology characterized by two phases: acute and chronic; both could be asymptomatic. The acute phase is present during the first weeks of infection and the chronic phase includes an indeterminate asymptomatic form and a chronic inflammation associated with myocarditis, heart failure, and megaviscera (megaesophagus and/or megacolon) [1, 2]. *T. cruzi* has infected millions of people in the world, mostly in Central and South America; the infection could be via triatomine insect vector, congenital transmission, organ transplantation, or blood transfusion [3, 4]. The life cycle of *T. cruzi* comprises several morphological transformations involving both mammalian and insect hosts, where three different major developmental stages are identified: epimastigotes, trypomastigotes, and amastigotes (Figure 1) [5, 6]. The developmental stages of *T. cruzi* alternate between noninfective and infective forms. Epimastigote and amastigote are noninfective but replicative stages in the gut of the insect vector and inside the mammalian cell, respectively. Trypomastigote stage is infective but nonreplicative and can

be also considered as two different developmental stages: the bloodstream trypomastigotes, found in the blood of the vertebrate host, and the metacyclic trypomastigotes, found in the rectum of the insect vector [6–9]. If one considers that the cycle starts with insect sucking the blood of mammalian host infected with the bloodstream trypomastigotes, the ingested trypomastigotes transform into epimastigotes inside the insect stomach and replicate intensely in the midgut. After that, metacyclic trypomastigotes arise from epimastigotes in the hindgut of the insect host which are eliminated with the faeces [6, 10]. When the insect vector takes a blood meal from a new noninfected host it subsequently defecates in the area near the puncture wound. The infection usually takes place through direct inoculation of excreted metacyclic trypomastigote which forms into the lesioned skin caused by the insect vector bite. Once inside the mammalian host, the metacyclic trypomastigote forms invade the host cells at the inoculation site and transform into the replicative amastigote form. Upon completion of a replicative cycle as intracellular amastigotes, they transform back into bloodstream trypomastigote forms which burst the eukaryotic cell host and are capable of infecting other cells or travel into the bloodstream

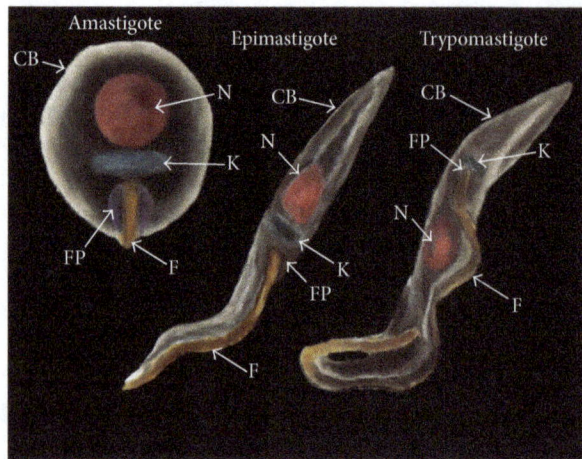

FIGURE 1: The different stages of *Trypanosoma cruzi*. The image depicted the amastigote, epimastigote, and trypomastigote stages from *T. cruzi* and their membrane domains: nucleus (N); kinetoplast (K); flagellum (F); flagellar pocket (FP); and cell body (CB).

[11]. The phenotypic and genotypic diversity of *T. cruzi* are well recognized. *T. cruzi* is partitioned into six discrete typing units (DTUs), TcI–TcVI. For a comprehensive review see [12, 13]. Despite the fact that many *T. cruzi* isolates have been described through the years, CL Brener was the reference organism used in the "*Trypanosoma cruzi* Genome Project." CL Brener is a clone derived from CL strain belonging to Tc VI lineage and has been thoroughly studied and well characterized [14–16]. The CL Brener genome sequence is now available and became the *T. cruzi* genome reference for other sequencing projects [13, 14, 17]. Moreover, new *T. cruzi* isolates are still being reported and features such as some of their surface membrane proteins are regularly compared with the CL Brener genome [17–20]. Although *T. cruzi* has several morphological transformations through its complex life cycle, studies on surface proteins have been mainly focused on the different stages through the infection process (Figure 1) [21–23]. Membrane proteins have been shown to play an important role in the biology of *T. cruzi*, including the interaction between parasite and host [2, 22, 24–28]. Scientists have made efforts to unravel the gaps on the structure and functions of these surface membrane proteins. However, despite their importance, the information is currently scattered. The aim of this review is to outline the families of surface membrane proteins from *T. cruzi* which are the most abundant and/or relevant during its life cycle (Table 1).

2. Mucin Family

Trypanosoma cruzi is covered by a dense layer of mucin-type molecules. Mucins are the major *T. cruzi* surface glycoproteins and their sugar residues are able to interact with mammalian cells [26, 63]. These proteins are characteristic and widely distributed over the cell body, flagellum, and flagellar pocket of the different developmental forms (Figure 1) [64]. Mucins play a key role in the parasite protection as

well as in the infectivity and modulation of the host immune response throughout the *T. cruzi* life cycle [25, 30, 60, 65, 66]. Based on sequence comparisons, *T. cruzi* mucins have been split into two gene families, termed TcMUC and TcSMUG (Figure 2, Table 1) [30, 67]. TcMUC expression seems to be restricted to the mammal-dwelling stages; these proteins are divided into three groups based on their central domains: TcMUC I to III [67, 68]. TcMUC I and II proteins are distributed on the amastigote and the bloodstream trypo-mastigote surface. TcMUC I is the major component in the amastigote form, whereas TcMUC II is predominant in membrane lipid rafts of the trypomastigote stage [31]. TcMUC I proteins show internal tandem repeats on their structure with a T_8KP_2 amino acid (aa) consensus sequence which are suitable targets for the O-glycosylation pathway in *T. cruzi*, flanked by an N-terminal signal peptide and a C-terminal glycosylphosphatidylinositol- (GPI-) anchor signal [30, 69]. TcMUC II genes encode proteins that share similar N- and C-termini with TcMUC I but without the T_8KP_2 motifs, although their central regions are still rich enough in threonine, serine, and proline residues [29, 70]. The single gene product of the TcMUC III group is termed trypomastigote small surface antigen (TSSA) and has been identified as a mucin-like glycoprotein (tGPI-mucins) [71]. TSSA are displayed on the surface of the trypomastigote forms of *Trypanosoma cruzi* and they are expressed in vivo as a ~20-kDa protein during the mammal-derived stages [72–74].

The second mucin family TcSMUG encodes for very small open reading frame containing a putative signal peptide at the N-terminus and a GPI-anchor signal in the C-terminus. This protein family is divided into two groups: small (S) and large (L) according to their encoded mRNA size [67, 71, 75]. The S group encodes for 35–50 kDa mucins N-glycosylated (Gp35/50 mucins) and they are the major acceptors of sialic acid on the parasite surface by parasite trans-sialidases in *T. cruzi*. This S group is found in the epimastigote and metacyclic trypomastigote forms [32, 33]. TcSMUG L group, in contrast, encodes for mucin-type glycoconjugates which are not sialic acid acceptors and they are only present in the surface of the epimastigote stage [34, 76]. Furthermore, depending on the origin of the encoding allele, TcSMUG L products contain one or two additional N-glycosylation signals between the N-terminal region and the threonine-rich region [34].

3. Trans-Sialidase Superfamily

Trypanosoma cruzi trans-sialidases (TS) genes are a large superfamily, which includes 1,430 gene members, including 693 pseudogenes [14, 60]. Similar to mucins, TS are dis-tributed along the cell body, flagellum, and flagellar pocket of *T. cruzi* [31, 77]. The TS superfamily is divided into four groups: Groups I to IV (Table 1) [78]. Their sequence similarity and functional properties were used as criteria for classification (Figure 3). Importantly, Group I comprises proteins with trans-sialidase (TS) and/or neuraminidase activities [79]. The TS activity involves the transfer of sialic acid from host glycoconjugates to mainly the parasite mucins

TABLE 1: Summary of surface protein families of *Trypanosoma cruzi* and their characteristics.

Protein Family	Group	Members	Host	Parasite stage	References
Mucin	TcMUC	TcMUC I	Mammal	Amastigote and bloodstream trypomastigote	[29]
		TcMUC II	Mammal	Amastigote and bloodstream trypomastigote	[30, 31]
		TcMUC III (TSSA)	Mammal	Bloodstream trypomastigote	[30]
	TcSMUG	TcSMUG S	Insect	Epimastigote and metacyclic trypomastigote	[30, 32, 33]
		TcSMUG L	Insect	Epimastigote	[34, 35]
Trans-sialidase	TS I	TCNA	Mammal	Bloodstream trypomastigote	[36–38]
		SAPA	Mammal	Bloodstream trypomastigote	[38]
		TS-epi	Insect	Epimastigote	[39]
		ASP-1 and ASP-2	Mammal	Amastigote	[40, 41]
	TS II	TSA-1, Tc85, and SA85	Mammal	Bloodstream trypomastigote	[42–44]
		GP82	Insect	Metacyclic trypomastigote	[45]
		GP90	Insect/mammal	Amastigote, bloodstream and metacyclic trypomastigote	[10, 46, 47]
	TS III	CRP, FLI60, CEA, and TESA	Mammal	Bloodstream trypomastigote	[48]
	TS IV	TsTc13	Insect	Metacyclic trypomastigote	[49]
TcGP63 family	TcGP63-I	61 kDa glycosylated isoform	Insect/Mammal	Epimastigote and amastigote	[50, 51]
		55 kDa nonglycosylated protein	Insect	Metacyclic trypomastigote	[50, 51]
	TcGP63-II	Two transcripts of 2.6 and 2.8 kb	Insect/mammal	Amastigotes, epimastigote, and bloodstream trypomastigote	[50, 52]
Amastin family	δ-Amastins	δ-Amastin and δ-ama40/50	Mammal	Present in all life cycle, up-regulated in amastigotes stage	[53]
	β-Amastin	β1-Amastin and β2-amastin	Insect/mammal	Present in all life cycle, upregulated in epimastigote stage	[53]
TcTASV family	TcTASV-A		Mammal	Bloodstream trypomastigote	[54]
	TcTASV-B		Mammal	Bloodstream trypomastigote	[54]
	TcTASV-C		Mammal	Bloodstream trypomastigote	[55]
MASP family			Insect/mammal	Amastigote, epimastigote, bloodstream and metacyclic trypomastigote	
Cruzipain family	N-cruzipain		Insect	Epimastigote	[56, 57]
	R-cruzipain 1 (cruzain)		Insect/mammal	Epimastigote, bloodstream trypomastigote and amastigote	[58, 59]
	R-cruzipain 2		Mammal	Bloodstream trypomastigote and amastigote	[56]

TcMUC in bloodstream trypomastigotes

I   MT$_2$LTM[M/V]TCRL$_2$[Y/C] - - - - | O-Glycosylated [T$_8$KP$_2$]$_{2-10}$ | ▓▓ | RAPS$_2$IR$_2$IDGSLG | ░░░

II   MT$_2$LTM$_2$[M/T]TCRL$_2$C - - - - - - - - - - | [T$_8$(K/Q)AP]$_{1-2}$ | ▓▓ | RAPSRLR[E/R]IDGSLS | ░░░

III MT$_2$CRL$_2$CAL$_2$ALALC$_3$LSACT$_2$AN | CORE 39-40 bp | | G(S/N)LS$_3$AWV(S/F)APLALA$_2$SALAYTALG | ░░░

TcSMUG in epimastigotes and metacyclic trypomastigotes

S   M$_2$LR$_2$VLCVLFLALC$_2$ACVC | ATAQE$_2$GQYDA$_2$V$_2$EAGEGQDQ | | [KNT$_7$ST$_3$S(S/K)AP]$_{1-3}$ | ▓▓   ░░░

L   M$_2$LR$_2$VLCVLFLALC$_2$ACVC | ATAQE$_2$GQYDA$_2$GFKA$_2$G$_2$DPK$_2$N | | DQT$_{17-20}$NAPAKDT$_{5-7}$NAPAK | ▓▓   ░░░

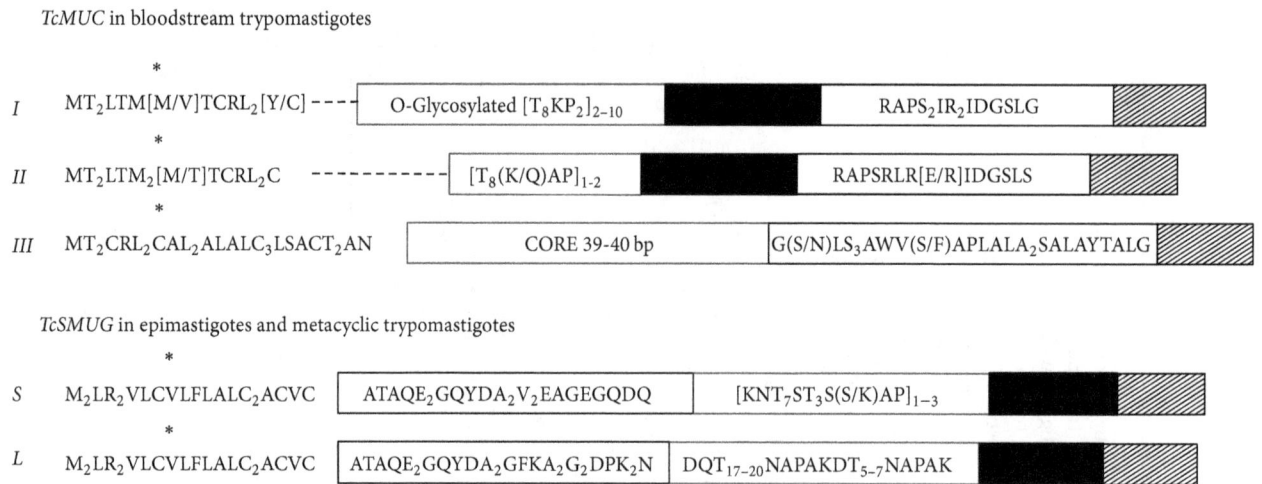

FIGURE 2: Mucin family. Schematic representation of mature proteins of mucin families TcMUC, found in bloodstream trypomastigotes, and TcSMUG, found in epimastigotes and metacyclic trypomastigotes. Signal peptide (*); protein fingerprints (white boxes); hypervariable region (---); threonine-rich region (black boxes) and glycosylphosphatidylinositol- (GPI-) anchor signal (shadowed boxes). Image based on Buscaglia and Frasch [30, 60].

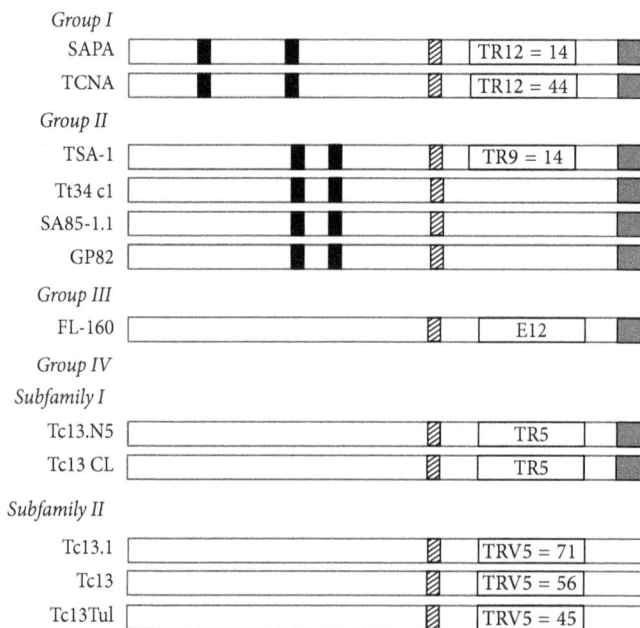

Group I
SAPA | ▓ ▓ ░ | TR12 = 14 |
TCNA | ▓ ▓ ░ | TR12 = 44 |

Group II
TSA-1 | ▓▓ ░ | TR9 = 14 |
Tt34 cl | ▓▓ ░ |
SA85-1.1 | ▓▓ ░ |
GP82 | ▓▓ ░ |

Group III
FL-160 | ░ | E12 |

Group IV
Subfamily I
Tc13.N5 | ░ | TR5 |
Tc13 CL | ░ | TR5 |

Subfamily II
Tc13.1 | ░ | TRV5 = 71 |
Tc13 | ░ | TRV5 = 56 |
Tc13Tul | ░ | TRV5 = 45 |

FIGURE 3: Trans-sialidase (TS) superfamily. Schematic representation of the four different groups of trans-sialidases (TS) from *Trypanosoma cruzi*. Characteristic motifs SXDXGTW and VTVXNVXLYNR for TS are depicted as black and shadowed boxes, respectively. The glycosylphosphatidylinositol- (GPI-) anchor signal in the C-terminus position is shown as grey boxes. Tandem repeats (TR) of 12 amino acid residues [DS$_2$AH(S/G)TPSTP(A/V)] are detected in SAPA and TCNA (TR12 inside an open box). Nine amino acid residue repeats [DK$_2$ESESGDSE] are identified in TSA-1 (TR9 inside an open box). A characteristic epitope [TPQRKT$_2$EDRPQ] is present in FL-160 (E12 inside an open box). The pentapeptide [EPKSA] is found once into subfamily I of Group IV (TR5 inside an open box) whereas, in subfamily II, EPKSA is repeatedly present (TRV5 inside an open box indicating the number of repeats). Image based on Colli and Schenkman [61, 62].

present in the plasma membrane of trypomastigotes [80–82]. On the other hand, neuraminidase activity occurs when nonsuitable acceptor molecules for sialic acid are present, and then sialic acid is transferred to water [83]. Trypanosomes are unable to synthesize the monosaccharide sialic acid; they need to scavenge it from the infected host using these TS activities. Therefore the sialylation process in *T. cruzi*

is crucial for its viability and propagation into the host [84–87]. Moreover, neuraminidase activity was proposed to be involved in the removal of sialic acid from parasites and/or host-cell molecules which is required for parasite internalization [84, 88]. TS Group I members are as follows: TCNA (neuraminidase), SAPA (shed acute-phase antigen), and TS-epi (Figure 3) [36, 37, 61]. SAPA and TCNA enzymes have active trans-sialidase and neuraminidase activities and are expressed during bloodstream trypomastigote stage [38]. Both enzymes are very close related; when compared they have 84% homology at the aa level. SAPA and TCNA have two main regions: an N-terminal catalytic region and a C-terminal extension, which repeats 12 amino acids (SAPA repeats) in tandem with the consensus sequence: D-S-S-A-H-[S/G]-T-P-S-T-P-[A/V] [89]. SAPA has only 14 tandem repeats compared to 44 for TCNA. The presence of SAPA repeats increases the half-life of the protein in the blood [90]. Both SAPA and TCNA proteins are anchored by glycosylphosphatidylinositol (GPI) to the parasite plasma membrane and can be found in serum from deeply infected mammals [38, 82]. Recently, Lantos and coworkers have shown that domains for mucins and TS are separated by about 150 nm, indicating that mucins do not pass through a TS-rich area for sialylation [31]. Moreover, they proposed a mechanism for the shedding of trans-sialidase into the extracellular space and/or bloodstream via microvesicles, where the phosphatidylinositol-phospholipase-C activity is actually not present in bloodstream trypomastigote stage [31]. TS-epi, the third member of Group I, is an active trans-sialidase expressed in the insect dwelling epimastigote form at the stationary phase and is different from the TS expressed of the blood trypomastigotes. TS-epi lacks SAPA repeats and is not anchored to the membrane by GPI; instead it is predicted that anchoring to the membrane is due to the presence of a transmembrane domain followed by a hydrophilic section in the C-terminus [39]. That last feature may explain why TS-epi is minimally secreted into the medium [91].

TS Group II comprises members of the GP85 surface glycoproteins: ASP-1, ASP-2, TSA-1, Tc85, SA85, GP82, and GP90. They all have been implicated in host-cell attachment and invasion [62, 92–94]. These proteins have complete or degenerate Asp box motifs (SxDxGxTW); the VTVxN-VxLYNR motif characteristic of all TS members; and a signal sequence for cleavage/addition of GPI anchor at the C-terminal region (Figure 3) [60, 62, 95]. ASP-1, ASP-2, and TSA-1 are targets of *T. cruzi*-specific CD8$^+$ cytotoxic T lymphocytes and they induce strong antibody responses in infected mice and humans [40, 41, 96, 97]. ASP-1 and ASP-2 are amastigote surface proteins, whereas TSA-1 is a trypomastigote surface antigen [40, 41]. SA85 glycoproteins are expressed by amastigote and bloodstream trypomastigote forms. However, only the amastigote form expresses the mannose-binding protein ligand which seems to be involved in the opsonization of the parasite enhancing its infection capability [98–100]. The Tc85 molecule is an 85 kDa glycoprotein and is found abundantly in bloodstream trypomastigotes. Tc85 is identified as a ligand capable of binding to different host receptor molecules (cytokeratin 18, fibronectin, and laminin) located on the cell surface of either monocytes, neutrophils, or fibroblasts [42, 92, 101, 102]. Furthermore, GP82 and GP90 are glycoproteins expressed on the surface of the metacyclic trypomastigote form [46, 103], and they are found mainly at the plasma membrane with opposite roles in mammalian cell invasion [33, 46]. GP82 is able to activate a Ca^{2+} signaling pathway in host cells following parasite adhesion, which is required for *T. cruzi* internalization [101,

104–106]. GP82 binds less efficiently to HeLa cells compared to GP90, but it is capable of triggering the Ca^{2+} signal in that host cell [105]. GP82 is also the signaling receptor that mediates protein tyrosine phosphorylation, which is necessary for host-cell invasion [106]. On the other hand, GP90 is a metacyclic stage-specific glycoprotein defined by its reactivity with monoclonal antibodies 1G7 and 5E7 [46]. GP90 expressed by metacyclic forms lacks any enzymatic activity [47]. GP90 is also present in the mammalian stages of *T. cruzi* (bloodstream trypomastigote and amastigotes stages) and has the antiphagocytic effect mediated by the removal of sugar residues necessary for parasite internalization. This surface glycoprotein appears to have glycosidase activity and downregulates host-cell invasion probably due to the fact that GP90 binds to mammalian cells in a receptor-mediated manner without triggering the Ca^{2+} signal-inducing activity [10, 47].

TS Group III is formed by surface proteins present in mammal-dwelling blood trypomastigotes which include the following: CRP, FL160, CEA, and TESA [48]. These proteins are recognized by sera from patients with Chagas' disease and they are able to inhibit the classical and the alternative pathways of complement activation, which could be a protection from lysis by the host in the trypomastigote form [48, 62, 95, 107–109]. TESA (trypomastigote excretory-secretory antigens) is distributed on the cell surface membrane of *T. cruzi* [107, 110] whereas CRP, FL160, and CEA are flagellum-associated membrane proteins [111–113]. Interestingly, the sequence of FL-160 contains an epitope which molecularly mimics a nervous tissue antigen from the mammalian host [114].

Finally, TS Group IV is composed of genes encoding trypomastigote surface antigens whose biological function is still unknown. This group is included in the TS superfamily because it contains the conserved motif VTVxNVxLYNR, which is shared by all known TS members [10, 37, 54, 61, 115]. However, the B5 peptide from TsTc13 protein, a representative of Group IV, has been shown to be highly antigenic and is present in the infective metacyclic trypomastigote form [49].

4. TcGP63 Family

Trypanosomes and *Leishmania* species express a family of cell surface-localized, zinc-dependent metalloproteases, which are also termed as GP63 proteins, major surface proteases, or leishmanolysins. Metallopeptidase activities have been described in trypanosomatids [116–118], but only the so-called GP63 from *Leishmania* spp. has been thoroughly characterized. *Trypanosoma cruzi* possesses GP63-like genes (*TcGP63*) and they are differentially regulated, which suggests its functional importance at multiple stages in the parasite life cycle [52, 119, 120]. The TcGP63 family has at least two groups of proteins: TcGP63-I and TcGP63-II (Figure 4, Table 1) [50]. It has been estimated that *TcGP63-I* has low (5–10) gene copies, whereas *TcGP63-II* has 62 gene copies into the *T. cruzi* genome [50, 51]. The TcGP63-I group is present in the three life-stages of *T. cruzi*. These proteins present metallopeptidase activity and are bound to the protozoan's membrane by a C-terminal glycosylphosphatidylinositol-

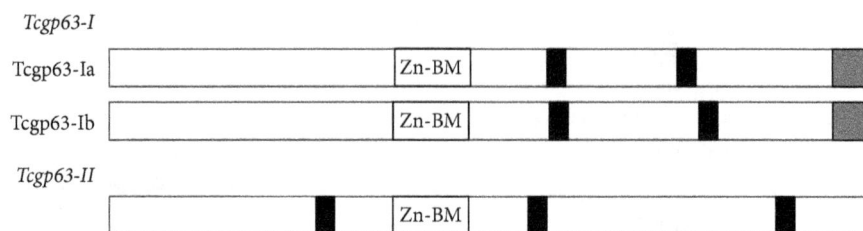

FIGURE 4: The TcGP63 family. The TcGP63 family consists of cell surface-localized, zinc-dependent metalloproteases also known as *T. cruzi* GP63-like proteins. This family has at least two groups: TcGP63-I and TcGP63-II. TcGP63-I members have two potential N-glycosylation sites, whereas TcGP63-II members have three [50]. The glycosylphosphatidylinositol- (GPI-) anchor signal in the C-terminus position is depicted as grey boxes; it is absent in the TcGP63-II members. Predicted N-glycosylation sites are shown in black boxes. Zn-BM: zinc-binding motifs [VXAHEX$_2$HA] associated with metalloprotease activity.

(GPI-) anchor signal [50]. Two isoforms are known of TcGP63-I in *T. cruzi*: a glycosylated and a nonglycosylated isoform. The 61 kDa glycosylated isoform is present in similar levels in both epimastigote and amastigote forms and is irregularly expressed on the surface membranes (cell body and flagellum) of the epimastigote. The second isoform is a 55 kDa TcGP63 nonglycosylated protein, which is located intracellularly near the kinetoplast and the flagellar pocket of the metacyclic trypomastigote [52]. TcGP63-II does not have GPI-anchor signal; instead its C-terminal sequence is replaced by a charged region containing three Asp and four Arg residues [50, 52].

5. Amastin Family

The amastin family is a group of transmembrane glycoproteins, which consists of small proteins of about 180 amino acids. Phylogenetic analysis of trypanosomatid amastins has defined four subfamilies named α-, β-, γ-, and δ-amastins, with distinct genomic organization as well as patterns of expression during the cell cycle of trypanosomatid [121, 122]. The *Trypanosoma cruzi* genome possesses two distinct subfamilies: β- and δ-amastins (Table 1), which have predicted the occurrence of four transmembrane regions (Figure 5) [53]. Genes encoding for the β1- and β2-amastin, belonging to the β- subfamily, are localized in the chromosome 32 of *T. cruzi*, whereas δ-amastin and δ-ama40/50 *loci* are found on chromosomes 34 and 26, respectively. β1- and δ-amastins are clearly located at the cell surface. Interestingly, β2-amastin shows a disperse distribution within the cytoplasm in addition to their surface localization [53]. The exact biological function of amastin is still unknown; however, as transmembrane proteins, amastins could play a role in proton or ion traffic across the membrane [123, 124]. Transcript levels of δ-amastins are upregulated in amastigotes from different *T. cruzi* strains, while β-amastin transcripts are more abundant in epimastigotes than in amastigotes or trypomastigotes; therefore β-amastins may be involved in the parasite adaptation to the insect vector [121, 125, 126]. Interestingly, Cruz and coworkers showed that δ-amastin plays a crucial role in the differentiation of *T. cruzi*; therefore it is a key molecule responsible for the parasite survival in the intracellular cell stage [127].

6. TcTASV Family

TcTASV (Trypomastigote Alanine Serine Valine-rich protein) is a family that comprises 40 members in *Trypanosoma cruzi*. They all have a C- and an N- terminus conserved with a variable central core. This variable core is rich in Ala, Ser, and Val residues, with a conserved Glu-Ala-Pro motif. It also has a high number of Ser and Thr susceptible to glycosylation and a signal peptide and a consensus sequence for the addition of a GPI anchor were predicted, suggesting that this family can be located at the parasite surface and/or be secreted to the milieu [54]. The TcTASV family is conserved across the genomes of *T. cruzi* strains and, to date, no orthologues in other trypanosomatids have been found [55].

TcTASV family was split into three subfamilies: A, B, and C apoproteins, based on their predicted molecular weights (18 kDa, 27 kDa, and 36 kDa, resp.) (Figure 1) [54]. Until now, only subfamilies A and C have been worked thoroughly. Subfamily B has presented experimental hurdles to overcome. Annotated genes identified as TcTASVs are present in 5 chromosomes; almost all annotated subfamily C on the chromosome 24 and a high proportion of subfamily A on the chromosome 16. A peptide entirely conserved in TcTASV-A is present in trypomastigote and amastigote extract. However, only the expression of TcTASV-A in bloodstream trypomastigotes was demonstrated, suggesting that the TcTASV population could undergo developmental regulation [54, 128]. The TcTASV-C subfamily is expressed mainly in the trypomastigote stage as a phosphorylated, heavily glycosylated protein with ca. 60 kDa. TcTASV-C is attached to the parasite surface by a GPI anchor on the cell body and flagellum, which may explain why it is shed spontaneously into the medium and is in contact with the immune system of the host during the course of the natural infection. The superficial localization and secretory nature of TcTASV-C suggest a possible role in the host-parasite interactions [55].

7. Mucin-Associated Surface Proteins (MASPs) Family

This family received its name because its members are located in close proximity of *Trypanosoma cruzi* mucins (TcMUC

δ-Amastin

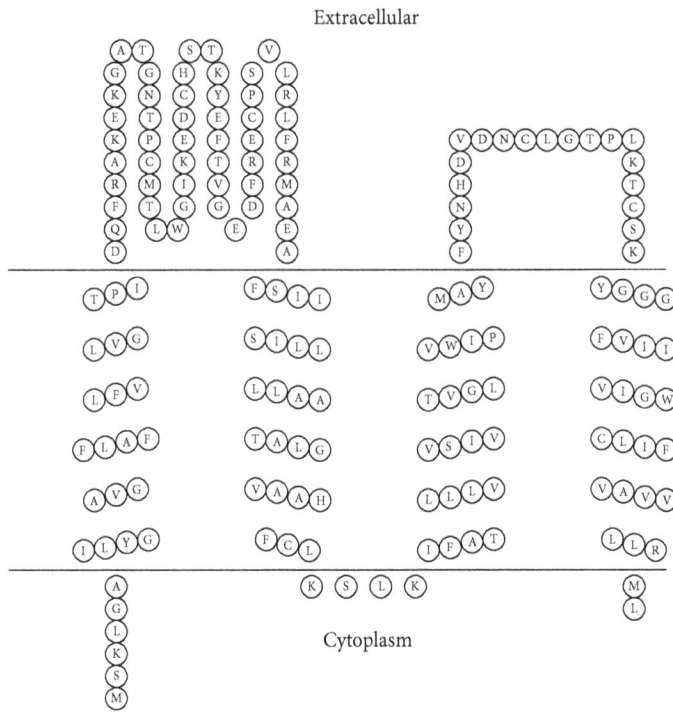

δ-Ama40/50

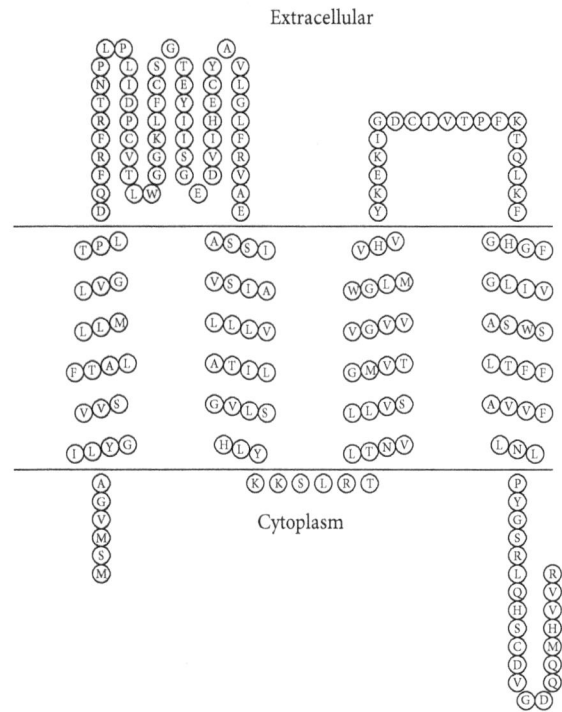

β1-Amastin

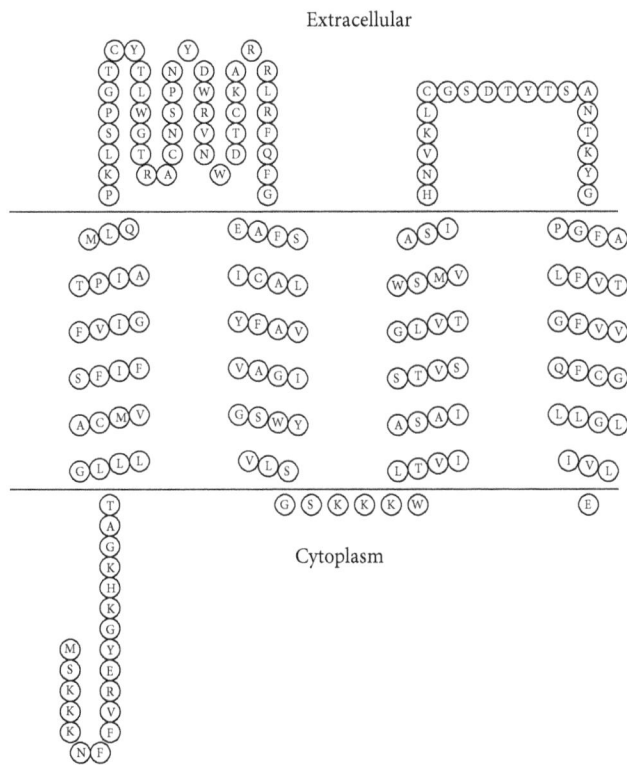

β2-Amastin

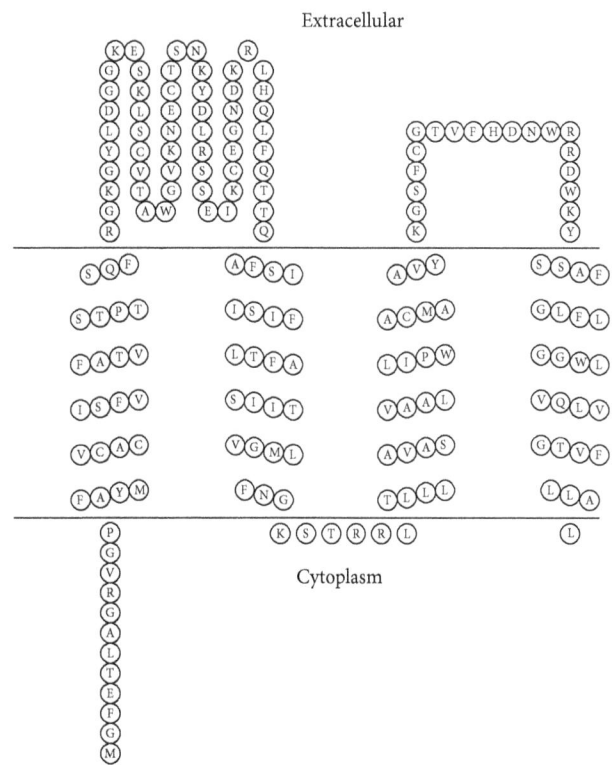

Figure 5: Amastin family. Topological model of subfamilies *β*- and *δ*-amastins from *Trypanosoma cruzi*. Amastins models share four predicted transmembrane helices and two extracellular hydrophilic loops. Although both N- and C-termini are predicted to be facing the cytoplasmic space, their length is variable among them. Topology predictions were performed using the "TOPO2, transmembrane protein display software" (http://www.sacs.ucsf.edu/TOPO2).

II); they are similar in structure, though not in sequence [129, 130]. MASP members contain N- and C-terminal conserved domains that encode a signal peptide and a GPI-anchor addition site, respectively. MASP is GPI anchored to the membrane and is preferentially expressed during the trypomastigote (bloodstream) stage (Table 1). Moreover, the central region is variable both in length (ranging 176–645 aa) and in sequence; its sequence also contains a large repertoire of repetitive motifs. Single aa residue repetitions are the most common, and those containing glutamic acid are more frequently being around 27% of the total of the identified repetitive motifs. Full-length MASP analysis revealed at least four potential O-glycosylation sites per sequence, 70% of which correspond to threonines [130]. The MASP expression was analyzed throughout the parasite life cycle and it was identified that they are expressed simultaneously in bloodstream trypomastigotes as well as in amastigotes and epimastigotes [45, 128]. MASP molecules are the most abundant antigens found on the surface of the infective trypomastigote stage of T. cruzi [130–133]. The overexpression of MASPs in the intracellular parasites prior to the division of the amastigotes located in the plasma membrane suggests that some of the proteins of this extensive family play a major biological role in the survival and multiplication of intracellular amastigotes [130, 134].

8. Cruzipain Family

Cruzipains are a papain-like cysteine proteases; cruzipain is expressed as a complex mixture of isoforms in all the *Trypanosoma cruzi* developmental stages (Figure 1) [135]. Despite the fact that cruzipain has high homology with other members of the papain proteases superfamily, this protein has a unique C-terminal region, which is retained in the mature protein [136, 137]. Cruzipains are expressed on all the body surface of epimastigotes and amastigotes. In contrast, on the trypomastigote form, cruzipain has only been present in the flagellar pocket region as well as within the pocket [138]. A specific, irreversible enzyme inhibitors for cruzipain GP57/51 was evaluated in heart muscle cells infected with trypomastigotes and proved to interfere with cell invasion and inhibit T. cruzi intracellular replication [139, 140]. The above suggests that cruzipain plays a role in the process of T. cruzi internalization into mammalian cells [138–140]. Additionally, cruzipain not only is essential for parasite survival but also generates a strong immune response in infected individuals [138, 141].

9. Concluding Remarks

A tangled mechanism is necessary for a "successful" host-pathogen interaction of *Trypanosoma cruzi* with its mammalian or insect host. The availability of the T. cruzi genome sequence made it possible to gain new insights into the parasite's biology and allowed the development of new powerful approaches to understand molecular pathogenesis and host-parasite interaction [14]. Proteome analysis was conducted in the different developmental stages of T. cruzi; as expected, surface proteins are part of the outstanding proteins

which were found differentially expressed among stages [128]. Another proteomic analysis has also been conducted in different organelles, on a specific developmental stage, or under certain stress conditions [8, 142–144]. Additionally, new technologies are now available to facilitate genome editing in T. cruzi, such as the Cre-recombinases and the CRISPR-Cas9 system. These genetic manipulation strategies have highly effective efficiency in different organisms and now were successfully adapted to disrupt genes from T. cruzi [145–147]. Furthermore the CRISPR-Cas9 system was recently used for endogenous tagging of proteins in T. cruzi which proved that this system is not limited to loss-of-function and made the localization/visualization of proteins from inside the parasite possible [148]. These new molecular strategies have now opened a new field of possibilities towards a more comprehensive functional analysis of the parasite biology and can be potentially used to move forward in the study of surface proteins of T. cruzi. As we present here, several studies show that surface membrane proteins are crucial for adaptation, differentiation, and survival of the parasite during its life cycle. Notably, some membrane protein families stand out during the host-parasite infection process, which make them potential targets to treat, or even prevent, the infection process. Altogether, these recent advancements can positively increase the current knowledge of host-parasite interactions and will help to accelerate the discovery of effective drugs against the Chagas disease. Despite all the research advances on these protein families on T. cruzi membrane, efforts to unravel their structure and function still have a long journey to be undertaken.

Acknowledgments

The authors acknowledges the support from the PROMEP/103.5/12/8515 and the National Council of Science and Technology from Mexico (CONACyT). They also thank David Ortega for his careful reading of the manuscript and helpful comments and suggestions. Ángel de la Cruz Pech-Canul and Rosa-Lidia Solís-Oviedo were supported by CONACyT with application nos. 250293 and 250294, respectively.

References

[1] C. Bern, "Chagas' disease," *New England Journal of Medicine*, vol. 373, no. 5, pp. 456–466, 2015.

[2] A. E. Balber, "The pellicle and the membrane of the flagellum, flagellar adhesion zone, and flagellar pocket: functionally discrete surface domains of the bloodstream form of African trypanosomes," *Critical Reviews in Immunology*, vol. 10, no. 3, pp. 177–201, 1990.

[3] L. V. Kirchhoff, "Epidemiology of American trypanosomiasis (Chagas Disease)," *Advances in Parasitology*, vol. 75, pp. 1–18, 2011.

[4] B. Y. Lee, K. M. Bacon, M. E. Bottazzi, and P. J. Hotez, "Global economic burden of Chagas disease: a computational

simulation model," *The Lancet Infectious Diseases*, vol. 13, no. 4, pp. 342–348, 2013.

[5] S. Goldenberg and A. R. Ávila, "Aspects of trypanosoma cruzi stage differentiation," in *Advances in Parasitology*, H. B. Tanowitz, L. M. Weiss, and L. V. Kirchhoff, Eds., pp. 285–305, Academic Press, Chapter 13 edition, 2011.

[6] W. De Souza, "Basic cell biology of Trypanosoma cruzi," *Current Pharmaceutical Design*, vol. 8, no. 4, pp. 269–285, 2002.

[7] M. L. Díaz, R. Torres, and C. I. González, "Expresión diferencial entre estadios de Trypanosoma cruzi I en el aislamiento de un paciente con cardiomiopatía chagásica crónica de zona endémica de Santander, Colombia," *Biomédica*, vol. 31, no. 4, pp. 503–513, 2011.

[8] A. d. C. M. d. Santos Júnior, D. E. Kalume, R. Camargo et al., "Unveiling the trypanosoma cruzi nuclear proteome," *PLoS ONE*, vol. 10, no. 9, Article ID e0138667, 2015.

[9] R. M. L. Queiroz, S. Charneau, I. M. D. Bastos et al., "Cell surface proteome analysis of human-hosted trypanosoma cruzi life stages," *Journal of Proteome Research*, vol. 13, no. 8, pp. 3530–3541, 2014.

[10] W. de Souza, T. M. de Carvalho, and E. S. Barrias, "Review on *trypanosoma cruzi*: host cell interaction," *International Journal of Cell Biology*, vol. 2010, article 295394, 18 pages, 2010.

[11] S. M. Teixeira, R. M. C. de Paiva, M. M. Kangussu-Marcolino, and W. D. DaRocha, "Trypanosomatid comparative genomics: contributions to the study of parasite biology and different parasitic diseases," *Genetics and Molecular Biology*, vol. 35, no. 1, pp. 1–17, 2012.

[12] S. F. Brenière, E. Waleckx, and C. Barnabé, "Over six thousand trypanosoma cruzi strains classified into discrete typing units (dtus): attempt at an inventory," *PLOS Neglected Tropical Diseases*, vol. 10, no. 8, Article ID e0004792, 2016.

[13] B. Zingales, M. A. Miles, D. A. Campbell et al., "The revised Trypanosoma cruzi subspecific nomenclature: rationale, epidemiological relevance and research applications," *Infection, Genetics and Evolution*, vol. 12, no. 2, pp. 240–253, 2012.

[14] N. M. El-Sayed, P. J. Myler, D. C. Bartholomeu et al., "The genome sequence of trypanosoma cruzi, etiologic agent of chagas disease," *Science*, vol. 309, no. 5733, pp. 409–415, 2005.

[15] M. R. M. Santos, M. I. Cano, A. Schijman et al., "The trypanosoma cruzi genome project: nuclear karyotype and gene mapping of clone cl brener," *Memórias do Instituto Oswaldo Cruz*, vol. 92, no. 6, pp. 821–828, 1997.

[16] W. Degrave, M. J. Levin, J. F. da Silveira et al., "Parasite genome projects and the trypanosoma cruzi genome initiative," *Memórias do Instituto Oswaldo Cruz*, vol. 92, no. 6, pp. 859–862, 1997.

[17] P. B. Hamilton and J. R. Stevens, "15—Classification and phylogeny of Trypanosoma cruzi A2—Telleria, Jenny," in *American Trypanosomiasis Chagas Disease*, M. Tibayrenc, Ed., pp. 321–344, Elsevier, London, 2nd edition, 2017.

[18] L. A. Shender, M. D. Lewis, D. Rejmanek, and J. A. K. Mazet, "Molecular diversity of trypanosoma cruzi detected in the vector triatoma protracta from california, USA," *PLoS Neglected Tropical Diseases*, vol. 10, no. 1, Article ID e0004291, 2016.

[19] R. Ruíz-Sánchez, M. P. de León, V. Matta et al., "Trypanosoma cruzi isolates from Mexican and Guatemalan acute and chronic chagasic cardiopathy patients belong to Trypanosoma cruzi I," *Memórias do Instituto Oswaldo Cruz*, vol. 100, no. 3, pp. 281–283, 2005.

[20] V. M. Monteón, R. López, A. A. Ramos-Ligonio, and K. Acosta-Viana, "Trans sialidase genes allow clustering of tci trypanosoma cruzi mexican isolates," *British Microbiology Research Journal*, vol. 4, no. 12, pp. 1299–1310, 2014.

[21] J. A. Perez-Molina, F. Norman, and R. Lopez-Velez, "Chagas disease in non-endemic countries: epidemiology, clinical presentation and treatment," *Current Infectious Disease Reports*, vol. 14, no. 3, pp. 263–274, 2012.

[22] K. Gull, "Host-parasite interactions and trypanosome morphogenesis: a flagellar pocketful of goodies," *Current Opinion in Microbiology*, vol. 6, no. 4, pp. 365–370, 2003.

[23] F. Noireau, P. Diosque, and A. M. Jansen, "*Trypanosoma cruzi*: adaptation to its vectors and its hosts," *Veterinary Research*, vol. 40, no. 2, article 26, pp. 1–23, 2009.

[24] C. Gadelha, S. Rothery, M. Morphew, J. R. McIntosh, N. J. Severs, and K. Gull, "Membrane domains and flagellar pocket boundaries are influenced by the cytoskeleton in African trypanosomes," *Proceedings of the National Academy of Sciences of the United States of America*, vol. 106, no. 41, pp. 17425–17430, 2009.

[25] S. M. Landfear and M. Ignatushchenko, "The flagellum and flagellar pocket of trypanosomatids," *Molecular and Biochemical Parasitology*, vol. 115, no. 1, pp. 1–17, 2001.

[26] F. Villalta and F. Kierszenbaum, "Host-cell invasion by *Trypanosoma cruzi*: role of cell surface galactose residues," *Biochemical and Biophysical Research Communications*, vol. 119, no. 1, pp. 228–235, 1984.

[27] S. Lacomble, S. Vaughan, M. Deghelt, F. F. Moreira-Leite, and K. Gull, "A trypanosoma brucei protein required for maintenance of the flagellum attachment zone and flagellar pocket ER domains," *Protist*, vol. 163, no. 4, pp. 602–615, 2012.

[28] C. L. Alcantara, J. C. Vidal, W. de Souza, and N. L. Cunha-e-Silva, "The three-dimensional structure of the cytostome-cytopharynx complex of Trypanosoma cruzi epimastigotes," *Journal of Cell Science*, vol. 127, no. 10, pp. 2227–2237, 2014.

[29] J. M. Di Noia, D. O. Sanchez, and A. C. C. Frasch, "The protozoan *Trypanosoma cruzi* has a family of genes resembling the mucin genes of mammalian cells," *The Journal of Biological Chemistry*, vol. 270, no. 41, pp. 24146–24149, 1995.

[30] C. A. Buscaglia, V. A. Campo, A. C. C. Frasch, and J. M. Di Noia, "*Trypanosoma cruzi* surface mucins: host-dependent coat diversity," *Nature Reviews Microbiology*, vol. 4, no. 3, pp. 229–236, 2006.

[31] A. B. Lantos, G. Carlevaro, B. Araoz et al., "Sialic acid glycobiology unveils trypanosoma cruzi trypomastigote membrane physiology," *PLoS Pathogens*, vol. 12, no. 4, Article ID e1005559, 2016.

[32] S. Schenkman, M. A. J. Ferguson, N. Heise, M. L. Cardoso de Almeida, R. A. Mortara, and N. Yoshida, "Mucin-like glycoproteins linked to the membrane by glycosylphosphatidylinositol anchor are the major acceptors of sialic acid in a reaction catalyzed by trans-sialidase in metacyclic forms of Trypanosoma cruzi," *Molecular and Biochemical Parasitology*, vol. 59, no. 2, pp. 293–303, 1993.

[33] N. Yoshida, "Molecular basis of mammalian cell invasion by Trypanosoma cruzi," *Anais da Academia Brasileira de Ciências*, vol. 78, no. 1, pp. 87–111, 2006.

[34] I. Urban, L. Boiani Santurio, A. Chidichimo et al., "Molecular diversity of the Trypanosoma cruzi TcSMUG family of mucin genes and proteins," *Biochemical Journal*, vol. 438, no. 2, pp. 303–313, 2011.

[35] M. S. Gonzalez, M. S. Souza, E. S. Garcia et al., "Trypanosoma cruzi tcsmug l-surface mucins promote development and infectivity in the triatomine vector rhodnius prolixus," *PLoS Neglected Tropical Diseases*, vol. 7, no. 11, Article ID e2552, 2013.

[36] P. Velge, M. A. Ouaissi, J. Cornette, D. Afchain, and A. Capron, "Identification and isolation of Trypanosoma cruzi trypomastigote collagen-binding proteins: possible role in cell-parasite interaction," *Parasitology*, vol. 97, no. 2, pp. 255–268, 1988.

[37] G. A. M. Cross and G. B. Takle, "The surface trans-sialidase family of Trypanosoma cruzi," *Annual Review of Microbiology*, vol. 47, pp. 385–411, 1993.

[38] S. Schenkman, L. P. De Carvalho, and V. Nussenzweig, "Trypanosoma cruzi trans-sialidase and neuraminidase activities can be mediated by the same enzymes," *Journal of Experimental Medicine*, vol. 175, no. 2, pp. 567–575, 1992.

[39] M. R. S. Briones, C. M. Egima, and S. Schenkman, "Trypanosoma cruzi trans-sialidase gene lacking C-terminal repeats and expressed in epimastigote forms," *Molecular and Biochemical Parasitology*, vol. 70, no. 1-2, pp. 9–17, 1995.

[40] H. P. Low and R. L. Tarleton, "Molecular cloning of the gene encoding the 83 kDa amastigote surface protein and its identification as a member of the Trypanosoma cruzi sialidase superfamily," *Molecular and Biochemical Parasitology*, vol. 88, no. 1-2, pp. 137–149, 1997.

[41] M. A. M. Santos, N. Garg, and R. L. Tarleton, "The identification and molecular characterization of Trypanosoma cruzi amastigote surface protein-1, a member of the trans-sialidase gene super-family," *Molecular and Biochemical Parasitology*, vol. 86, no. 1, pp. 1–11, 1997.

[42] R. Giordano, D. L. Fouts, D. Tewari, W. Colli, J. E. Manning, and M. J. M. Alves, "Cloning of a surface membrane glycoprotein specific for the infective form of Trypanosoma cruzi having adhesive properties to laminin," *Journal of Biological Chemistry*, vol. 274, no. 6, pp. 3461–3468, 1999.

[43] S. Kahn, T. G. Colbert, J. C. Wallace et al., "The major 85-kDa surface antigen of the mammalian-stage forms of Trypanosoma cruzi is a family of sialidases," *Proceedings of the National Academy of Sciences of the United States of America*, vol. 88, no. 10, pp. 4481–4485, 1991.

[44] D. L. Fouts, B. J. Ruef, P. T. Ridley, R. A. Wrightsman, D. S. Peterson, and J. E. Manning, "Nucleotide sequence and transcription of a trypomastigote surface antigen gene of Trypanosoma cruzi," *Molecular and Biochemical Parasitology*, vol. 46, no. 2, pp. 189–200, 1991.

[45] E. Bayer-Santos, C. Aguilar-Bonavides, S. P. Rodrigues et al., "Proteomic analysis of trypanosoma cruzi secretome: characterization of two populations of extracellular vesicles and soluble proteins," *Journal of Proteome Research*, vol. 12, no. 2, pp. 883–897, 2013.

[46] M. M. G. Teixeira and N. Yoshida, "Stage-specific surface antigens of metacyclic trypomastigotes of Trypanosoma cruzi identified by monoclonal antibodies," *Molecular and Biochemical Parasitology*, vol. 18, no. 3, pp. 271–282, 1986.

[47] N. Nogueira, "Host and parasite factors affecting the invasion of mononuclear phagocytes by Trypanosoma cruzi," *Ciba Foundation symposium*, vol. 99, pp. 52–73, 1983.

[48] P. R. C. Correa, E. M. Cordero, L. G. Gentil, E. Bayer-Santos, and J. F. D. Silveira, "Genetic structure and expression of the surface glycoprotein GP82, the main adhesin of Trypanosoma cruzi metacyclic trypomastigotes," *The Scientific World Journal*, vol. 2013, Article ID 156734, 11 pages, 2013.

[49] L. M. Freitas, S. L. dos Santos, G. F. Rodrigues-Luiz et al., "Genomic analyses, gene expression and antigenic profile of the trans-sialidase superfamily of trypanosoma cruzi reveal an undetected level of complexity," *PLoS ONE*, vol. 6, no. 10, Article ID e25914, 2011.

[50] I. C. Cuevas, J. J. Cazzulo, and D. O. Sánchez, "gp63 homologues in Trypanosoma cruzi: surface antigens with metalloprotease activity and a possible role in host cell infection," *Infection and Immunity*, vol. 71, no. 10, pp. 5739–5749, 2003.

[51] M. S. Llewellyn, L. A. Messenger, A. O. Luquetti et al., "Deep sequencing of the trypanosoma cruzi GP63 surface proteases reveals diversity and diversifying selection among chronic and congenital chagas disease patients," *PLoS Neglected Tropical Diseases*, vol. 9, no. 4, Article ID e0003458, 2015.

[52] M. M. Kulkarni, C. L. Olson, D. M. Engman, and B. S. McGwire, "Trypanosoma cruzi GP63 proteins undergo stage-specific differential posttranslational modification and are important for host cell infection," *Infection and Immunity*, vol. 77, no. 5, pp. 2193–2200, 2009.

[53] M. M. Kangussu-Marcolino, R. M. C. De Paiva, P. R. Araújo et al., "Distinct genomic organization, mRNA expression and cellular localization of members of two amastin sub-families present in Trypanosoma cruzi," *BMC Microbiology*, vol. 13, no. 1, article 10, 2013.

[54] E. A. García, M. Ziliani, F. Agüero, G. Bernabó, D. O. Sánchez, and V. Tekiel, "TcTASV: a novel protein family in Trypanosoma cruzi identified from a subtractive trypomastigote cDNA library," *PLoS Neglected Tropical Diseases*, vol. 4, no. 10, article e841, 2010.

[55] G. Bernabó, G. Levy, M. Ziliani, L. D. Caeiro, D. O. Sánchez, and V. Tekiel, "TcTASV-C, a protein family in trypanosoma cruzi that is predominantly trypomastigote-stage specific and secreted to the medium," *PLoS ONE*, vol. 8, no. 7, Article ID e71192, 2013.

[56] F. C. G. Dos Reis, W. A. S. Júdice, M. A. Juliano, L. Juliano, J. Scharfstein, and A. P. C. De A. Lima, "The substrate specificity of cruzipain 2, a cysteine protease isoform from Trypanosoma cruzi," *FEMS Microbiology Letters*, vol. 259, no. 2, pp. 215–220, 2006.

[57] E. D. Nery, M. A. Juliano, M. Meldal et al., "Characterization of the substrate specificity of the major cysteine protease (cruzipain) from *Trypanosoma cruzi* using a portion-mixing combinatorial library and fluorogenic peptides," *Biochemical Journal*, vol. 323, no. 2, pp. 427–433, 1997.

[58] A. E. Eakin, A. A. Mills, G. Harth, J. H. Mckerrow, and C. S. Craik, "The sequence, organization, and expression of the major cysteine protease (cruzain) from Trypanosoma cruzi," *The Journal of Biological Chemistry*, vol. 267, no. 11, pp. 7411-20, 1992.

[59] W. A. S. Judice, M. H. S. Cezari, A. P. C. A. Lima et al., "Comparison of the specificity, stability and individual rate constants with respective activation parameters for the peptidase activity of cruzipain and its recombinant form, cruzain, from Trypanosoma cruzi," *European Journal of Biochemistry*, vol. 268, no. 24, pp. 6578–6586, 2001.

[60] A. C. C. Frasch, "Functional diversity in the trans-sialidase and mucin families in Trypanosoma cruzi," *Parasitology Today*, vol. 16, no. 7, pp. 282–286, 2000.

[61] S. Schenkman, D. Eichinger, M. E. A. Pereira, and V. Nussenzweig, "Structural and functional properties of *Trypanosoma* trans-sialidase," *Annual Review of Microbiology*, vol. 48, pp. 499–523, 1994.

[62] W. Colli, "Trans-sialidase: a unique enzyme activity discovered in the protozoan Trypanosoma cruzi," *FASEB Journal*, vol. 7, no. 13, pp. 1257–1264, 1993.

[63] N. Yoshida, R. A. Mortara, M. F. Araguth, J. C. Gonzalez, and M. Russo, "Metacyclic neutralizing effect of monoclonal antibody 10D8 directed to the 35- and 50-kilodalton surface glycoconjugates of Trypanosoma cruzi," *Infection and Immunity*, vol. 57, no. 6, pp. 1663–1667, 1989.

[64] G. E. Cánepa, A. C. Mesías, H. Yu, X. Chen, and C. A. Buscaglia, "Structural features affecting trafficking, processing, and secretion of Trypanosoma cruzi mucins," *Journal of Biological Chemistry*, vol. 287, no. 31, pp. 26365–26376, 2012.

[65] I. C. Almeida and R. T. Gazzinelli, "Proinflammatory activity of glycosylphosphatidylinositol anchors derived from *Trypanosoma cruzi*: structural and functional analyses," *Journal of Leukocyte Biology*, vol. 70, no. 4, pp. 467–477, 2001.

[66] A. Acosta-Serrano, I. C. Almeida, L. H. Freitas-Junior, N. Yoshida, and S. Schenkman, "The mucin-like glycoprotein super-family of *Trypanosoma cruzi*: structure and biological roles," *Molecular and Biochemical Parasitology*, vol. 114, no. 2, pp. 143–150, 2001.

[67] J. M. Di Noia, I. D'Orso, D. O. Sánchez, and A. C. C. Frasch, "AU-rich elements in the 3'-untranslated region of a new mucin-type gene family of Trypanosoma cruzi confers mRNA instability and modulates translation efficiency," *Journal of Biological Chemistry*, vol. 275, no. 14, pp. 10218–10227, 2000.

[68] V. A. Campo, C. A. Buscaglia, J. M. Di Noia, and A. C. C. Frasch, "Immunocharacterization of the mucin-type proteins from the intracellular stage of Trypanosoma cruzi," *Microbes and Infection*, vol. 8, no. 2, pp. 401–409, 2006.

[69] M. V. Han and C. M. Zmasek, "PhyloXML: XML for evolutionary biology and comparative genomics," *BMC Bioinformatics*, vol. 10, article 356, 2009.

[70] J. M. Di Noia, G. D. Pollevick, M. T. Xavier et al., "High diversity in mucin genes and mucin molecules in *Trypanosoma cruzi*," *The Journal of Biological Chemistry*, vol. 271, no. 50, pp. 32078–32083, 1996.

[71] E. Barreto-Bergter and A. B. Vermelho, "Structures of glycolipids found in trypanosomatids: contribution to parasite functions," *Open Parasitology Journal*, vol. 4, no. 1, pp. 84–97, 2010.

[72] J. M. Di Noia, C. A. Buscaglia, C. R. De Marchi, I. C. Almeida, and A. C. C. Frasch, "A Trypanosoma cruzi small surface molecule provides the first immunological evidence that Chagas' disease is due to a single parasite lineage," *Journal of Experimental Medicine*, vol. 195, no. 4, pp. 401–413, 2002.

[73] C. A. Buscaglia and J. M. Di Noia, "Trypanosoma cruzi clonal diversity and the epidemiology of Chagas' disease," *Microbes and Infection*, vol. 5, no. 5, pp. 419–427, 2003.

[74] C. R. De Marchi, J. M. Di Noia, A. C. C. Frasch, V. A. Neto, I. C. Almeid, and C. A. Buscaglia, "Evaluation of a recombinant Trypanosoma cruzi mucin-like antigen for serodiagnosis of Chagas' disease," *Clinical and Vaccine Immunology*, vol. 18, no. 11, pp. 1850–1855, 2011.

[75] E. S. Nakayasu, D. V. Yashunsky, L. L. Nohara, A. C. T. Torrecilhas, A. V. Nikolaev, and I. C. Almeida, "GPIomics: global analysis of glycosylphosphatidylinositol-anchored molecules of Trypanosoma cruzi," *Molecular Systems Biology*, vol. 5, article 261, 2009.

[76] L. M. De Pablos and A. Osuna, "Conserved regions as markers of different patterns of expression and distribution of the mucin-associated surface proteins of Trypanosoma cruzi," *Infection and Immunity*, vol. 80, no. 1, pp. 169–174, 2012.

[77] U. Frevert, S. Schenkman, and V. Nussenzweig, "Stage-specific expression and intracellular shedding of the cell surface trans-sialidase of Trypanosoma cruzi," *Infection and Immunity*, vol. 60, no. 6, pp. 2349–2360, 1992.

[78] R. R. Moraes Barros, M. M. Marini, C. R. Antônio et al., "Anatomy and evolution of telomeric and subtelomeric regions in the human protozoan parasite Trypanosoma cruzi," *BMC Genomics*, vol. 13, no. 1, article 229, 2012.

[79] S. Schenkman, M.-S. Jiang, G. W. Hart, and V. Nussenzweig, "A novel cell surface trans-sialidase of *Trypanosoma cruzi* generates a stage-specific epitope required for invasion of mammalian cells," *Cell*, vol. 65, no. 7, pp. 1117–1125, 1991.

[80] M. E. A. Pereira, M. A. Loures, F. Villalta, and A. F. B. Andrade, "Lectin receptors as markers for *Trypanosoma cruzi*. Developmental stages and a study of the interaction of wheat germ agglutinin with sialic acid residues on epimastigote cells," *Journal of Experimental Medicine*, vol. 152, no. 5, pp. 1375–1392, 1980.

[81] J. O. Previato, A. F. B. Andrade, M. C. V. Pessolani, and L. Mendonca-Previato, "Incorporation of sialic acid into *Trypanosoma cruzi* macromolecules: a proposal for a new metabolic route," *Molecular and Biochemical Parasitology*, vol. 16, no. 1, pp. 85–96, 1985.

[82] B. Zingales, C. Carniol, R. M. de Lederkremer, and W. Colli, "Direct sialic acid transfer from a protein donor to glycolipids of trypomastigote forms of Trypanosoma cruzi," *Molecular and Biochemical Parasitology*, vol. 26, no. 1-2, pp. 135–144, 1987.

[83] R. P. Prioli, I. Rosenberg, and M. E. A. Pereira, "High- and low-density lipoproteins enhance infection of *Trypanosoma cruzi* in vitro," *Molecular and Biochemical Parasitology*, vol. 38, no. 2, pp. 191–198, 1990.

[84] S. Tomlinson, L. C. Pontes de Carvalho, F. Vandekerckhove, and V. Nussenzweig, "Role of sialic acid in the resistance of Trypanosoma cruzi trypomastigotes to complement," *Journal of Immunology*, vol. 153, no. 7, pp. 3141–3147, 1994.

[85] V. L. Pereira-Chioccola, A. Acosta-Serrano, I. Correia de Almeida et al., "Mucin-like molecules form a negatively charged coat that protects Trypanosoma cruzi trypomastigotes from killing by human anti-α-galactosyl antibodies," *Journal of Cell Science*, vol. 113, part 7, pp. 1299–1307, 2000.

[86] K. Nagamune, A. Acosta-Serrano, H. Uemura et al., "Surface sialic acids taken from the host allow trypanosome survival in Tsetse fly vectors," *Journal of Experimental Medicine*, vol. 199, no. 10, pp. 1445–1450, 2004.

[87] T. Jacobs, H. Erdmann, and B. Fleischer, "Molecular interaction of Siglecs (sialic acid-binding Ig-like lectins) with sialylated ligands on *Trypanosoma cruzi*," *European Journal of Cell Biology*, vol. 89, no. 1, pp. 113–116, 2010.

[88] N. Yoshida, M. L. Dorta, A. T. Ferreira et al., "Removal of sialic acid from mucin-like surface molecules of Trypanosoma cruzi metacyclic trypomastigotes enhances parasite-host cell interaction," *Molecular and Biochemical Parasitology*, vol. 84, no. 1, pp. 57–67, 1997.

[89] G. D. Pollevick, J. Affranchino, A. C. C. Frasch, and D. O. Sánchez, "The complete sequence of a shed acute-phase antigen of Trypanosoma cruzi," *Molecular and Biochemical Parasitology*, vol. 47, no. 2, pp. 247–250, 1991.

[90] C. A. Buscaglia, J. Alfonso, O. Campetella et al., "Tandem amino acid repeats from Trypanosoma cruzi shed antigens increase the

half-life of proteins in blood," *Blood*, vol. 93, no. 6, pp. 2025–2032, 1999.

[91] S. S. C. Rubin-de-Celis, H. Uemura, N. Yoshida, and S. Schenkman, "Expression of trypomastigote trans-sialidase in metacyclic forms of Trypanosoma cruzi increases parasite escape from its parasitophorous vacuole," *Cellular Microbiology*, vol. 8, no. 12, pp. 1888–1898, 2006.

[92] M. A. Ouaissi, J. P. Kusnierz, H. Gras-masse et al., "Fluorescence-activated cell-sorting analysis of fibronectin peptides binding to trypanosoma cruzi trypomastigotes," *The Journal of Protozoology*, vol. 35, no. 1, pp. 111–114, 1988.

[93] C. Claser, N. M. Espíndola, G. Sasso, A. J. Vaz, S. B. Boscardin, and M. M. Rodrigues, "Immunologically relevant strain polymorphism in the Amastigote Surface Protein 2 of Trypanosoma cruzi," *Microbes and Infection*, vol. 9, no. 8, pp. 1011–1019, 2007.

[94] W. Colli and M. J. M. Alves, "Relevant glycoconjugates on the surface of trypanosoma cruzi," *Memórias do Instituto Oswaldo Cruz*, vol. 94, no. 1, pp. 37–49, 1999.

[95] O. Campetella, D. Sánchez, J. J. Cazzulo, and A. C. C. Frasch, "A superfamily of trypanosoma cruzi surface antigens," *Parasitology Today*, vol. 8, no. 11, pp. 378–381, 1992.

[96] B. Wizel, M. Nunes, and R. L. Tarleton, "Identification of Trypanosoma cruzi trans-sialidase family members as targets of protective CD8+ TC1 responses," *Journal of Immunology*, vol. 159, no. 12, pp. 6120–6130, 1997.

[97] B. Wizel, M. Palmieri, C. Mendoza et al., "Human infection with Trypanosoma cruzi induces parasite antigen-specific cytotoxic T lymphocyte responses," *Journal of Clinical Investigation*, vol. 102, no. 5, pp. 1062–1071, 1998.

[98] S. Kahn, M. Wleklinski, A. Aruffo, A. Farr, D. Coder, and M. Kahn, "Trypanosoma cruzi amastigote adhesion to macrophages is facilitated by the mannose receptor," *Journal of Experimental Medicine*, vol. 182, no. 5, pp. 1243–1258, 1995.

[99] N. A. Peterslund, C. Koch, J. C. Jensenius, and S. Thiel, "Association between deficiency of mannose-binding lectin and severe infections after chemotherapy," *Lancet*, vol. 358, no. 9282, pp. 637–638, 2001.

[100] I. D. S. Cestari, I. Evans-Osses, J. C. Freitas, J. M. Inal, and M. I. Ramirez, "Complement C2 receptor inhibitor trispanning confers an increased ability to resist complement-mediated lysis in Trypanosoma cruzi," *Journal of Infectious Diseases*, vol. 198, no. 9, pp. 1276–1283, 2008.

[101] M. H. Magdesian, R. Giordano, H. Ulrich et al., "Infection by *Trypanosoma cruzi*: identification of a parasite ligand and its host cell receptor," *The Journal of Biological Chemistry*, vol. 276, no. 22, pp. 19382–19389, 2001.

[102] R. R. Tonelli, R. J. Giordano, E. M. Barbu et al., "Role of the gp85/trans-sialidases in *Trypanosoma cruzi* tissue tropism: preferential binding of a conserved peptide motif to the vasculature in vivo," *PLoS Neglected Tropical Diseases*, vol. 4, no. 11, article e864, 2010.

[103] E. M. Cordero, L. G. Gentil, G. Crisante et al., "Expression of GP82 and GP90 surface glycoprotein genes of Trypanosoma cruzi during in vivo metacyclogenesis in the insect vector Rhodnius prolixus," *Acta Tropica*, vol. 105, no. 1, pp. 87–91, 2008.

[104] M. I. Ramirez, R. De Cassia Ruiz, J. E. Araya et al., "Involvement of the stage-specific 82-kilodalton adhesion molecule of Trypanosoma cruzi metacyclic trypomastigotes in host cell invasion," *Infection and Immunity*, vol. 61, no. 9, pp. 3636–3641, 1993.

[105] R. C. Ruiz, J. Favoreto S., M. L. Dorta et al., "Infectivity of Trypanosoma cruzi strains is associated with differential expression of surface glycoproteins with differential Ca2+ signalling activity," *Biochemical Journal*, vol. 330, no. 1, pp. 505–511, 1998.

[106] S. Favoreto Jr., M. L. Dorta, and N. Yoshida, "Trypanosoma cruzi 175-kDa protein tyrosine phosphorylation is associated with host cell invasion," *Experimental Parasitology*, vol. 89, no. 2, pp. 188–194, 1998.

[107] T. K. Matsumoto, P. C. Cotrim, J. F. Da Silveira, A. M. S. Stolf, and E. S. Umezawa, "Trypanosoma cruzi: Isolation of an immunodominant peptide of TESA (trypomastigote excreted-secreted antigens) by gene cloning," *Diagnostic Microbiology and Infectious Disease*, vol. 42, no. 3, pp. 187–192, 2002.

[108] M. Beucher and K. A. Norris, "Sequence diversity of the Trypanosoma cruzi complement regulatory protein family," *Infection and Immunity*, vol. 76, no. 2, pp. 750–758, 2008.

[109] T. L. Kipnis, J. R. David, C. A. Alper, A. Sher, and W. D. da Silva, "Enzymatic treatment transforms trypomastigotes of *Trypanosoma cruzi* into activators of alternative complement pathway and potentiates their uptake by macrophages," *Proceedings of the National Academy of Sciences of the United States of America*, vol. 78, no. 1, pp. 602–605, 1981.

[110] M. Berrizbeitia, M. Ndao, J. Bubis et al., "Purified excreted-secreted antigens from *Trypanosoma cruzi* trypomastigotes as tools for diagnosis of Chagas' disease," *Journal of Clinical Microbiology*, vol. 44, no. 2, pp. 291–296, 2006.

[111] M. S. Cetron, R. Hoff, S. Kahn, H. Eisen, and W. C. Van Voorhis, "Evaluation of recombinant trypomastigote surface antigens of Trypanosoma cruzi in screening sera from a population in rural Northeastern Brazil endemic for chagas' disease," *Acta Tropica*, vol. 50, no. 3, pp. 259–266, 1992.

[112] E. E. Jazin, L. Åslund, J. Henriksson, and U. Pettersson, "Trypanosoma cruzi exoantigen is a member of a 160 kDa gene family," *Parasitology*, vol. 110, no. 1, pp. 61–69, 1995.

[113] K. A. Norris, J. E. Schrimpf, and M. J. Szabo, "Identification of the gene family encoding the 160-kilodalton Trypanosoma cruzi complement regulatory protein," *Infection and Immunity*, vol. 65, no. 2, pp. 349–357, 1997.

[114] W. C. Van Voorhis, L. Schlekewy, and H. L. Trong, "Molecular mimicry by Trypanosoma cruzi: the Fl-160 epitope that mimics mammalian nerve can be mapped to a 12-amino acid peptide," *Proceedings of the National Academy of Sciences of the United States of America*, vol. 88, no. 14, pp. 5993–5997, 1991.

[115] L. Freire-De-Lima, L. M. Fonseca, T. Oeltmann, L. Mendonça-Previato, and J. O. Previato, "The trans-sialidase, the major Trypanosoma cruzi virulence factor: three decades of studies," *Glycobiology*, vol. 25, no. 11, pp. 1142–1149, 2015.

[116] M. H. Branquinha, A. B. Vermelho, S. Goldenberg, and M. C. Bonaldo, "Ubiquity of cysteine- and metalloproteinase activities in a wide range of trypanosomatids," *Journal of Eukaryotic Microbiology*, vol. 43, no. 2, pp. 131–135, 1996.

[117] F. B. Nogueira, M. A. Krieger, P. Nirdé, S. Goldenberg, A. J. Romanha, and S. M. F. Murta, "Increased expression of iron-containing superoxide dismutase-A (TcFeSOD-A) enzyme in Trypanosoma cruzi population with in vitro-induced resistance to benznidazole," *Acta Tropica*, vol. 100, no. 1-2, pp. 119–132, 2006.

[118] A. L. S. D. Santos, R. M. D. A. Soares, C. S. Alviano, and L. F. Kneipp, "Heterogeneous production of metallo-type peptidases in parasites belonging to the family Trypanosomatidae," *European Journal of Protistology*, vol. 44, no. 2, pp. 103–113, 2008.

[119] N. M. A. El-Sayed and J. E. Donelson, "African trypanosomes have differentially expressed genes encoding homologues of

the Leishmania GP63 surface protease," *Journal of Biological Chemistry*, vol. 272, no. 42, pp. 26742–26748, 1997.

[120] P. M. Grandgenett, B. C. Coughlin, L. V. Kirchhoff, and J. E. Donelson, "Differential expression of GP63 genes in Trypanosoma cruzi," *Molecular and Biochemical Parasitology*, vol. 110, no. 2, pp. 409–415, 2000.

[121] S. M. R. Teixeira, D. G. Russell, L. V. Kirchhoff et al., "A differentially expressed gene family encoding 'amastin,' a surface protein of Trypanosoma cruzi amastigotes," *Journal of Biological Chemistry*, vol. 269, no. 32, pp. 20509–20516, 1994.

[122] A. P. Jackson, "The evolution of amastin surface glycoproteins in trypanosomatid parasites," *Molecular Biology and Evolution*, vol. 27, no. 1, pp. 33–45, 2010.

[123] A. Rochette, F. McNicoll, J. Girard et al., "Characterization and developmental gene regulation of a large gene family encoding amastin surface proteins in Leishmania spp.," *Molecular and Biochemical Parasitology*, vol. 140, no. 2, pp. 205–220, 2005.

[124] H. Azizi, K. Hassani, Y. Taslimi, H. S. Najafabadi, B. Papadopoulou, and S. Rafati, "Searching for virulence factors in the non-pathogenic parasite to humans Leishmania tarentolae," *Parasitology*, vol. 136, no. 7, pp. 723–735, 2009.

[125] G. C. Cerqueira, D. C. Bartholomeu, W. D. DaRocha et al., "Sequence diversity and evolution of multigene families in Trypanosoma cruzi," *Molecular and Biochemical Parasitology*, vol. 157, no. 1, pp. 65–72, 2008.

[126] T. A. Minning, D. B. Weatherly, J. Atwood III, R. Orlando, and R. L. Tarleton, "The steady-state transcriptome of the four major life-cycle stages of Trypanosoma cruzi," *BMC Genomics*, vol. 10, article 370, 2009.

[127] M. C. Cruz, N. Souza-Melo, C. V. da Silva et al., "Trypanosoma cruzi: role of δ-amastin on extracellular amastigote cell invasion and differentiation," *PLoS ONE*, vol. 7, no. 12, Article ID e51804, 2012.

[128] J. A. Atwood III, D. B. Weatherly, T. A. Minning et al., "Microbiology: the Trypanosoma cruzi proteome," *Science*, vol. 309, no. 5733, pp. 473–476, 2005.

[129] N. M. El-Sayed, P. J. Myler, G. Blandin et al., "Comparative genomics of trypanosomatid parasitic protozoa," *Science*, vol. 309, no. 5733, pp. 404–409, 2005.

[130] D. C. Bartholomeu, G. C. Cerqueira, A. C. A. Leão et al., "Genomic organization and expression profile of the mucin-associated surface protein (masp) family of the human pathogen Trypanosoma cruzi," *Nucleic Acids Research*, vol. 37, no. 10, pp. 3407–3417, 2009.

[131] L. M. De Pablos, G. G. González, J. S. Parada et al., "Differential expression and characterization of a member of the mucin-associated surface protein family secreted by Trypanosoma cruzi," *Infection and Immunity*, vol. 79, no. 10, pp. 3993–4001, 2011.

[132] L. M. De Pablos, I. M. Díaz Lozano, M. I. Jercic et al., "The C-terminal region of Trypanosoma cruzi MASPs is antigenic and secreted via exovesicles," *Scientific Reports*, vol. 6, Article ID 27293, 2016.

[133] C. Serna, J. A. Lara, S. P. Rodrigues, A. F. Marques, I. C. Almeida, and R. A. Maldonado, "A synthetic peptide from Trypanosoma cruzi mucin-like associated surface protein as candidate for a vaccine against Chagas disease," *Vaccine*, vol. 32, no. 28, pp. 3525–3532, 2014.

[134] M. V. Chuenkova and M. PereiraPerrin, "Chagas' disease parasite promotes neuron survival and differentiation through TrkA nerve growth factor receptor," *Journal of Neurochemistry*, vol. 91, no. 2, pp. 385–394, 2004.

[135] L. Lima, P. A. Ortiz, F. M. da Silva et al., "Repertoire, genealogy and genomic organization of cruzipain and homologous genes in Trypanosoma cruzi, T. cruzi-like and other Trypanosome species," *PLoS ONE*, vol. 7, no. 6, Article ID e38385, 2012.

[136] L. Åslund, J. Henriksson, O. Campetella, A. C. C. Frasch, U. Pettersson, and J. J. Cazzulo, "The C-terminal extension of the major cysteine proteinase (cruzipain) from Trypanosoma cruzi," *Molecular and Biochemical Parasitology*, vol. 45, no. 2, pp. 345–347, 1991.

[137] V. Alvarez, F. Parussini, L. Åslund, and J. J. Cazzulo, "Expression in insect cells of active mature cruzipain from trypanosoma cruzi, containing its c-terminal domain," *Protein Expression and Purification*, vol. 26, no. 3, pp. 467–475, 2002.

[138] T. Souto-Padron, O. E. Campetella, J. J. Cazzulo et al., "Cysteine proteinase in *Trypanosoma cruzi*: immunocytochemical localization and involvement in parasite-host cell interaction," *Journal of Cell Science*, vol. 96, part 3, pp. 485–490, 1990.

[139] M. N. L. Meirelles, L. Juliano, E. Carmona et al., "Inhibitors of the major cysteinyl proteinase (GP57/51) impair host cell invasion and arrest the intracellular development of *Trypanosoma cruzi* in vitro," *Molecular and Biochemical Parasitology*, vol. 52, no. 2, pp. 175–184, 1992.

[140] S. Gea, N. Guiñazu, A. Pellegrini et al., "Cruzipain, a major Trypanosoma cruzi cystein protease in the host-parasite interplay," *Inmunologia*, vol. 25, no. 4, pp. 225–238, 2006.

[141] P. M. Ferrão, C. M. D'Avila-Levy, T. C. Araujo-Jorge et al., "Cruzipain activates latent TGF-β from host cells during T. cruzi invasion," *PLoS ONE*, vol. 10, no. 5, Article ID e0124832, 2015.

[142] G. V. F. Brunoro, M. A. Caminha, A. T. D. S. Ferreira et al., "Reevaluating the Trypanosoma cruzi proteomic map: the shotgun description of bloodstream trypomastigotes," *Journal of Proteomics*, vol. 115, pp. 58–65, 2015.

[143] S. B. Roberts, J. L. Robichaux, A. K. Chavali et al., "Proteomic and network analysis characterize stage-specific metabolism in Trypanosoma cruzi," *BMC Systems Biology*, vol. 3, article 52, 2009.

[144] D. Pérez-Morales, H. Lanz-Mendoza, G. Hurtado, R. Martínez-Espinosa, and B. Espinoza, "Proteomic analysis of Trypanosoma cruzi epimastigotes subjected to heat shock," *Journal of Biomedicine and Biotechnology*, vol. 2012, Article ID 902803, 9 pages, 2012.

[145] M. M. Kangussu-Marcolino, A. P. Cunha, A. R. Avila, J.-P. Herman, and W. D. Darocha, "Conditional removal of selectable markers in Trypanosoma cruzi using a site-specific recombination tool: proof of concept," *Molecular and Biochemical Parasitology*, vol. 198, no. 2, pp. 71–74, 2014.

[146] D. Peng, S. P. Kurup, P. Y. Yao, T. A. Minning, and R. L. Tarleton, "CRISPR-Cas9-mediated single-gene and gene family disruption in Trypanosoma cruzi," *mBio*, vol. 6, no. 1, Article ID e02097-14, 2015.

[147] N. Lander, Z.-H. Li, S. Niyogi, and R. Docampo, "CRISPR/Cas9-induced disruption of paraflagellar rod protein 1 and 2 genes in Trypanosoma cruzi reveals their role in flagellar attachment," *mBio*, vol. 6, no. 4, Article ID e01012-15, 2015.

[148] N. Lander, M. A. Chiurillo, M. Storey et al., "CRISPR/Cas9-mediated endogenous C-terminal tagging of Trypanosoma cruzi genes reveals the acidocalcisome localization of the inositol 1,4,5-Trisphosphate receptor," *Journal of Biological Chemistry*, vol. 291, no. 49, pp. 25505–25515, 2016.

11

New Scenarios of Chagas Disease Transmission in Northern Colombia

Catalina Tovar Acero,[1] Jorge Negrete Peñata,[2] Camila González,[3]
Cielo León,[3] Mario Ortiz,[3] Julio Chacón Pacheco,[4,5] Elkin Monterrosa,[6]
Abraham Luna,[7] Dina Ricardo Caldera,[1] and Lyda Espitia-Pérez[8]

[1]Grupo de Investigación en Enfermedades Tropicales y Resistencia Bacteriana, Facultad de Ciencias de la Salud, Universidad del Sinú, Montería, Colombia
[2]Laboratorio de Investigaciones Biomédicas, Universidad del Sinú, Montería, Colombia
[3]Departamento de Ciencias Biológicas, Centro de Investigaciones en Microbiología y Parasitología Tropical (CIMPAT), Universidad de los Andes, Bogotá, Colombia
[4]Fundación Colombia Mia, Montería, Colombia
[5]Grupo de Investigación Biodiversidad Unicordoba, Universidad de Córdoba, Montería, Colombia
[6]Área de Entomología, Laboratorio de Salud Pública de Córdoba, Montería, Colombia
[7]Hospital San Juan de Sahagún, Sahagún, Colombia
[8]Grupo de Investigación Biomédica y Biología Molecular, Facultad de Ciencias de la Salud, Universidad del Sinú, Montería, Colombia

Correspondence should be addressed to Catalina Tovar Acero; catalina@unisinu.edu.co

Academic Editor: Bernard Marchand

Chagas disease (CD) is a systemic parasitic infection caused by the flagellated form of *Trypanosoma cruzi*. Córdoba department, located in the Colombian Caribbean Coast, was not considered as a region at risk of *T. cruzi* transmission. In this article, we describe the first acute CD case in Salitral village in Sahagún, Córdoba, confirmed by microscopy and serological tests. Our results draw attention to a new scenario of transmission of acute CD in nonendemic areas of Colombia and highlight the need to include CD in the differential diagnosis of febrile syndromes in this region.

Chagas disease (CD) also known as American Trypanosomiasis is a systemic parasitic infection caused by the protozoan parasite *Trypanosoma cruzi (T. cruzi)*, which affects six to seven million people worldwide with an annual incidence of 28.000 cases in the Americas [1]. Transmission to humans as well as to domestic and sylvatic mammals occurs mainly through the introduction of the parasite present in triatomine bug feces during its blood meals; however, alternate transmission routes include blood transfusion, organ transplants, laboratory accidents, congenital and oral ingestion of contaminated food [2, 3]. During the acute phase of the disease up to 30% of patients suffer from cardiac disorders and up to 10% suffer from digestive (typical enlargement of the esophagus, spleen, or colon), neurological, or mixed alterations [4].

Colombia has been one of the Latin American countries with a considerable number of acute Chagas disease outbreaks where oral transmission of *T. cruzi* has been recorded specially in endemic areas of Santander, Norte de Santander, Cundinamarca, Boyacá, Casanare, Meta, Arauca, and some areas of the Sierra Nevada of Santa Marta [5]. Córdoba department located in the Colombian Caribbean Coast is not considered as an endemic region for *T. cruzi* transmission; therefore CD is not included in the diagnosis of febrile diseases in hospitals and health centres. Additionally, due to the lack of knowledge about CD clinical symptoms, diagnosis in a setting where multiple infectious tropical diseases are present, such as tuberculosis, malaria, dengue, chikungunya, and Zika, is a challenge; therefore annual reports considered

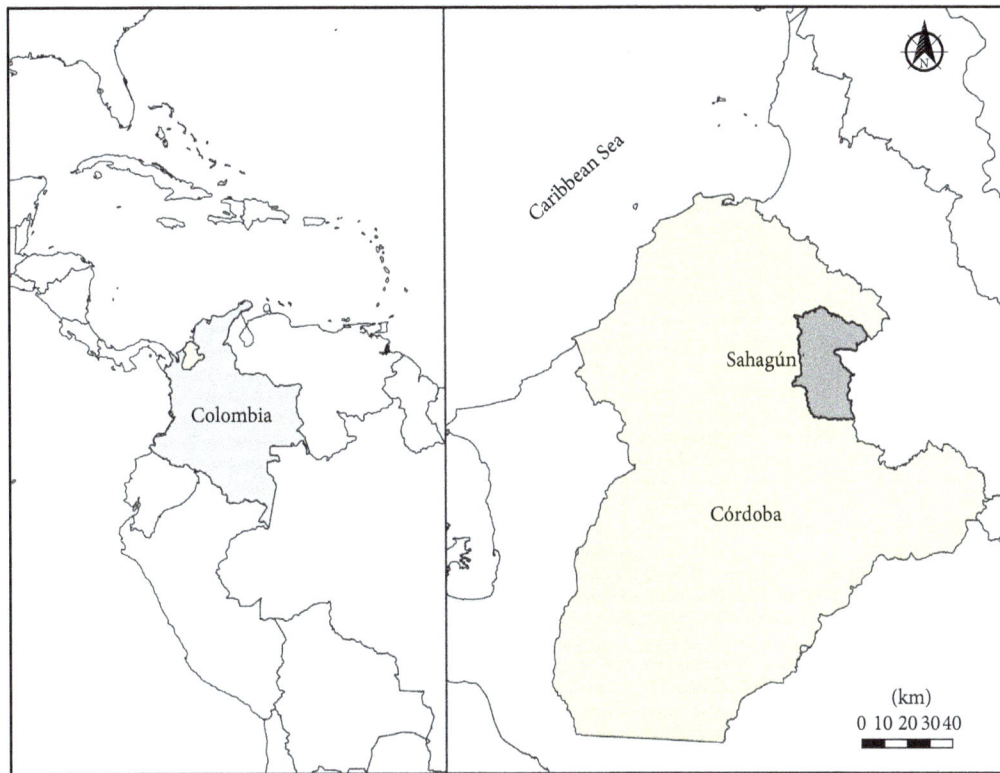

FIGURE 1: Sampling area (Salitral village, Córdoba, Colombia).

CD cases to be imported from other departments. In this article, we describe the first acute CD case in Salitral village in Sahagún, Córdoba, confirmed by microscopy and serological tests. Sampling area was located in Salitral village in the Sahagún municipality of Córdoba department, located in the northwest part of Colombia (8°49′47.9″N, 75°31′31.5″W, and 75 m.a.s.l.) (Figure 1). This area has a tropical climate with an annual mean temperature between 27°C and 30°C and a relative humidity of 84%. Main economic activities are related to mixed extensive crop-livestock systems, usually linked to the proliferation of wild small mammals (rodents and marsupials) described as *T. cruzi* reservoirs.

The 16-year-old young male patient born and resident in Salitral village was referred to the emergency service of San Juan de Sahagún Hospital (ESE HSJS) with eight days of headache and high fever (38°C) associated with chills, generalized myalgia, asthenia, adynamia, choluria, severe epigastralgia without epistaxis, and gingivorrhagia. On his epidemiological history, the patient denied knowing triatomine bugs, having received any blood transfusion or organ transplant, or traveling outside Córdoba prior to the beginning of the symptoms. No inoculation point either in skin or periocular region suggesting vectorial transmission could be detected. The patient was alert, without respiratory distress or cardiovascular involvement (electrocardiogram). Abdominal palpation and ultrasound examination confirmed symptoms of a moderate spleen enlargement (splenomegaly). Thick blood smear examination was negative for *Plasmodium* spp. but positive for *T. cruzi* trypomastigotes (Figure 2).

Serological analysis by enzyme linked immunosorbent assay (ELISA) for CD was negative at the seventh day of hospitalization.

All experimental and sampling protocols were approved by the Ethics Committee of Universidad del Sinú according to national normativity for human populations studies and the NIH Guide for the Care and Use of Animals [6].

In order to determine the presence of triatomine bugs, active manual search was carried out by the professional staff and community members in walls, cracks in the walls and ceiling, mattresses, and floor of the patient's house and other 24 houses in the neighborhood area according to OMS recommended methodology [7]. Additionally, live-baited traps [8] and Gómez-Nuñez boxes [9] were also placed in the intra and peridomicile of each selected household. Taxonomic identification of captured specimens was performed based on external morphology, according to Lent and Wygodzinsky [10]. Detection of *T. cruzi* infection in captured triatomines was confirmed by direct and molecular techniques examining intestinal contents and rectal ampulla. Small- and mid-sized mammals were also captured using 5 mist nets for bats and 20 Tomahawks and 40 Sherman traps. Captured mammals were taxonomically identified according to Emmons and Feer [11], Linares [12], Tirira [13], Gardner [14], and Patton et al. [15] and whole blood samples were taken for molecular identification of *T. cruzi*. Sampling was performed on 80 volunteers selected from the entire population. All participants filled out a clinical-epidemiological survey including identification variables and

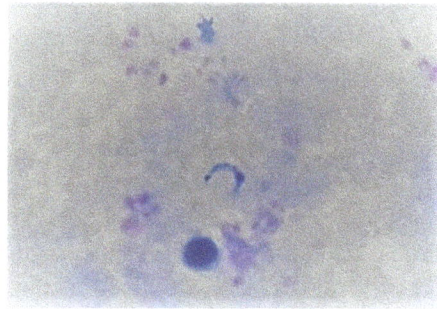

FIGURE 2: *T. cruzi* trypomastigotes detected in thick blood smears of the infected patient (1000x).

FIGURE 3: Patient's house infrastructure and peridomicile.

evidence of signs or symptoms according to case definition. All family members and other individuals related to the acute case patient were also analyzed. All samples were collected after obtaining the corresponding informed consent. The serological analysis included detection of IgG antibodies by ELISA and indirect immunofluorescence (IFI). For *T. cruzi* detection, human blood samples and triatomine bugs rectal ampulla were collected in a volume solution containing EDTA and guanidine 6M and stored at room temperature. A spin column-based nucleic acid purification kit was used to perform DNA extraction (High Pure PCR Template Preparation, de Roche®). Molecular detection was carried out through amplification of the variable region of kinetoplast DNA (kDNA) according to the methodology previously described by [16, 17] and tandem repeat satellite region from *T. cruzi* using the *cruzi1* and *cruzi2* primers described by [18]. Amplification cycles for kDNA were performed using a two-step procedure using an initial denaturation step at 94°C for 3 min; 5 cycles of denaturation at 94°C for 1 min, annealing at 68°C for 1 min, and extension at 72°C for 1 min, followed by 35 cycles at 94°C for 45 sec; annealing at 64°C for 45 sec, extension at 72°C for 45 sec, and final extension at 72°C for 10 min. Cycling conditions for *cruzi1* and *cruzi2* were initial denaturation at 94°C for 5 min and 40 cycles of denaturation at 94°C for 1 min, annealing at 64°C for 30 sec, extension at 72°C for 1 min, and final extension at 72°C for 10 min. Patient's house was built with wood walls, palm roofs, and dirt floors surrounded by dense vegetation consisting of trees and palms (Figure 3). Among the 24 houses included in the study 32% were constructed with wooden walls, 40% with dirt floors, and 56% with thatch palm roof and 32% had unplastered walls. Conventional parasitological methods,

serological screening, and molecular testing for detection of *T. cruzi* infection performed on blood samples of 80 voluntary patients showed negative results. In this particular community, serological test showed positive results only for the acute case described in this work. During the entomological sampling, seven individuals identified as *Rhodnius pallescens* and two classified as *Panstrongylus geniculatus* were captured. Most captured insects were collected by members of the community. Analysis of intestinal content and rectal ampulla confirmed the presence of *T. cruzi* in one specimen of each species. Analysis of blood samples of 29 specimens of small- and mid-sized mammals using molecular methods confirmed the presence of *T. cruzi* DNA in two specimens of *Didelphis marsupialis* and two specimens of *Heteromys anomalus*. The transmission scenario of CD in Córdoba still remains a challenge and must be addressed through clinical and ecoepidemiological studies since, as our results showed, a sylvatic cycle exists and accidental human cases might be occurring. In this particular case, signs and symptoms presented by the patient including prolonged febrile illness, epigastralgia, and absence of lesions in either the skin or the periocular region indicating the insect bite, together with patient statement of never being bitten by triatomines and never leaving Córdoba department, would suggest oral transmission as the most likely pathway of infection [19, 20].

Considering the *T. cruzi* detection in specimens of the triatomine bugs *Panstrongylus geniculatus* and *Rhodnius pallescens* and the mammals *Didelphis marsupialis* and *Heteromys anomalus*, there is an evident risk of infection to humans either. In Córdoba department previous studies reported the presence of several triatomine species [21]. In line with our findings, *E. cuspidatus*, *P. geniculatus*, and *R.*

pallescens had previously been reported in Sahagún municipality [22]. However, no evidence of domiciliated triatomines was found. Even when no evidence of domiciliated *P. geniculatus* has been reported for Colombia, recent reports about the increasing frequency of *Panstrongylus* species displaying ability to invade and colonize human habitats are focusing the interest of entomologists and CD control managers throughout Latin America [23]. Current land use changes in Colombia and particularly in Córdoba, where vast forested areas have been cleared for livestock and agricultural activities, may favor triatomine domiciliation. This new situation could impose necessary changes in the strategy of CD control programs in Colombia, which until now have been limited to vector control activities in rural communities in endemic areas. Community engagement in sampling activities constituted a very effective approach for triatomine collection [24]. In our case, triatomine collection by community members was 3.5 times more effective compared to conventional sampling. These data are of key importance for the successful implementation of vector control in Córdoba and community participation could be a method of choice for sustained monitoring of triatomines in this area. Considering that there were no previous reports of *T. cruzi* infected reservoirs in Córdoba, our study represents the first report of *T. cruzi* infection detected in small mammals (*Didelphis marsupialis* and *Heteromys anomalus*) from this particular region. As documented by Cantillo-Barraza et al. [25], *D. marsupialis* may play an active role in the amplification of *T. cruzi* transmission in peridomestic areas mediating enzootic cycle or acting as a link between the enzootic and domestic cycles. In previous studies, *Heteromys anomalus* has been associated with sylvatic transmission cycles of CD in Colombia [26]. Our study also confirmed *T. cruzi* transmission in Salitral municipality evidenced by the presence of the parasite in different actors involved in the transmission cycle (human, reservoir, and vector). Even when no domiciliated vectors were found, our findings suggest the existence of autochthonous human cases in Córdoba and highlight the need to include CD in the differential diagnosis of febrile syndromes and diseases of this region. Despite the improvements in building materials and construction conditions of human dwellings in Córdoba, in some rural areas contact with natural and sylvatic environments persists, thereby creating the constant presence of potential vectors and reservoirs like triatomines, marsupials, and small mammals around the peridomicile. This close contact presumably could enable the emergence of CD cases [27]. Similarly, in rural areas some practices related to food preparation and storage may also constitute a potential risk factor increasing the contact with triatomine feces and small mammals dejections [28]. This study draws attention to new scenarios transmission of CD in nonendemic areas of Córdoba department in Colombia and highlights the need to include CD in the differential diagnosis of febrile syndromes and diseases of this region.

Acknowledgments

The authors gratefully wish to acknowledge the support of all the members of the Biomedical and Molecular Biology Laboratory of Universidad del Sinú for logistic cooperation during the sampling period. This research was supported by Gobernación de Córdoba and Sistema General de Regalías (SGR), Colombia (Grant no. 754/2013), and Universidad del Sinú-Elías Bechara Zainúm (UNISINU), Colombia.

References

[1] J. R. Coura and P. A. Vĩas, "Chagas disease: a new worldwide challenge," *Nature*, vol. 465, no. 7301, pp. S6–S7, 2010.

[2] K. Rueda, J. E. Trujillo, J. C. Carranza, and G. A. Vallejo, "Transmisión oral de Trypanosoma cruzi: un nuevo escenario epidemiológico de la enfermedad de Chagas en Colombia y otros países suramericanos," *Biomédica*, vol. 34, no. 4, 2014.

[3] J. F. Ríos, M. Arboleda, A. N. Montoya, E. P. Alarcón, and G. J. Parra-Henao, "Probable brote de transmisión oral de enfermedad de Chagas en Turbo, Antioquia," *Biomédica*, vol. 31, no. 2, p. 185, 2011.

[4] W.H.O., Chagas disease (American trypanosomiasis): Fact sheet N°340. 2014.

[5] Ministerio de la Protección Social, I.N.d.S., Organización Panamericana de la Salud OPS/OMS, Guía para la Atención Clínica Integral del paciente con enfermedad de Chagas. 2010: p. 1–81.

[6] Care, I.o.L.A.R.C.o., U.o.L. Animals, and N.I.o.H.D.o.R. Resources, Guide for the care and use of laboratory animals. 1985: National Academies.

[7] M. D. Feliciangeli, M. Hernández, B. Suarez et al., "Comparación de métodos de captura intradoméstica de triatominos vectores de la enfermedad de Chagas en Venezuela," *Boletín de Malariología y Salud Ambiental*, vol. 47, pp. 103–117, 2007.

[8] V. M. Angulo and L. Esteban, "Nueva trampa para la captura de triatominos en hábitats silvestres y peridomésticos," *Biomédica*, vol. 31, no. 2, p. 264, 2011.

[9] A. Longa and J. V. Scorza, "Acrocomia aculeata (Palmae), hábitat silvestre de Rhodnius robustus en el Estado Trujillo, Venezuela," *Boletín de Malariología y Salud Ambiental*, vol. 47, no. 1-2, pp. 213–220, 2007.

[10] H. Lent and P. Wygodzinsky, "Revision of the Triatominae (Hemiptera, Reduviidae), and their significance as vectors of Chagas disease," *Bulletin of the American Museum of Natural History*, vol. 163, pp. 123–520, 1979.

[11] L. Emmons and F. Feer, *Neotropical Rainforest Mammals: A Field Guide*, University of Chicago Press, 1997.

[12] O. J. Linares, *Mamíferos de Venezuela*, Sociedad Conservacionista Audubon de Venezuela, 1998.

[13] D. Tirira, "Guía de campo de los Mamíferos del Ecuador. Ediciones ed. 2007, Quito, Ecuador: Publicación especial sobre los mamíferos del Ecuador 6. 576-576".

[14] A. L. Gardner, *Mammals of South America, Volume 1: Marsupials, Xenarthrans, Shrews, and Bats*, University of Chicago Press, 2008.

[15] J. L. Patton, U. F. Pardiñas, and G. D'Elía, *Mammals of South America, Volume 2: Rodents*, University of Chicago Press, 2015.

[16] J. M. Burgos, J. Altcheh, M. Bisio et al., "Direct molecular profiling of minicircle signatures and lineages of Trypanosoma cruzi bloodstream populations causing congenital Chagas disease," *International Journal for Parasitology*, vol. 37, no. 12, pp. 1319–1327, 2007.

[17] A. G. Schijman, M. Bisio, L. Orellana et al., "International study to evaluate PCR methods for detection of Trypanosoma cruzi DNA in blood samples from Chagas disease patients," *PLoS neglected tropical diseases*, vol. 5, no. 1, e931 pages, 2011.

[18] M. Bisio, C. Cura, T. Duffy et al., "Trypanosoma cruzi discrete typing units in Chagas disease patients with HIV co-infection," *Revista Biomédica*, vol. 20, pp. 166–178, 2009.

[19] B. Alarcón de Noya, J. Veas, R. Ruiz-Guevara et al., "Evaluación clínica y de laboratorio de pacientes hospitalizados durante el primer brote urbano de enfermedad de chagas de transmisión oral en venezuela," *Revista de Patologia Tropical*, vol. 42, no. 2, 2013.

[20] L. Zuleta, "Enfermedad de Chagas: posible transmisión oral en trabajadores del sector hidrocarburos, Casanare, Colombia," *Biomédica Revista del Instituto Nacional de Salud*, vol. 37, no. 2, pp. 2–12, 2014.

[21] F. Guhl, G. Aguilera, N. Pinto, and D. Vergara, "Updated geographical distribution and ecoepidemiology of the triatomine fauna (Reduviidae: Triatominae) in Colombia," *Biomedica*, vol. 27, 2007.

[22] F. Guhl, G. Aguilera, N. Pinto, and D. Vergara, "Actualización de la distribución geográfica y ecoepidemiología de la fauna de triatominos (Reduviidae: Triatominae) en Colombia," *Biomedica*, vol. 27, no. 1, pp. 143–162, 2007.

[23] M. Reyes-Lugo and A. Rodriguez-Acosta, "Domiciliation of the sylvatic Chagas disease vector Panstrongylus geniculatus Latreille, 1811 (Triatominae: Reduviidae) in Venezuela," *Transactions of the Royal Society of Tropical Medicine and Hygiene*, vol. 94, no. 5, p. 508, 2000.

[24] E. Dumonteil, M. J. Ramirez-Sierra, J. Ferral, M. Euan-Garcia, and L. Chavez-Nuñez, "Usefulness of community participation for the fine temporal monitoring of house infestation by non-domiciliated triatomines," *Journal of Parasitology*, vol. 95, no. 2, pp. 469–471, 2009.

[25] O. Cantillo-Barraza, E. Garcés, A. Gómez-Palacio et al., "Eco-epidemiological study of an endemic Chagas disease region in northern Colombia reveals the importance of Triatoma maculata (Hemiptera: Reduviidae), dogs and Didelphis marsupialis in Trypanosoma cruzi maintenance," *Parasites and Vectors*, vol. 8, no. 1, article no. 482, 2015.

[26] A. M. Mejía-Jaramillo, L. A. Agudelo-Uribe, J. C. Dib, S. Ortiz, A. Solari, and O. Triana-Chávez, "Genotyping of Trypanosoma cruzi in a hyper-endemic area of Colombia reveals an overlap among domestic and sylvatic cycles of Chagas disease," *Parasites and Vectors*, vol. 7, no. 1, article no. 108, 2014.

[27] F. Guhl, M. Restrepo, V. M. Angulo, C. M. F. Antunes, D. Campbell-Lendrum, and C. R. Davies, "Lessons from a national survey of Chagas disease transmission risk in Colombia," *Trends in Parasitology*, vol. 21, no. 6, pp. 259–262, 2005.

[28] J. Buendía, Guía de atención de la enfermedad de Chagas. Guías promoción la salud y prevención enfermedades en la salud pública, 2005: p. 1–48.

12

Therapeutic and Safety Evaluation of Combined Aqueous Extracts of *Azadirachta indica* and *Khaya senegalensis* in Chickens Experimentally Infected with *Eimeria* Oocysts

J. G. Gotep,[1] J. T. Tanko,[2] G. E. Forcados,[1] I. A. Muraina,[1] N. Ozele,[1] B. B. Dogonyaro,[3] O. O. Oladipo,[1] M. S. Makoshi,[1] O. B. Akanbi,[4] H. Kinjir,[5] A. L. Samuel,[1] T. E. Onyiche,[1,6] G. O. Ochigbo,[1,7] O. B. Aladelokun,[1,8] H. A. Ozoani,[1,9] V. Z. Viyoff,[1,10] C. C. Dapuliga,[1,11] A. A. Atiku,[1] P. A. Okewole,[4] D. Shamaki,[12] M. S. Ahmed,[12] and C. I. Nduaka[13]

[1] *Biochemistry Division, National Veterinary Research Institute, PMB 01, Vom, Nigeria*
[2] *Parasitology Division, National Veterinary Research Institute, PMB 01, Vom, Nigeria*
[3] *Virology Division, National Veterinary Research Institute, PMB 01, Vom, Nigeria*
[4] *Central Diagnostics Laboratory, National Veterinary Research Institute, PMB 01, Vom, Nigeria*
[5] *Haematology Department, Federal College of Veterinary and Medical Laboratory Technology, PMB 01, Vom, Nigeria*
[6] *Department of Veterinary Microbiology and Parasitology, University of Maiduguri, Bama Road, Maiduguri, Borno State, Nigeria*
[7] *Department of Veterinary Physiology, Pharmacology and Biochemistry, University of Ibadan, PMB 0248, Ibadan, Nigeria*
[8] *Department of Biochemistry, University of Ibadan, PMB 0248, Ibadan, Nigeria*
[9] *Department of Medical Laboratory Science, Rivers State University of Science and Technology, PMB 5080, Port Harcourt, Nigeria*
[10] *Department of Epidemiology, University of Buea, P.O. Box 63, Buea, Cameroon*
[11] *Microbiology Department, Kwame Nkrumah University of Science and Technology, Kumasi, Ghana*
[12] *National Veterinary Research Institute, Vom, Nigeria*
[13] *Africa Education Initiative (NEF), 9401 Sentinel Ridge, Eagleville, PA 19403, USA*

Correspondence should be addressed to J. G. Gotep; jurbe4u@yahoo.com

Academic Editor: Emmanuel Serrano Ferron

Coccidiosis is a disease of economic importance in poultry causing morbidity and mortality. Reports show that *Azadirachta indica* and *Khaya senegalensis* have been used individually in the treatment of avian coccidiosis. We thus investigated the efficacy and safety of the combined aqueous extracts of these plants for the treatment of experimentally induced coccidiosis in broiler chickens using oocyst count, oxidative stress biomarkers, serum biochemistry, histology, and haematological parameters. The phytochemical screening revealed the presence of tannins, saponins, cardiac glycosides, and steroids in both extracts. In addition, alkaloids and flavonoids were present in *Azadirachta indica*. There was significant ($p < 0.05$) dose dependent decrease in oocyst count across the treatment groups with 400 mg/kg of the combined extract being the most efficacious dose. Immunomodulatory and erythropoietic activity was observed. There were decreased intestinal lesions and enhanced antioxidant activity across the treatment groups compared to the negative control. Administration of the combined extract did not cause damage to the liver as ALT, AST, and ALP levels were significantly reduced in the uninfected chickens treated with the extracts compared to control suggesting safety at the doses used. The combined aqueous extracts of *K. senegalensis* stem bark and *Azadirachta indica* leaves were ameliorative in chickens infected with coccidiosis.

1. Introduction

Coccidiosis is a major parasitic disease of poultry caused by an Apicomplexan protozoan belonging to the subclass Coccidia, family Eimeriidae, and genus *Eimeria* [1]. The disease has significant economic impact on the poultry industry causing high mortality, poor growth, decreased productivity, and high medical cost [2]. Anticoccidial drugs are commonly used to prevent and treat coccidiosis. However, indiscriminate and long-time use of anticoccidial drugs has led to the emergence of drug resistant parasites and presence of residual drugs in chicken products raising concerns about public health and food safety [3, 4]. According to Yang et al. [5], anticoccidial vaccines are an alternative means to prevent coccidiosis. However, efficacy, safety, and cost-effectiveness are still challenges for anticoccidial vaccine use in poultry [6]. Consumers and poultry farmers around the world have voiced concerns about the use of present anticoccidial agents [5]. Therefore, there is an expedient need for an alternative approach to prevent and treat avian coccidiosis necessitating an examination of the potential of natural products from plant extracts.

In recent time, various researchers have tested several plants for anticoccidial activity in chickens [4, 5, 7–10]. *Azadirachta indica (AI)* and *Khaya senegalensis (KS)* both belonging to the family Meliaceae have been reported to possess anticoccidial properties and have been used individually to combat avian coccidiosis. This property has been demonstrated by their ability to reduce oocyst count [11, 12], inhibit inflammation [13], and enhance erythropoiesis [14, 15]. In addition, *KS* has antidiarrhoeal properties [16]. The pathogenesis of coccidiosis is associated with oxidative stress caused by increased generation of reactive oxygen species due to activities of the parasite as well as the host immune system causing a depletion of antioxidant enzymes and GSH level and increased lipid peroxidation of cells in the intestinal linings and surrounding tissues. *AI* and *KS* have free radical scavenging ability as well as cellular immune-modulatory properties in mice [17] and human colorectal cancer [18].

Tipu et al. [19] showed that combinations of herbs used against coccidiosis are effective and an economical alternative for prophylactic anticoccidial medication. *AI* administered at dosages of 200, 400, 800, and 1600 mg/kg in broiler chickens for 4 days showed 800 mg/kg to be the most effective dosage [11]. In another study by our group using *KS* and *AI* as single extracts, 800 mg/kg was found to be most effective for each extract [unpublished data]. We therefore hypothesized that combining the extracts should be more effective at a lower dosage. This study was, therefore, aimed at evaluating the therapeutic efficacy of the combined aqueous extracts of leaves of *AI* and stem bark of *KS* in chickens experimentally infected with *Eimeria* oocyst using oocysts count, oxidative stress markers, intestinal histopathology, and haematological parameters and also determining the safety of the doses used on serum levels of liver and kidney function parameters and biomarkers of oxidative stress in liver homogenates of uninfected chickens treated with the extracts.

2. Materials and Methods

2.1. Plant Collection and Preparation. The stem bark of *KS* and leaves of *AI* were collected from the environs of the National Veterinary Research Institute (NVRI), Vom, Nigeria. The plants were identified and authenticated in the herbarium at Federal College of Forestry, Jos, Nigeria, and assigned voucher numbers FHJ 198 and FHJ 199 for *AI* and

KS, respectively. After collection, the samples were washed, air-dried, and ground into powder under aseptic conditions. Eight hundred grams (800 g) of the individual pulverized plant was macerated with distilled water for 72 hours. At the end of the extraction, the mixture was sieved and filtered. The filtrate was concentrated by drying in the oven at 40°C [20] with modifications. The dried extracts were stored at 4°C until needed.

2.2. Phytochemical Screening. The extracts of both *KS* and *AI* were screened individually to detect the presence of some phytochemicals according to the methods described [21].

2.3. Source of Oocyst. Mixed *Eimeria* oocyst suspension (*Eimeria tenella*, *E. necatrix*, and *E. brunetti*) was obtained from the Parasitology Division of the National Veterinary Research Institute (NVRI), and each 1 mL of the oocyst suspension contained a total of 2185 mixed *Eimeria* oocysts which was determined by the Mac Master technique [22].

2.4. Experimental Animals. Apparently healthy day-old broiler chicks were obtained from a hatchery in Jos, Nigeria, and brooded under standard conditions for three weeks before commencement of the study. The chicks were fed standard pelletized broiler starter feed (Vital Feed® Grand Cereals, Nigeria, Plc., Jos, Nigeria) and water *ad libitum*. Birds were housed in individual cages with proper lighting and heat. The birds were vaccinated against infectious bursal disease (IBD) and Newcastle disease virus with IBD and La Sota vaccines, respectively, using NVRI, Vom, vaccines. All experiments were conducted in accordance with the Principles and Guide for the Care and Use of Laboratory Animals [23] and approved by the Animal Ethics Committee of NVRI, Vom.

2.5. Experimental Design

2.5.1. Efficacy Study. Twenty-five (25) experimental birds were weighed and each infected with 2185 sporulated oocysts (0.1 mL) as a single oral gavage according to Biu et al. [11]. Daily collection and screening of faeces for oocyst presence and count were carried out. Birds were also monitored for clinical signs of coccidiosis. After establishment of the infection (7 days after inoculation), treatment commenced by oral gavage of the extract. The combined aqueous extracts of *AI* and *KS* were administered at a dose ratio of 1 : 1 at 100 mg/kg, 200 mg/kg, and 400 mg/kg. The experiment included two control groups, negative and positive, treated with distilled water and amprolium (Amprolium 250 WSP, Kepro® B.V., Holland), respectively. All treatments lasted for five (5) days which is the usual period of chemotherapy for coccidiosis using the standard drug amprolium. At the end of the experiment, chicks were sacrificed by cervical dislocation and blood was collected from the jugular vein into EDTA containers for haematological analysis and processed immediately. Tissues were collected in 10% buffered formalin and physiological saline solution for histopathological and oxidative stress examinations, respectively.

2.5.2. Safety Study. Twenty (20) apparently healthy birds were divided into four groups of five (5) birds each. The combined aqueous extracts of *AI* and *KS* were administered at the dose ratio of 1:1 at 100 mg/kg, 200 mg/kg, and 400 mg/kg for 5 days while the control group received distilled water. The experiment was terminated 24 hours after the last administration. Blood was collected for serum biochemical analysis and liver harvested for oxidative stress assays.

2.6. Oocyst Estimation. Evaluation of faeces for the oocyst per gram (OPG) counts was performed using modified McMaster's technique [22].

2.7. Tissue Homogenization. The harvested tissues (intestine for therapeutic efficacy and liver for the safety study) were rinsed with phosphate buffered saline (PBS) and blotted with filter paper and weighed. They were then chopped into bits and homogenized in ice using homogenizing buffer (0.1 M phosphate buffer, pH 7.4) at ratio of 1 : 4 w/v. The resulting homogenate was centrifuged at 10,000 g for 15 minutes at $-4°C$ to obtain the postmitochondrial fraction. The supernatant was collected, stored at $-4°C$, and used for oxidative stress assays. All samples were analyzed within 7 days after termination of experiment.

2.8. Oxidative Stress Assays. Superoxide dismutase (SOD), reduced glutathione (GSH) levels, malondialdehyde (MDA), catalase, and total protein were determined [24].

2.9. Haematological Evaluation. Red blood cell (RBC), white blood cell (WBC), packed cell volume (PCV), haemoglobin concentration (Hb), mean corpuscular volume (MCV), mean corpuscular haemoglobin (MCH), and mean corpuscular haemoglobin concentration (MCHC) together with absolute count of heterophils and lymphocytes as well as H/L ratio were determined [25].

2.10. Serum Biochemical Analysis. Alkaline phosphatase (ALP), alanine aminotransferase (ALT) and aspartate aminotransferase (AST), blood urea nitrogen (BUN), serum creatinine (CRE), total protein (TP), albumin (ALB), total bilirubin (TBIL), and direct bilirubin (DBIL) were analysed using Randox Diagnostic Test kits according to manufacturer's instructions.

2.11. Histopathological Examination. Tissue sample (intestine) was harvested from the infected and treated birds immediately after sacrifice, fixed in 10% buffered formal saline, embedded in paraffin wax, sectioned at 5 μ thickness, stained with haemotoxylin and eosin (H&E) stain, cleared in xylene, and mounted in a mountant [26, 27].

2.12. Statistical Analysis. Data obtained from the study were summarized as means ± standard error of mean and differences between the means determined at 5% level of significance using the one-way analysis of variance [28].

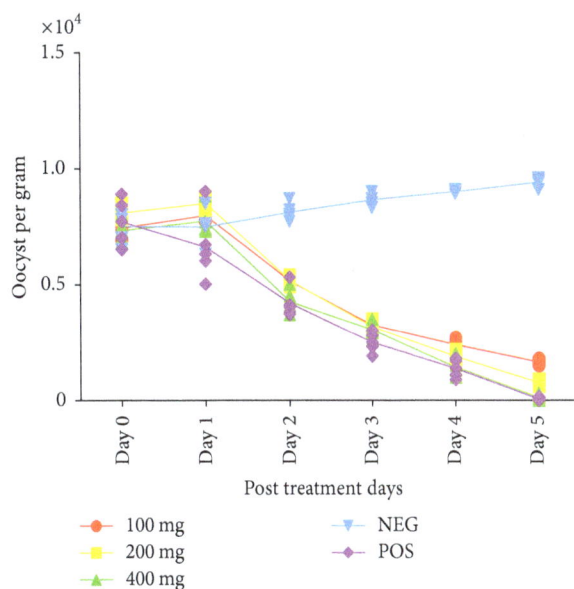

FIGURE 1: Therapeutic effect of combined extract of *Azadirachta indica* and *Khaya senegalensis* on oocyst count of chicken infected with *Eimeria* oocysts. NEG = negative control (infected, not treated). POS = positive control (infected and treated with amprolium).

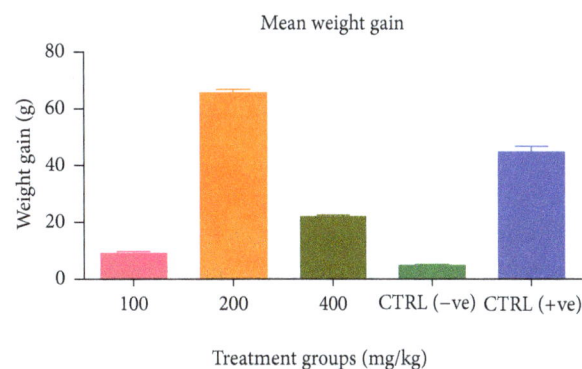

FIGURE 2: Effect of the combined extracts on weight gain of chickens infected with *Eimeria* oocysts.

3. Results

3.1. Extraction and Phytochemical Screening. The yield of the plants was 4.90% for *AI* and 7.02% for *KS*. The phytochemical screening revealed the presence of tannins, saponins, cardiac glycosides, and steroids in both extracts. Alkaloids and flavonoids were present in *AI* while resins, terpenes, and anthraquinones were not detected in both extracts.

3.2. Effect of Combined Extracts on Oocyst Count and Weight Gain (Efficacy Study). A dose dependent reduction in oocyst count was observed in birds treated with the extracts (Figure 1). A significant ($p < 0.05$) increase in weight gain was also recorded among treated groups (Figure 2).

3.3. Effect of Combined Extracts on Haematological Parameters (Efficacy Study). There was no significant difference ($p >$

TABLE 1: Effect of combined extracts of *Azadirachta indica* and *Khaya senegalensis* on red blood cell count, haemoglobin concentration, total white blood cell/differential counts of chickens infected with *Eimeria* oocysts.

Treatment	RBC $\times 10^{12}$	Hb (g/dL)	WBC $\times 10^9$	Heterophils (%)	Lymphocyte (%)
Control (−ve)	1.944 ± 0.121	11.00 ± 1.266	12.32 ± 0.476	25.80 ± 3.137	78.33 ± 2.906
Control (+ve)	2.466 ± 0.370	12.20 ± 0.938	10.44 ± 0.33[a]	23.00 ± 4.438	73.50 ± 3.524
100 mg/kg	2.123 ± 0.358	14.04 ± 0.248	9.02 ± 0.233[a]	27.80 ± 4.067	42.40 ± 3.894[a,b]
200 mg/kg	2.107 ± 0.066	12.82 ± 1.352	8.448 ± 0.354[a,b]	25.20 ± 3.527	40.20 ± 5.417[a,b]
400 mg/kg	2.040 ± 0.050	11.17 ± 0.667	10.56 ± 0.640[a]	20.00 ± 2.983	76.33 ± 3.283

Values in the same column with different letters superscripts are significantly different.

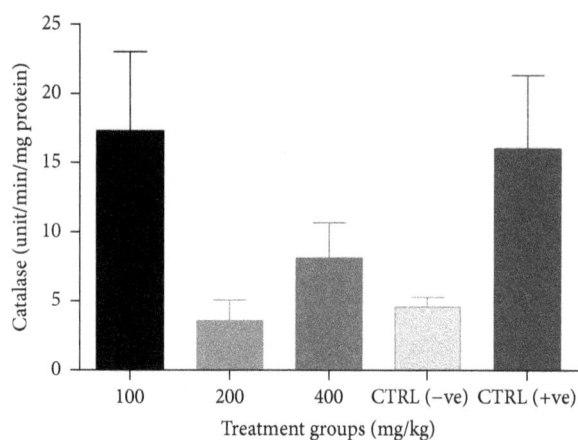

FIGURE 3: Effect of combined extracts of *Azadirachta indica* and *Khaya senegalensis* on catalase activity in the intestine of chickens infected with *Eimeria* oocysts.

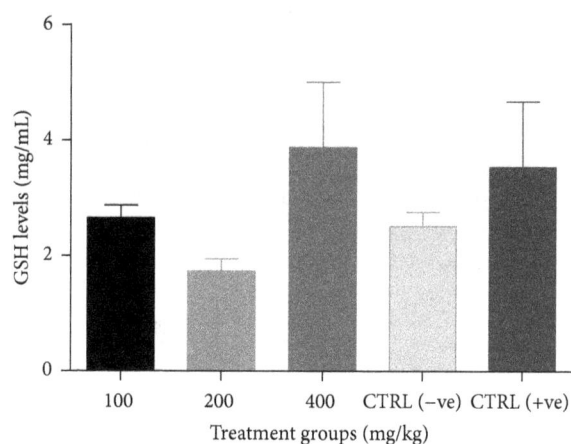

FIGURE 4: Effect of combined extracts of *Azadirachta indica* and *Khaya senegalensis* on glutathione (GSH) levels of the intestine of chickens infected with *Eimeria* oocysts.

0.05) in RBC count and haemoglobin concentration within treatment groups but an increase in mean RBC count was observed when compared with the negative control. There was a significant ($p < 0.05$) decrease in WBC count in the treated groups compared to negative control. There was no significant ($p > 0.05$) change in heterophils between the negative control and the treated groups but a dose dependent decrease in heterophils was observed (Table 1).

3.4. Effect of Combined Extract on Oxidative Stress Markers in the Intestine (Efficacy Study).
There was a significant increase ($p < 0.05$) in catalase activity of all groups treated with the combined extract when compared to the negative control (Figure 3). No significant change ($p > 0.05$) in GSH levels (Figure 4) and SOD activity (Figure 5) was recorded between the negative control and the treated groups. However, there was a general decrease in MDA levels in all treatment groups with significant reduction ($p < 0.05$) observed at lower doses of the combined extract (Figure 6).

3.5. Histological Findings from the Chicken Intestine Infected with *Eimeria* Oocyst (Efficacy Study).
For the 100 mg/kg and 200 mg/kg of the combined extract, moderate cryptic destruction as well as intracryptic developmental stages was observed (Figures 7(a) and 7(b)). The architecture of the intestine was moderately preserved in the 400 mg/kg

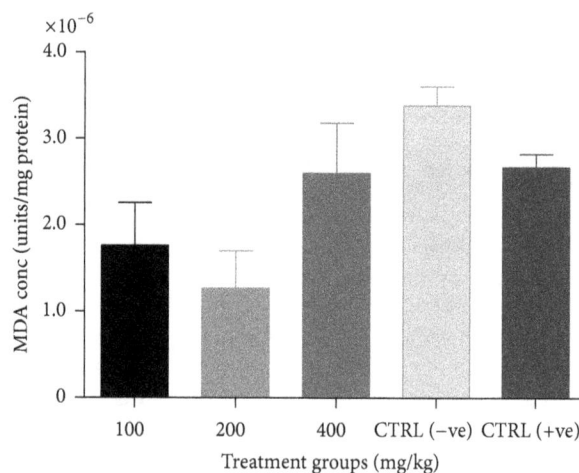

FIGURE 5: Effect of combined extracts of *Azadirachta indica* and *Khaya senegalensis* on lipid peroxidation (malondialdehyde, MDA) of chickens infected with *Eimeria* oocysts.

treated group. The crypts and the lamina propria were preserved (Figure 7(c)) similar to the positive control group (Figure 7(d)). Intracryptic developmental stages of the parasite were observed in the negative control, causing severe cryptic destruction and intestinal fibrosis (Figure 7(e)).

TABLE 2: Clinico-chemical parameters of liver function of healthy chickens exposed to combined aqueous extract of *Azadirachta indica* and *Khaya senegalensis*.

Treatment	Parameter						
	ALT (U/L)	AST (U/L)	ALP (U/L)	TP (g/L)	TBIL (mg/dL)	DBIL (mg/dL)	ALB (mg/dL)
Control	8.320 ± 0.489^b	21.32 ± 0.598^b	447.80 ± 7.055^b	35.32 ± 2.71^b	0.444 ± 0.005^b	0.233 ± 0.024^b	9.083 ± 1.124^b
100 mg/kg	7.560 ± 0.796^b	19.84 ± 0.379^b	420.70 ± 3.143^a	32.89 ± 1.459^b	0.406 ± 0.072^b	0.130 ± 0.017^a	14.16 ± 1.089^a
200 mg/kg	6.927 ± 0.058^b	18.89 ± 0.741^a	412.70 ± 6.374^a	26.51 ± 0.685^a	0.399 ± 0.032^b	0.087 ± 0.048^a	14.83 ± 0.724^a
400 mg/kg	5.600 ± 0.489^a	16.10 ± 2.330^a	312.40 ± 4.347^a	28.00 ± 0.907^a	0.358 ± 0.035^a	0.120 ± 0.032^a	13.88 ± 0.409^a

Values are expressed as mean ± SEM.
Columns with different superscript show significant difference ($p < 0.05$).

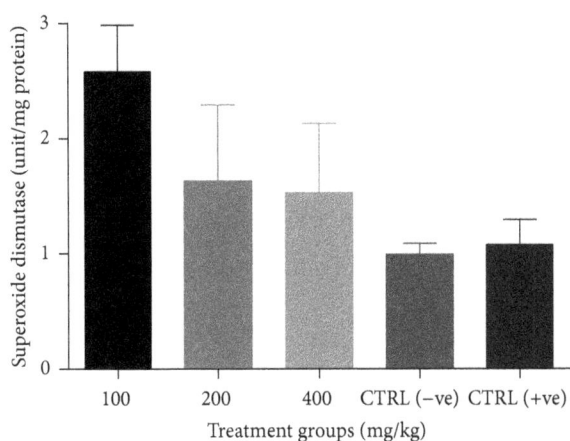

FIGURE 6: Effect of combined extracts of *Azadirachta indica* and *Khaya senegalensis* on superoxide dismutase (SOD) activity of the intestine of chickens infected with *Eimeria* oocysts.

3.6. Effect of Combined Extract on the Serum Biochemical Markers of Healthy Birds (Safety Study). Administration of the combined extract of *KS* and *AI* to apparently healthy birds significantly reduced ($p < 0.05$) ALT, AST, ALP, TP, DBIL, and TBIL across the groups treated with the extract, with the 400 mg/kg group showing a marked reduction in the levels of these serum markers compared to the control (Table 2). The albumin levels increased significantly ($p < 0.05$) across the groups treated with the extract compared to the control (Table 2).

There was a significant decrease ($p < 0.05$) in the creatinine and urea levels when compared to the control (Table 3). In addition, the combined extract did not significantly ($p > 0.05$) affect the sodium and potassium levels when compared to the control (Table 3).

3.7. Effect of Combined Aqueous Extract on Oxidative Stress Markers in Liver of Healthy Chickens (Safety Study). There was a dose dependent increase in GSH levels which was significant ($p < 0.05$) at 400 mg/kg when compared to the control. Similarly, there was a significant dose dependent increase ($p < 0.05$) in SOD and catalase activities when compared to the control (Figures 8, 9, and 10).

4. Discussion

The dose dependent reduction in oocyst count observed in the treated groups was comparable with amprolium which could be attributed to the presence of some bioactive compounds in the plant extracts. Saponin, for example, is known to bind membrane cholesterol, altering the integrity of the parasite membrane, resulting in loss of homeostasis and eventual death of the parasite [29]. Also, limonoids contained in AI inhibit protein digestion and uptake of vitamins and minerals by the parasites in the gut [17]. This action results in impaired nutrient utilization, reduced growth, and multiplication of the parasite which could contribute to the reduced oocyst count observed.

Extracts of neem and mahogany when used individually have been reported to reduce oocyst count in avian coccidiosis [11, 12]. In a separate study carried out by our group, similar observation was made when each plant extract was administered singly (data not shown). However, in this study, the oocyst count was more reduced when the combined extract was used.

The significant increase in mean weight gain in treated birds when compared to the negative control is possibly due to the inhibition of inflammation in the intestinal mucosa which is suggestive of an increased nutrient absorption across the intestinal wall and enhanced feed conversion ratio compared to the negative control. Nwosu et al. [12] and Biu et al. [11] reported an increased weight gain and feed conversion ratio in birds treated with only KS extracts and AI, respectively.

The observed increase in RBC and haemoglobin concentration is indicative of the erythropoietic ability of the combined extracts, which is beneficial since the *Eimeria* parasite in the epithelia of the intestines causes bloody diarrhoea and consequently anaemia. This finding is in consonance with that of Sanni et al., [15] who reported an antianaemic effect of KS on phenylhydrazine-induced anaemia in rats.

AI has been shown to possess antianaemic properties in rats [14]. The dose dependent decrease in white blood cells count and heterophil-lymphocyte ratio with a concomitant reduction in heterophils is suggestive of decreased inflammation. It can be extrapolated that the decrease in parasitic load downregulates the activity of the immune system leading to decrease in inflammation and consequently a decrease in

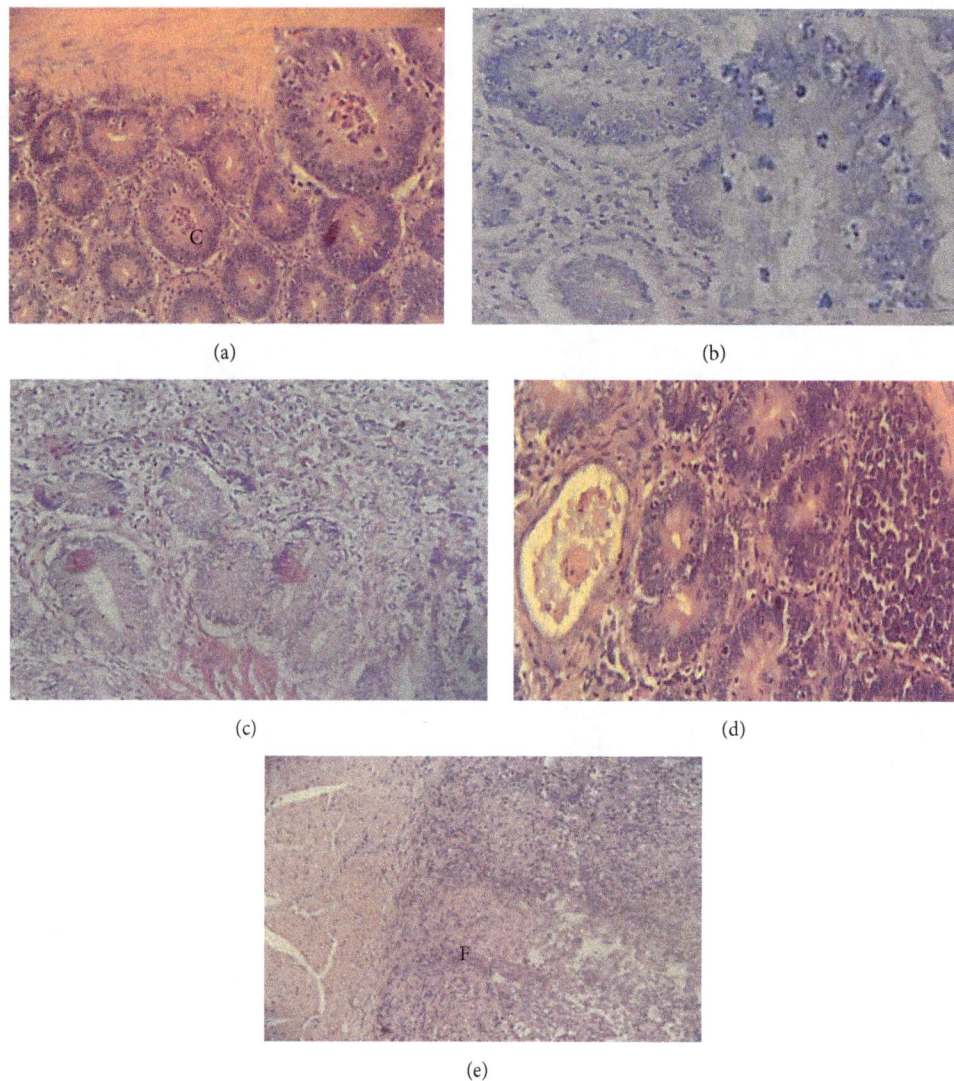

FIGURE 7: Photomicrograph of the intestine of chickens infected with *Eimeria* oocysts and treated with graded doses of the combined extract and control. (a) 100 mg/kg: combined extract of *KS* and *AI* treatment; moderate cryptic destruction and intracryptic developmental stages of *Eimeria* H&E: ×400. (b) 200 mg/kg: intestine infected with *Eimeria* spp.; moderate cryptic destruction and intracryptic developmental stages of *Eimeria* and intestinal fibrosis H&E: ×400. (c) Amprolium 10 mg/L: chicken, intestine, moderate cryptic destruction and ectasia, intracryptic developmental stages of *Eimeria*, and intestinal fibrosis H&E: ×400. (d) 400 mg/kg extract: moderate cryptic destruction and intracryptic developmental stages of Eimeria H&E: ×: 400. (e) Control (−ve) intestine infected with *Eimeria* spp.; severe cryptic destruction and intracryptic developmental stages of *Eimeria* and intestinal fibrosis H&E: ×400.

TABLE 3: Clinicochemical parameters of kidney function of chickens exposed to combined aqueous extract of *Azadirachta indica* and *Khaya senegalensis*.

Treatment	Parameter			
	CRE (mg/dL)	URE (mg/dL)	K (mEq/L)	Na (mEq/L)
Control	0.418 ± 0.024^b	9.179 ± 0.540^c	5.613 ± 0.248^a	147.43 ± 1.003^b
100 mg	0.288 ± 0.001^a	5.847 ± 0.540^a	5.013 ± 0.622^a	150.89 ± 1.028^b
200 mg	0.285 ± 0.039^a	5.439 ± 0.490^a	5.403 ± 0.315^a	151.33 ± 4.290^b
400 mg	0.285 ± 0.039^a	7.207 ± 0.544^a	5.957 ± 0.915^a	147.17 ± 0.301^b

Values are expressed as mean ± SEM.
Rows with different superscript show significant difference ($p < 0.05$).

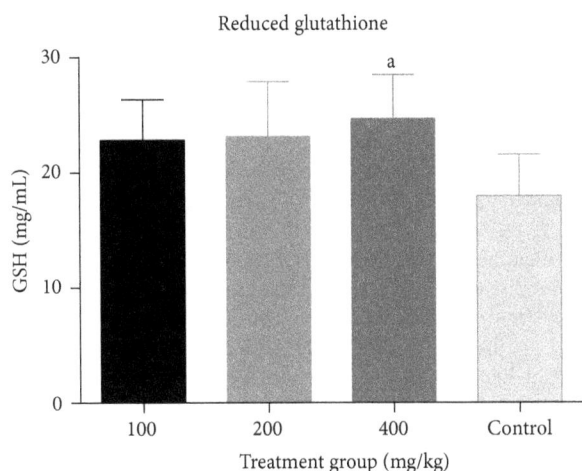

FIGURE 8: Effect of combined aqueous extracts of *Azadirachta indica* and *Khaya senegalensis* on reduced glutathione (GSH) levels of the liver of healthy chickens. Values are expressed as mean ± SEM where $n = 5$. [a]Significant as compared with control ($p < 0.05$).

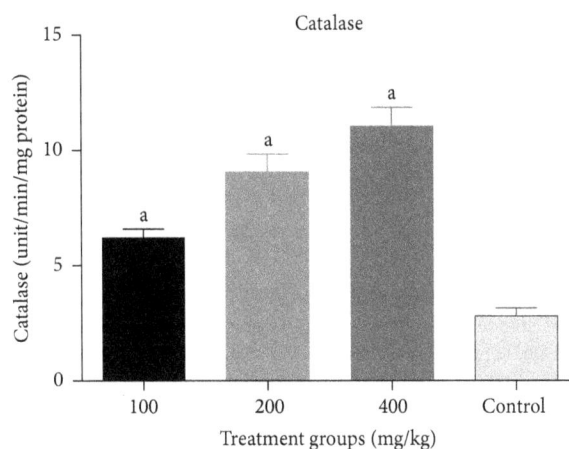

FIGURE 10: Effect of combined aqueous extracts of *Azadirachta indica* and *Khaya senegalensis* on catalase activity of the liver of healthy chickens. Values are expressed as mean ± SEM where $n = 5$. [a]Significant as compared with control ($p < 0.05$).

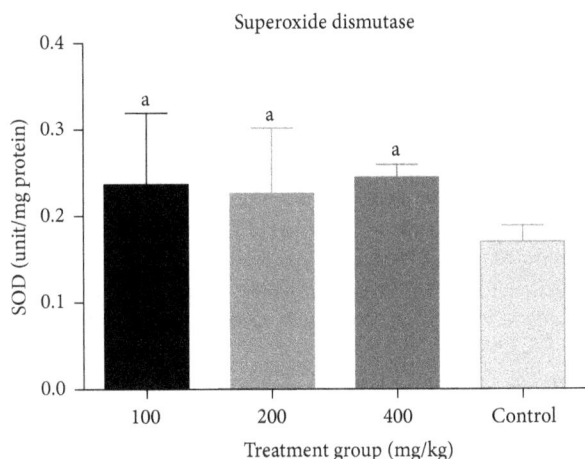

FIGURE 9: Effect of combined aqueous extracts of *Azadirachta indica* and *Khaya senegalensis* on superoxide dismutase activity (SOD) of the liver of healthy chickens. Values are expressed as mean ± SEM where $n = 5$. [a]Significant as compared with control ($p < 0.05$).

heterophils, tending towards the normal blood picture of a greater ratio of lymphocytes to heterophils in avian species. In addition, *Khaya senegalensis* has been reported to inhibit inflammation in rats [13].

The increase in intestinal glutathione level, catalase, and superoxide dismutase activity as well as decrease in malondialdehyde levels in the treated groups is suggestive of the *in vivo* antioxidant enhancing capacity of the combined extract. During *Eimeria* infection in chickens, the innate immune system of these chickens protects them by producing reactive oxygen species (ROS) in a process termed "oxidative burst" in an attempt to destroy the *Eimeria* pathogens [30]. Unfortunately, since such ROS generated are not target specific, the reactive species also damage cells of the gastrointestinal tract resulting in ulcers [31]. Flavonoids, limonoids, and saponins, amongst other active phytocompounds, present in the plant

extracts possess antioxidant properties which aid in free radical scavenging and reactive oxygen quenching activities, thereby ameliorating oxidative stress mediated damage [13, 17, 18]. AI aqueous extract has been reported to reduce MDA levels in mice infected with *Eimeria* species [17]. In addition, KS was reported to also reduce MDA levels in rats with ethanol-induced ulcers [13].

From the result of the safety study, the reduction in AST, ALT, and ALP activities compared to control shows that the combination of the extracts is not hepatotoxic. The elevated catalase and superoxide dismutase activities in the liver of the chicken may suggest that administration of combined extracts of KS and AI could be hepatoprotective. It has been previously reported that the methanolic extract of AI leaves (500 mg/kg) and aqueous extract of KS stem bark (250 and 500 mg/kg, resp.) have hepatoprotective activity [17, 32]. The increase in albumin levels observed in the different dose levels is suggestive of proper maintenance of the integrity of the liver and other extrahepatic tissues involved in protein synthesis [33].

The significant decrease in serum creatinine and urea levels across the dose levels when compared to the control shows that the extract combination was not nephrotoxic. In addition, the sodium and potassium levels were not affected, thus giving further credence to the safety of the extracts on the kidney. Decreased urea level with no marked changes in serum potassium and sodium levels following supplementation of broiler feed with neem leaves has been reported [33].

5. Conclusion

This study shows novel findings with respect to the possible synergistic efficacy of the combination of aqueous extracts of *K. senegalensis* stem bark and *A. indica* leaves lending further credence to the folkloric use of these plants in the treatment of coccidiosis.

Acknowledgments

The authors wish to acknowledge the African Education Initiative (NEF), USA, in partnership with the National Veterinary Research Institute, Vom, Nigeria, for providing grant for the research.

References

[1] J. U. Gararawa, N. G. Usman, V. K. Ayi et al., "Anticoccidial resistance in poultry: a review," *New York Science Journal*, vol. 4, no. 8, pp. 102–109, 2011.

[2] R. B. Williams, "Epidemiological aspects of the use of live anticoccidial vaccines for chickens," *International Journal for Parasitology*, vol. 28, no. 7, pp. 1089–1098, 1998.

[3] H. D. Chapman, "Biochemical, genetic and applied aspects of drug resistance in *Eimeria* parasites of the fowl," *Avian Pathology*, vol. 26, no. 2, pp. 221–244, 1997.

[4] J. Orengo, A. J. Buendía, M. R. Ruiz-Ibáñez et al., "Evaluating the efficacy of cinnamaldehyde and Echinacea purpurea plant extract in broilers against *Eimeria acervulina*," *Veterinary Parasitology*, vol. 185, no. 2-4, pp. 158–163, 2012.

[5] W. C. Yang, Y. J. Tien, C. Y. Chung et al., "Effect of *Bidens pilosa* on infection and drug resistance of *Eimeria* in chickens," *Research in Veterinary Science*, vol. 98, pp. 74–81, 2015.

[6] P. A. Sharman, N. C. Smith, M. G. Wallach, and M. Katrib, "Chasing the golden egg: vaccination against poultry coccidiosis," *Parasite Immunology*, vol. 32, no. 8, pp. 590–598, 2010.

[7] T. O. Oyagbemi and J. O. Adejinmi, "Supplementation of broiler feed with leaves of *Vernonia amygdalina* and *Azadirachta indica* protected birds naturally infected with *Eimeria* sp," *African Journal of Biotechnology*, vol. 11, no. 33, pp. 8407–8413, 2012.

[8] S. D. Ola-Fadunsin and I. O. Ademola, "Direct effects of *Moringa oleifera Lam* (Moringaceae) acetone leaf extract on broiler chickens naturally infected with *Eimeria* species," *Tropical Animal Health and Production*, vol. 45, no. 6, pp. 1423–1428, 2013.

[9] V. Naidoo, L. J. McGaw, S. P. R. Bisschop, N. Duncan, and J. N. Eloff, "The value of plant extracts with antioxidant activity in attenuating coccidiosis in broiler chickens," *Veterinary Parasitology*, vol. 153, no. 3-4, pp. 214–219, 2008.

[10] S. H. Lee, H. S. Lillehoj, S. I. Jang, K. W. Lee, D. Bravo, and E. P. Lillehoj, "Effects of dietary supplementation with phytonutrients on vaccine-stimulated immunity against infection with *Eimeria tenella*," *Veterinary Parasitology*, vol. 181, no. 2–4, pp. 97–105, 2011.

[11] A. A. Biu, S. D. Yusuf, and J. S. Rabo, "Use of neem (*Azadirachta indica*) aqueous extract as a treatment for poultry coccidiosis in Borno State, Nigeria," *African Scientist*, vol. 7, no. 3, pp. 47–153, 2006.

[12] C. U. Nwosu, S. W. Hassan, M. G. Abubakar, and A. A. Ebbo, "Anti-diarrhoeal and toxicological studies of leaf extracts of *Khaya senegalensis*," *Journal of Pharmacology and Toxicology*, vol. 7, no. 1, pp. 1–10, 2012.

[13] F. N. Ishaq, A. U. Zezi, and T. O. Olurishe, "*Khaya senegalensis* inhibits piroxicam mediated gastro-toxicity in wistar rats," *Avicenna Journal of Phytomedicine*, vol. 4, no. 6, pp. 377–384, 2014.

[14] E. E. Iyare and N. N. Obaji, "Effects of aqueous leaf extract of *Azadirachta indica* on some haematological parameters and blood glucose level in female rats," *Nigerian Journal of Experimental and Clinical Biosciences*, vol. 2, no. 1, pp. 54–58, 2014.

[15] F. S. Sanni, S. Ibrahim, K. A. N. Esievo, and S. Sanni, "Effect of oral administration of aqueous extract of *Khaya senegalensis* stem bark on phenylhydrazine-induced anaemia in rats," *Pakistan Journal of Biological Sciences*, vol. 8, no. 2, pp. 255–258, 2005.

[16] I. L. Elisha, M. S. Makoshi, S. Makama et al., "Antidiarrheal evaluation of aqueous and ethanolic stem bark extracts of *Khaya senegalensis* A. Juss (Meliaceace) in albino rats," *Pakistan Veterinary Journal*, vol. 33, no. 1, pp. 32–36, 2013.

[17] M. A. Dhkil, A. E. Abdel Moneim, and S. Al-Quraishy, "Antioxidant, hepatoprotective, and ameliorative effects of *Azadirachta indica* on *Eimeria papillata*-induced infection in mice," *Journal of Medicinal Plants Research*, vol. 6, no. 20, pp. 3640–3647, 2012.

[18] X. M. Androulakis, S. J. Muga, F. Chen, Y. Koita, B. Toure, and M. J. Wargovich, "Chemopreventive effects of *Khaya senegalensis* bark extract on human colorectal cancer," *Anticancer Research*, vol. 26, no. 3, pp. 2397–2406, 2006.

[19] A. M. Tipu, M. S. Akhtar, M. I. Anjum, and M. L. Raja, "New dimension of medicinal plants as animal feed," *Pakistan Veterinary Journal*, vol. 26, no. 3, pp. 144–148, 2006.

[20] J. Singh, "Maceration, percolation and infusion techniques for the extraction of medicinal and aromatic plants," in *Extraction Technologies for Medicinal and Aromatic plants*, United Nations Industrial Development Organization and the International Center for Science and High Technology, pp. 67–82, 2008.

[21] A. Sofowora, *Medicinal Plants and Traditional Medicine in Africa*, Spectrum Books, Ibadan, Nigeria, 1989.

[22] J. Kaufmann, *Parasitic Infections of Domestic Animals: A Diagnostic Manual*, Birkhäuser, Basel, Switzerland, 1996.

[23] National Research Council, *Guide for the Care and Use of Laboratory Animals*, National Academy Press, Washington, DC, USA, 1996.

[24] M. Del Carmen Contini, N. Millen, L. Riera, and S. Mahieu, "Kidney and liver functions and stress oxidative markers of monosodium glutamate-induced obese rats," *Food and Public Health*, vol. 2, no. 5, pp. 168–177, 2012.

[25] T. W. Campbell, *Avian Haematology and Cytology*, Iowa State University Press, Ames, Iowa, USA, 1988.

[26] O. B. Akanbi and V. O. Taiwo, "Mortality and pathology associated with highly pathogenic avian influenza H5N1 outbreaks in commercial poultry production systems in Nigeria," *International Scholarly Research Notices*, vol. 2014, Article ID 415418, 7 pages, 2014.

[27] R. A. B. Drury and E. A. Wallington, *Carleton's Histological Techniques*, Oxford University Press, London, UK, 4th edition, 1976.

[28] Graphpad Software, *InStat Guide to Choosing and Interpreting Statistical Tests*, Graphpad Software, San Diego, Calif, USA, 1998.

[29] Y. Wang, T. A. McAllister, C. J. Newbold, L. M. Rode, P. R. Cheeke, and K.-J. Cheng, "Effects of *Yucca schidigera* extract on fermentation and degradation of steroidal saponins in the rumen simulation technique (RUSITEC)," *Animal Feed Science and Technology*, vol. 74, no. 2, pp. 143–153, 1998.

[30] H. S. Lillehoj and J. M. Trout, "Avian gut-associated lymphoid tissues and intestinal immune responses to *Eimeria* parasites," *Clinical Microbiology Reviews*, vol. 9, no. 3, pp. 349–360, 1996.

13

Selecting PCR for the Diagnosis of Intestinal Parasitosis: Choice of Targets, Evaluation of In-House Assays, and Comparison with Commercial Kits

G. N. Hartmeyer,[1,2] S. V. Hoegh,[2] M. N. Skov,[1,2] R. B. Dessau,[3] and M. Kemp[1,2]

[1]Research Unit of Clinical Microbiology, Institute of Clinical Research Faculty of Health Science,
 University of Southern Denmark, Odense, Denmark
[2]Department of Clinical Microbiology, Odense University Hospital, Odense, Denmark
[3]Department of Clinical Microbiology, Slagelse Hospital, Slagelse, Denmark

Correspondence should be addressed to G. N. Hartmeyer; gitte.hartmeyer@rsyd.dk

Academic Editor: Bernard Marchand

Microscopy of stool samples is a labour-intensive and inaccurate technique for detection of intestinal parasites causing diarrhoea and replacement by PCR is attractive. Almost all cases of diarrhoea induced by parasites over a nine-year period in our laboratory were due to *Giardia lamblia*, *Cryptosporidium* species, or *Entamoeba histolytica* detected by microscopy. We evaluated and selected in-house singleplex real-time PCR (RT-PCR) assays for these pathogens in 99 stool samples from patients suspected of having intestinal parasitosis tested by microscopy. The strategy included a genus-specific PCR assay for *C. parvum* and *C. hominis*, with subsequent identification by a PCR that distinguishes between the two species. *G. lamblia* was detected in five and *C. parvum* in one out of 68 microscopy-negative samples. The performance of the in-house RT-PCR assays was compared to three commercially available multiplex test (MT-PCR) kit systems in 81 stool samples, collected in 28 microscopy-positive and 27 microscopy-negative samples from individuals suspected of intestinal parasitosis and in 26 samples from individuals without suspicion of parasitic infection. The in-house assays detected parasites in more samples from patients suspected of having parasitosis than did any of the kits. We conclude that commercial kits are targeting relevant parasites, but their performance may vary.

1. Background

Correct identification of microbial agents causing diarrhoea in humans is crucial for optimal treatment. Detection of disease-causing intestinal parasites is traditionally done by microscopic examination of stool samples. Over the last years this has been changed in favour of using PCR. Studies have shown that both sensitivity and specificity of PCR are better compared to microscopy [1–5]. Moreover, microscopy can lead to false conclusions, with harmless parasites being interpreted as disease-causing, while life-threatening parasites may not be detected. This has in particular been demonstrated for intestinal amoeba [6–10]. For estimating the true impact of parasitic intestinal infections, it is important to establish valid and reliable laboratory techniques for testing stool samples from patients. Use of optimized laboratory methods will improve patient safety through rapid and correct diagnosis, which leads to timely start of appropriate treatment.

The aim of this study was to evaluate the consequences of replacing microscopy by real-time PCR (RT-PCR) for detection of intestinal parasites causing diarrhoea. In order to do so, we first established which parasites were detected by microscopy in our laboratory over a period of nine years, to determine which parasites were relevant in our patient population. We determined which previously detected parasites would be missed by introducing a limited number of species-specific PCR assays and how many cases they represented. We then evaluated the performance of in-house singleplex RT-PCR assays for the three most important intestinal parasitic pathogens. Finally, the performance of three selected in-house RT-PCR assays for detection of *Giardia lamblia, Cryptosporidium parvum/Cryptosporidium hominis,* and *Entamoeba histolytica* was compared to those of three commercial multiplex real-time PCR (MT-PCR) kits.

Two specific objectives were defined: (1) evaluation of performance of species-specific in-house RT-PCR assays for detection of *G. lamblia, C. parvum/C. hominis,* and *E. histolytica* in stool samples submitted for examination for parasites; (2) comparison of the performance of the in-house RT-PCR assays with the performance of three commercial MT-PCR kits for detection of the same parasites.

2. Methods

2.1. Data Collection from the Laboratory Information System (LIS). Data on faecal samples examined for parasites from October 2005 to January 2015 was extracted from the electronic LIS. The total number of samples and patients and results of microscopy were registered.

2.2. Stool Samples. In total 125 stool samples, of which 99 were examined by microscopy on suspicion of parasitosis, were randomly collected from individuals with gastrointestinal complaints between June 2010 and January 2015. Ninety-nine of these samples were included for objective one (31 microscopy-positive and 68 microscopy-negative) and eighty-one (28 microscopy-positive and 27 microscopy-negative) were included for testing objective two. In addition 26 samples from individuals without suspicion of parasitosis were included without microscopy for objective two.

For objective 1, a total of 99 samples were analysed by in-house RT-PCR. For objective 2, a total of 81 samples were analysed by in-house RT-PCR and by three commercial MT-PCR kits. All samples were kept at −80°C until PCR were performed.

2.3. Microscopy for Intestinal Parasites. Microscopic examination for the presence of ova and cysts was routinely performed by examination of iodine-stained wet-mount preparations after formalin-ethyl acetate concentration, at a magnification of ×400 [11]. On specific request and when *Cryptosporidium* species, *Cyclospora* species, or *Cystoisospora* species was suspected from routine microscopy a smear stained by modified Ziehl-Neelsen technique was also examined [12].

2.4. In-House RT-PCR. For the in-house PCR assays, DNA was extracted by using NucliSENS easyMAG system (bioMérieux, France) in accordance with the manufacturer's instructions. Prior to DNA extraction, a cotton swab was submerged into the stool sample and suspended in 4 ml physiological NaCl solution. An internal extraction and PCR control, phocine herpesvirus (PhHV laboratory strain) was added to each sample prior to DNA extraction [13]. The nucleic acids were eluted in 100 μl and processed for PCR immediately.

Detection of the three intestinal parasites (*G. lamblia, C. parvum/C. hominis,* and *E. histolytica*) was performed

as singleplex RT-PCR in 99 samples analysed as duplicate. One assay was tested for detection of *G. lamblia,* three were tested for *C. parvum/C. hominis,* and one assay (assay 3) was included to distinguish between *C. parvum* and *C. hominis.* Finally one assay for *E. histolytica* was tested, using primers and probes (Table 1) described previously [1, 4, 13–16].

The 25 μl reactions mixture contained 1x TaqMan® Fast Universal PCR Master Mix, 2x No AmpErase® UNG (Thermo Fisher Scientific, Waltham, MA, USA), 1000 nM of the primers, and 200 nM of the probes and 5 μl DNA eluate.

The real-time PCR was performed using an Applied Biosystems 7500 Fast Real-Time PCR Thermocycler (Thermo Fisher Scientific) with the following cycling conditions: 95°C for 20 sec, followed by 45 cycles of 95°C for 3 sec and 60°C for 30 sec.

The PCR products were analysed using Sequence Detection Software v.1.4 (Thermo Fisher Scientific). A manual cycle threshold was set to 0,1 with an automatic baseline. The sample was regarded as positive if the Ct-value was ≤42 and had an exponential curve. Negative and positive extraction and PCR controls were included in all PCR analysis.

The singleplex in-house RT-PCR assays for *G. lamblia, C. parvum/C. hominis* (assay 1), and *E. histolytica* were collective called kit A and compared to three commercial MT-PCR kits, used in objective 2 on 79 samples.

2.5. Diagnostic Test Kits. Three different commercial kits available at the market were tested: RIDA®GENE Parasitic Stool Panel (PG1705) from R-Biopharm AG, Darmstadt, Germany (kit B), LightMix® Modular Gastroenteritis Assays from TIB MOLBIOL, Berlin, Germany (kit C), and BD MAX™ Enteric Parasite Panel from BD Diagnostic, Franklin Lakes, NJ, USA (kit D). DNA extraction for kits B and C was done as described for the in-house assays. For kit D, DNA extraction was done according BD MAX enteric parasite panel instructions on the BD MAX system. In all commercial kits, we used the internal control DNA, which was recommended and included in the kits. Seventy-nine samples were tested in duplicate in all three kits, in accordance with the manufacturer's instructions. For kit B the PCR assays were carried out using an Applied Biosystems 7500 Fast Real-Time PCR Thermocycler with the following cycling conditions: 1 min at 95°C, followed by 45 cycles of 95°C for 15 sec and 60°C for 30 sec. For kit C the PCR assays were done on a Roche LightCycler 480® II real-time instrument with the following cycling conditions: 10 min at 95°C, followed by 50 cycles of 95°C for 5 sec, 62°C for 5 sec, and 72°C for 15 sec. For kit D analysing was done according BD MAX enteric parasite panel instructions on the BD MAX system. A positive result in kits was regarded positive, if one out of two duplicates was positive, used in objective 2.

2.6. Analysis. McNemar's test was used for the statistical comparison of the paired data in objective 1.

2.7. Ethics, Biobank, and Data Storage. The study is part of a Ph.D. project and approved by the Danish Data Protection Agency (j.nr. 2008-58-0035). All samples were stored at −80°C in an approved research biobank established for the

TABLE 1: Primer and probes used for in-house real-time PCR assays in study.

Parasites	Forward primer sequence, 5′ →3′ Reverse primer sequence, 5′ →3′ Probe sequence (FAM), 5′ →3′	Target	Reference
Giardia lamblia	GAC GGC TCA GGA CAA CGG TT TTG CCA GCG GTG TCC G CCC GCG GCG GTC CCT GCT AG	ssu-rRNA	Verweij et al. (2004) [1]
C. parvum/C. hominis (assay 1)	CTT TTT ACC AAT CAC AGA ATC ATC AGA TGT GTT TGC CAA TGC ATA TGA A TCG ACT GGT ATC CCT ATA A	DNA J-like protein gene	Bruijnesteijn van Coppenraet et al. (2009) [4]
C. parvum/C. hominis (assay 2)	CGC TTC TCT AGC CTT TCA TGA CTT CAC GTG TGT TTG CCA AT CCA ATC ACA GAA TCA TCA GAA TCG ACT GGT ATC	DNA J-like protein gene	Fontaine and Guillot (2002) [14]
C. parvum/C. hominis (assay 3)	GAA CTG TAC AGA TGC TTG GGA GAA T CCT T CGT TAG TTG AAT CCT CTT TCC A *C. hominis* probe CTT GGA GCT CGT ATC AG *C. parvum* probe TTG GAG CTC ATA TCA G	Specific protein-coding gene	Yang et al. (2013) [15]
Entamoeba histolytica	ATT GTC GTG GCA TCC TAA CTC A GCG GAC GGC TCA TTA TAA CA CAT TGA ATG AAT TGG CCA TT	ssu-rRNA	Verweij et al. (2004) [1] Stensvold et al. (2010) [16]
Internal control			
Phocine herpesvirus (PhHV)	GGG CGA ATC ACA GAT TGA ATC GCG GTT CCA AAC GTA CCA A TTT TTA TGT GTC CGC CAC CAT CTG GAT C	Glycoprotein B	Niesters (2001) [13]

TABLE 2: Samples with discordant results obtained from different test kits.

Sample number	Mic.	Giardia				Cryptosporidium			
		Kit A	Kit B	Kit C	Kit D	Kit A	Kit B	Kit C	Kit D
6	Giardia	0/0	0/0	0/0	0/0	0/0	0/0	0/0	0/0
7	Giardia	+/+	0/0	**+/0**	+/+	0/0	0/0	0/0	0/0
8	Giardia	+/+	+/+	+/+	0/0	0/0	0/0	0/0	**+/0**
9	Giardia	+/+	+/+	+/+	**+/0**	0/0	0/0	0/0	0/0
12	Giardia	+/+	**+/0**	**+/0**	0/0	0/0	0/0	0/0	0/0
15	Crypto	0/0	0/0	0/0	0/0	+/+	0/0	+/+	+/+
18	Crypto	0/0	0/0	0/0	0/0	+/+	+/+	**+/0**	+/+
23	Crypto	0/0	0/0	0/0	0/0	0/0	0/0	0/0	0/0
24	Crypto	0/0	0/0	0/0	0/0	+/+	+/+	0/0	+/+
32	undet	+/+	0/0	0/0	0/0	0/0	0/0	0/0	0/0
33	undet	**+/0**	0/0	0/0	0/0	0/0	0/0	0/0	0/0
34	undet	+/+	+/+	+/+	+/+	0/0	0/0	0/0	0/0
35	undet	+/+	+/+	+/+	+/+	0/0	0/0	0/0	0/0
42	undet	0/0	0/0	0/0	0/0	+/+	+/+	+/+	+/+
49	undet	+/+	+/+	+/+	**+/0**	0/0	0/0	0/0	0/0
Total positive		**9**	**6**	**7**	**5**	**4**	**3**	**3**	**5**

Mic. = microscopy; undet = undetected; + = positive; 0 = negative. Results from duplicate tests are shown as +/+, +/0, and 0/0. Discordant results are in bold. Kit A = in-house RT-PCR assays, kit B = RIDA GENE Parasitic Stool Panel (PG1705), kit C = LightMix Modular Gastroenteritis Assays, and kit D = BD MAX Enteric Parasite Panel.

study. Data were stored in an approved and secure portal belonging to Region of Southern Denmark.

3. Results

3.1. LIS Data. In the period of October 2005 to January 2015, 10593 samples from 4887 patients were examined for intestinal parasites. Based on microscopy the reported diarrhoea-causing parasites were *G. lamblia* (500 samples/237 patients), *Cryptosporidium* species (78 samples/41 patients), and *E. histolytica* or *E. histolytica/dispar* (159 samples/62 patients) and *Cyclospora* species (25 samples/12 patients). As previously described [10] the majority of parasites reported as *E. histolytica* after microscopy were most probably in fact *E. dispar*, which is considered nonpathogenic. On the basis of these data we decided in the study to test for *G. lamblia* and *C. parvum/C. hominis* because of the frequencies by which they cause disease. For *Cryptosporidium* sp. the severe illness it causes in immunocompromised patients and the importance to public health as an agent of food and water borne outbreaks also contributed to the decision. In addition a test for *E. histolytica* was included because amoebic dysentery and amoebic liver abscess are severe conditions requiring immediate and appropriate treatment. Thus only the rare cases of cyclosporiasis (approximately one per year in our diagnostic laboratory) would be missed because of the selection of targets.

3.2. Objective 1. The in-house RT-PCR assay for *G. lamblia* was positive in 13 of the 14 microscopy-positive samples

and in 5 microscopy-negative samples. All three tested RT-PCR assays for *C. parvum/C. hominis* were positive in ten of the eleven microscopy-positive samples as well as in one microscopy-negative sample. The selected species-specific in-house RT-PCR assays detected the expected pathogen in all the microscopy-positive samples except for one sample (number 6, Table 2) which based on microscopy was reported with *G. lamblia* and one sample (number 23, Table 2) with *Cryptosporidium* spp. As none of the RT-PCR assays or MT-PCR kits detected the expected parasites in these samples they may represent false positive microscopy reporting (Table 2). The *Cryptosporidium* assay 3 identified *C. hominis* in four and *C. parvum* in eight samples. The *E. histolytica* RT-PCR assay was only positive in 1 of 8 microscopy-positive samples and none of the microscopy-negative samples. The seven microscopy-positive samples that were negative for *E. histolytica* by PCR were all positive when analysed for *E. dispar* by species-specific PCR.

There was no statistical difference in the number of positive samples when tested by microscopy and PCR for the three selected parasites ($p = 0,22$).

3.3. Objective 2. Kit A detected *G. lamblia* in three samples more than kit B, two samples more than kit C, and four samples more than kit D. For *C. parvum/C. hominis,* kit A detected one positive sample more than kit B and one positive sample more than kit C but one less than kit D. Only kit D was positive for *C. parvum/C. hominis* in sample (number 8) in which the other kits detected *G. lamblia* (Table 2). All kits detected *E. histolytica* in the same sample.

Of all samples tested, discordant results were obtained from duplicate determinations in one sample using in-house assays (kit A), two using kit B, two using kit C, and three using kit D (Table 2).

None of the MT-PCR kits confirmed the presence of *G. lamblia* in sample numbers 32 and 33, which were positive in two out of two and one of two replicates, respectively, by the in-house RT-PCR. These two samples were from the same patient. A sample three days later was positive when tested at the National Reference Laboratory at Statens Serum Institut. Sample numbers 32 and 33 were therefore considered to be true positive but weak.

Sample number 86 (not shown in Table 2) was only tested in kits A, B, and D and therefore not included in the total number. In this sample, kit B was positive for both *G. lamblia* and *E. histolytica,* in one out of two duplicates. This was not confirmed by any of the other test kits.

Inhibition by faecal constituents was not a problem in this study, as it has been reported previously [17].

4. Discussion

In this study we have evaluated PCR assays for replacement of microscopy for routine detection of diarrhoea-causing parasites. In contrast to microscopy PCR only detects specific parasites. Careful selection of targets for the PCR assays is therefore mandatory. It is also important to be aware of which diarrhoea-causing parasites present in the population that are not targeted by the selected PCR assays and the number of patients affected by exclusion of assays for particular rare parasites [18, 19].

Based on previous frequencies of detection by microscopy and severity of disease we decided to establish PCR assays for *G. lamblia, C. parvum/C. hominis,* and *E. histolytica.* The only diarrhoea-causing parasite previously detected and not targeted by the PCR assays was *Cyclospora* spp. with a little more than one case on average each year.

The three different assays for *C. parvum/C. hominis* performed equally but assay 1 resulted in the lowest CT values and was used for objective 2. Assay 3 distinguished between *C. hominis* and *C. parvum* and was used for subsequent species identification in positive samples. Rapid species identification is valuable for epidemiological investigations.

The in-house RT-PCR assays detected *G. lamblia* and *Cryptosporidium* spp. in microscopy-negative samples from patients suspected of suffering from intestinal parasitosis and thus appeared more sensitive than microscopy. PCR has previously been reported to be more sensitive than microscopy for detection of specific parasites. In ten Hove's study in 2007, PCR showed 3.6% better sensitivity than microscopy for *Giardia* in clinical stool samples [3], and, in Starks study in 2011, PCR had 2.9% better sensitivity for *Giardia* and 2% better sensitivity for *Cryptosporidium* than microscopy [5]. We found PCR detected 4.1% more *Giardia* than microscopy but did not find any difference for *Cryptosporidium* in this study. A major advantage of PCR over microscopy is the specificity obtained from the discrimination between *E. histolytica* and *E. dispar* [7–10]. Seven of the eight samples originally reported with *E. histolytica* by microscopy were

negative in the in-house RT-PCR and were subsequently identified as *E. dispar.*

The comparison of four test kits, including the in-house assay based kit A, showed varying results from replicate tests. Testing in single determinations may lead to false results in a minority of cases. Future use of these assays may be improved by running tests in duplicate.

The limitation of this study was first of all the sample size, which does not allow statistical analysis of differences in performance of microscopy and RT-PCR in-house assays. Tendencies in favour of the in-house assays were seen when comparing variation of replicates and sensitivities to commercial test kits.

As indicated by the numbers of cases of intestinal parasitosis registered in our LIS over the years, collection of larger number of samples will take time. The detection of parasites in microscopy-negative samples suggests that replacement of microscopy with PCR will increase the positive rates and thereby shorten the time needed to establish large sample collections.

5. Conclusion

In our setting it is relevant to test for *G. lamblia, C. parvum/C. hominis,* and *E. histolytica.* We expect that replacement of microscopy with in-house RT-PCR assays for these parasites will result in higher positive rates for *G. lamblia* and *C. parvum/C. hominis,* while false positive results for *E. histolytica* will be avoided. Addition of a secondary test differentiating between *C. parvum* and *C. hominis* will be of value for early discovery of outbreaks.

All the commercial MT-PCR kits evaluated here tested for the relevant targets. However, some variation in performance was seen when using the kits. The choice of method for detection of intestinal protozoa may depend on the setting. Compared to PCR microscopy is less sensitive and less specific, more time consuming, and more dependent on individual skills. Use of commercial PCR kits may be attractive in laboratories handling moderate numbers of samples, while in-house PCR assays can be established and maintained for large-scale throughput analyses, mainly due to lower costs.

Disclosure

The sponsors did not have any role in the study design, collection, analysis, and interpretation of data.

Acknowledgments

The study is a part of a Ph.D. project and supported by grant from following Danish organizations: A.P. Møller Foundation for the Advancement of Medical Science, "Fonden for Læge Else Poulsens Mindelegat," Beckett-Fonden, Region Syddanmarks and Region Sjællands Fælles Forskningspulje, University of Southern Denmark, and the Department for Clinical Microbiology at Odense University Hospital. The authors thank the laboratory staff of the Sections of Parasitology in the Department for Clinical Microbiology at Odense University Hospital, Odense, Denmark, for their help in collecting the stool samples and the Department for Clinical Microbiology at Slagelse Hospital, Slagelse, Denmark, for use of their BD MAX system, specially Tina Vasehus Madsen, Ph.D., for assisting with the testing. They also thank Dr. Ming Chen in the Department for Clinical Microbiology, Hospital of Southern Jutland, Sønderborg, Denmark, for critically revising the paper.

References

[1] J. J. Verweij, R. A. Blangé, K. Templeton et al., "Simultaneous detection of *Entamoeba histolytica, Giardia lamblia*, and *Cryptosporidium parvum* in fecal samples by using multiplex real-time PCR," *Journal of Clinical Microbiology*, vol. 42, no. 3, pp. 1220–1223, 2004.

[2] T. Schuurman, P. Lankamp, A. van Belkum, M. Kooistra-Smid, and A. van Zwet, "Comparison of microscopy, real-time PCR and a rapid immunoassay for the detection of Giardia lamblia in human stool specimens," *Clinical Microbiology and Infection*, vol. 13, no. 12, pp. 1187–1191, 2007.

[3] R. ten Hove, T. Schuurman, M. Kooistra, L. Möller, L. Van Lieshout, and J. J. Verweij, "Detection of diarrhoea-causing protozoa in general practice patients in The Netherlands by multiplex real-time PCR," *Clinical Microbiology and Infection*, vol. 13, no. 10, pp. 1001–1007, 2007.

[4] L. E. S. Bruijnesteijn van Coppenraet, J. A. Wallinga, G. J. H. M. Ruijs, M. J. Bruins, and J. J. Verweij, "Parasitological diagnosis combining an internally controlled real-time PCR assay for the detection of four protozoa in stool samples with a testing algorithm for microscopy," *Clinical Microbiology and Infection*, vol. 15, no. 9, pp. 869–874, 2009.

[5] D. Stark, S. E. Al-Qassab, J. L. N. Barratt et al., "Evaluation of multiplex tandem real-time PCR for detection of cryptosporidium spp., Dientamoeba fragilis, Entamoeba histolytica, and Giardia intestinalis in clinical stool samples," *Journal of Clinical Microbiology*, vol. 49, no. 1, pp. 257–262, 2011.

[6] A. Kebede, J. J. Verweij, B. Petros, and A. M. Polderman, "Short communication: misleading microscopy in amoebiasis," *Tropical Medicine and International Health*, vol. 9, no. 5, pp. 651-652, 2004.

[7] B. S. Ayed, B. R. Abdallah, K. Mousli, M. Aolin, M. Thellier, and A. Bouratbine, "Molecular differentiation of entamoeba histolytica and entamoeba dispar from Tunisian food handlers with amoeba infection initially diagnosed by microscopy," *Parasite*, vol. 15, no. 1, pp. 65–68, 2008.

[8] K. Khairnar, S. C. Parija, and R. Palaniappan, "Diagnosis of intestinal amoebiasis by using nested polymerase chain reaction-restriction fragment length polymorphism assay," *Journal of Gastroenterology*, vol. 42, no. 8, pp. 631–640, 2007.

[9] J. J. Verweij, F. Oostvogel, E. A. T. Brienen, A. Nang-Beifubah, J. Ziem, and A. M. Polderman, "Short communication: prevalence of entamoeba histolytica and entamoeba dispar in northern Ghana," *Tropical Medicine and International Health*, vol. 8, no. 12, pp. 1153–1156, 2003.

[10] G. N. Hartmeyer, S. V. Høgh, M. Chen, H. Holt, M. N. Skov, and M. Kemp, "Need for species-specific detection for the diagnosis of amoebiasis in a non-endemic setting," *Scandinavian Journal of Infectious Diseases*, vol. 45, no. 11, pp. 868–871, 2013.

[11] A. V. Allen and D. S. Ridley, "Further observations on the formol-ether concentration technique for faecal parasites.," *Journal of Clinical Pathology*, vol. 23, no. 6, pp. 545-546, 1970.

[12] S. A. Henriksen and J. F. Pohlenz, "Staining of cryptosporidia by a modified Ziehl-Neelsen technique," *Acta veterinaria Scandinavica*, vol. 22, no. 3-4, pp. 594–596, 1981.

[13] H. G. M. Niesters, "Quantitation of viral load using real-time amplification techniques," *Methods*, vol. 25, no. 4, pp. 419–429, 2001.

[14] M. Fontaine and E. Guillot, "Development of a TaqMan quantitative PCR assay specific for Cryptosporidium parvum," *FEMS Microbiology Letters*, vol. 214, no. 1, pp. 13–17, 2002.

[15] R. Yang, C. Murphy, Y. Song et al., "Specific and quantitative detection and identification of Cryptosporidium hominis and C. parvum in clinical and environmental samples," *Experimental Parasitology*, vol. 135, no. 1, pp. 142–147, 2013.

[16] C. R. Stensvold, M. Lebbad, J. J. Verweij et al., "Identification and delineation of members of the entamoeba complex by pyrosequencing," *Molecular and Cellular Probes*, vol. 24, no. 6, pp. 403–406, 2010.

[17] L. Monteiro, D. Bonnemaison, A. Vekris et al., "Complex polysaccharides as PCR inhibitors in feces: helicobacter pylori model," *Journal of Clinical Microbiology*, vol. 35, no. 4, pp. 995–998, 1997.

[18] J. J. Verweij, "Application of PCR-based methods for diagnosis of intestinal parasitic infections in the clinical laboratory," *Parasitology*, vol. 141, no. 14, pp. 1863–1872, 2014.

[19] L. van Lieshout and J. J. Verweij, "Newer diagnostic approaches to intestinal protozoa," *Current Opinion in Infectious Diseases*, vol. 23, no. 5, pp. 488–493, 2010.

14

TLR Specific Immune Responses against Helminth Infections

Sivaprakasam Rajasekaran,[1] Rajamanickam Anuradha,[2] and Ramalingam Bethunaickan[1]

[1]*Department of Immunology, National Institute for Research in Tuberculosis, Chennai, India*
[2]*International Center for Excellence in Research, National Institutes of Health, National Institute for Research in Tuberculosis, Chennai, India*

Correspondence should be addressed to Ramalingam Bethunaickan; bramalingam@gmail.com

Academic Editor: José F. Silveira

Despite marked improvement in the quality of lives across the globe, more than 2 million individuals in socioeconomically disadvantaged environments remain infected by helminth (worm) parasites. Owing to the longevity of the worms and paucity of immunologic controls, these parasites survive for long periods within the bloodstream, lymphatics, and gastrointestinal tract resulting in pathologic conditions such as anemia, cirrhosis, and lymphatic filariasis. Despite infection, an asymptomatic state may be maintained by the host immunoregulatory environment, which involves multiple levels of regulatory cells and cytokines; a breakdown of this regulation is observed in pathological disease. The role of TLR expression and function in relation to intracellular parasites has been documented but limited studies are available for multicellular helminth parasites. In this review, we discuss the unique and shared host effector mechanisms elicited by systemic helminth parasites and their derived products, including the role of TLRs and sphingolipids. Understanding and exploiting the interactions between these parasites and the host regulatory network are likely to highlight new strategies to control both infectious and immunological diseases.

1. Introduction

Helminth parasites (worms) include an array of metazoan organisms. Over 60% of the world populations are at the risk of helminth infections in tropical and subtropical regions [1]. Parasitic infections are major public health problems that have impact on socioeconomic influence. Chronic infection may lead to physical disabilities, anemia, and malnourishment [2]. Helminth parasitic infection has been largely eliminated in developed countries due to control of the insect vector population by the safe disposal of human waste and the availability of efficient drugs. Nevertheless, in developing countries, these types of parasite control methods are often not yet practical and helminths persist as a significant biomedical problem.

Helminth parasites have evolved to survive and reproduce within immune-exposed niches, such as the blood, lymphatics, and gastrointestinal tract [2, 3]. Several parasites have a complex multistage lifecycle, which requires numerous intermediate hosts for completion. Inside the mammalian host, parasites undergo extensive growth and differentiation to produce developmental stages ready for transmission to the next intermediate host. Larva migrates within the host to its suitable niche that supports its growth and reproduction. The resulting offspring are capable of transmission from one host to another; this process varies among helminths [4]. The social and medical impact of the global parasitic worm burden necessitates more attention and research focus on modulation of immune responses to helminth infection and factors that influence disease pathogenesis [5]. The host immune response to helminths includes multiple strategies for induction of regulatory networks and immune responses that involve both the innate and adaptive immune system [6]. Helminth parasites have evolved immune evasion strategies necessary for their continued transmission. This immune evasion is achieved at the expense of both antigen-presenting cells (APCs) and T cells. Similar to intracellular parasitic infections, pattern-recognition receptors (PRR) play a pivotal role in initiating the host immune response against multicellular helminth parasites [6]. Most of the pathogen-associated molecular patterns (PAMPs) from these parasites are recognized by Toll-like receptors (TLRs) [7]. TLRs are expressed on many cell types, for example, epithelial cells of the gastrointestinal and respiratory tracts, myofibroblasts, enteroendocrine cells, astrocytes, and immune cells such as T cells, B cells, and dendritic cells (DCs) [8]. TLRs dictate the downstream pathways involved in adaptive immune responses by influencing multiple antigen-presenting cell (APC) functions [9]. The potential contribution of TLRs to fighting parasitic infections has gained much attention in the last decade [10]. In addition to TLR, NOD (nucleotide oligomerization domain-like receptor) recognizes intracellular PAMPS and initiates signaling pathways that induce production of inflammatory cytokines [11, 12].

2. Overview of TLR

TLRs are central players in many aspects of microbial elimination, including recruitment of phagocytes to infected tissue, following microbial killing. TLRs are expressed by macrophages and dendritic cells (DCs); T and B lymphocytes also express TLRs [13]. TLRs are membrane spanning and noncatalytic receptors, which are capable of recognizing structurally conserved molecules derived from pathogens and directing the downstream immune response [14]. Currently thirteen TLRs have been identified, TLR1–TLR13; of these, TLR1–TLR9 are conserved between humans and mice. In mice TLR10 is nonfunctional due to a retrovirus insertion whereas TLR11–TLR13 are present within endosomal compartments of mice but are lost from the human genome [15]. TLR1, TLR2, TLR4, TLR5, and TLR6 are expressed on the plasma membrane and TLR3, TLR7, TLR8, and TLR9 are present in endosome of leukocytes. These receptors are expressed on various immune and nonimmune cells in a variety of combinations in order to recognize most of the pathogen-associated molecular patterns (PAMPs), thereby providing a link between the innate and adaptive immune systems [16, 17]. TLR signals through MyD88 pathway leading to activation of MAPK and induces the translocation of nuclear factor kappa B (NF-kB) to the nucleus. NF-kB promotes the transcription and synthesis of proinflammatory cytokines [18].

3. TLR Pathway

Immune and nonimmune cells with unique combinations of TLR expression patterns have been identified within mammals [8]. They can recognize PAMPs derived from parasites or microbes, including proteins, lipoproteins, lipids, and nucleic acids. In addition, endogenous ligands (heat-shock proteins, fragments from extracellular matrix, fibrinogen, and end products of cellular apoptosis like DNA and RNA) can also bind to TLRs and trigger inflammatory cascades [19]. TLRs are important in the recognition of *Leishmania* species [20]. They sequentially trigger the innate and then adaptive immune responses required for controlling *Leishmania* parasite [21]. Purified *Leishmania* lipophosphoglycan stimulates upregulation of TLR2 on human NK cells and elicits leishmanicidal reactions via release of inflammatory mediators, for example, TNF-α, IFN-γ, nitric oxide (NO), and reactive oxygen species (Th1 response). TLR2 can also induce the

antileishmanial immune response through decreased expression of TLR9. The PAMP-dependent DC activation could be based on TLR expression. Plasmacytoid dendritic cells (PDC) express TLR9 and recognize CpG motifs in mice while myeloid DC (mDC) expresses TLR4 and its reacts to LPS [22]. Endogenous RNA and DNA activate TLR7 and TLR9, by entering into the endosomal compartment, thereby inducing production of proinflammatory cytokines by plasmacytoid (pDC) and conventional DCs (cDC) [23]. Endogenous TLRs have been crucial for resistance to *Leishmania major* [18]. Within mammalian genomes, the CpG motif occurs much less frequently and remains highly methylated; as a result, TLR9 does have a limited activation role in eliciting innate immunity. CpG DNA induces a conformational change in TLR9 that is required for its activation [24]. TLR9 recognizes unmethylated CpG motifs as a conserved molecular pattern in pathogen DNA and abnormal composition, structure, or chemical features in any kind of DNA [25]. TLR9 signaling is essential for NK cell activation and production of IL-12 by bone marrow-derived DC, which can reduce the parasite burden [10, 26]. Fakher et al. have shown that TLR9-deficient mice have increased *Leishmania* burden which indicates that TLR9 plays an important role in reducing parasite burden [27]. Steady-state production of IL-12 by migratory CD103(+) DCs, independent of signals from commensals or TLR-initiated events, was necessary and sufficient to exert the suppressive effects on Th2 response development in *S. mansoni* [28].

Downregulation of TLRs is a strategy used by protozoa to evade immune responses. Protozoan parasites such as *Trypanosome* spp. and *Entamoeba histolytica* were shown to inhibit the immune response particularly, by downregulating TLR2 expression [29]. Similarly, mRNA expression of TLR3, TLR4, TLR5, and TLR7 from monocyte derived DC was significantly downregulated by live microfilariae (Mf) of *B. malayi* (BmA) [30]. Filarial infected individuals have shown decreased mRNA and protein expression of TLR1, TLR2, TLR4, and TLR9 in B cells [31]. T cells play an additional role in TLR signaling, because T cells express many of the TLRs. In lymphatic filarial infected patients, T cells express lower levels of TLR1, TLR3, and TLR4 after stimulation by B cells and monocytes [32].

TLR plays the major role in intestinal homeostasis [33]. The MUC2 genes are associated with TLR pathways. Helminth and it products may stimulate physical barrier function of IECs by TLR [34]. Control of inflammation by helminths in the TLR pathway is highly possible for efficient host protection through TLR-dependent proinflammatory cascades elicited by parasitic infections, which must be firmly controlled to evade severe pathology, reviewed by [35].

4. Immune Response to Helminths

Innate and adaptive immune systems are crucial for the induction of type 2 immunity, which distinguishes the response to helminth infection. The key players in T helper (Th) type 2 immunity are CD4$^+$ Th2 cells and involve the cytokines IL-4, IL-5, IL-9, IL-10, and IL-13 and immunoglobulin (Ig)E. Th2-type immune responses are comprised of three features:

inflammation, wound repair, and resistance to helminths [36]. A diverse range of multicellular parasites dwelling inside humans elicit a stereotypical immune response in order to protect themselves from immune attack [6, 37, 38]. Helminth influence of DC may also be facilitated by method by the enzymatic activities of helminth-derived products. For example, helminth parasites inhibit the host innate immune response, initially by releasing many types of enzymatically active products which are assumed to play a key role in determining and supporting infection by contributing to the degrading the soluble antiparasitic molecules or the weakening of innate immune cells [39]. This results in the production of Th2 associated cytokines, particularly, IL-5, IL-13, and IL-4 together with IgE elicited through mast cell and eosinophil mobilization. Th2 immune responses are not however sufficient to expel the parasite [40]. IL-4 and IL-13 together with apoptotic cells provoke host protection against helminth infections and the anti-inflammatory and tissue repair phenotype in macrophages [41, 42]. Several animal studies, carried out with *Schistosoma mansoni*, have shown that parasites are capable of attenuating Th1 responses (decreased IFN-γ, TNF-α, IL-12, and NO) and promoting Th2 immune responses (IL-10 and TGF-β) [43, 44]. The filarial endosymbiont *Wolbachia* is known to elicit immune responses through TLR2 and TLR4 and is known to be the major mediator of inflammatory responses in lymphatic filariasis and onchocerciasis [45, 46]. Immunologists are often intrigued by the way the host tolerates helminth infection by the immunomodulation, even after exhibiting severe immunopathological condition [47]. Gao et al. suggested that TLR4 might play a role in the protection against infection, whereas TLR2 was favorable for the parasite [48]. The expression of TLR2 and subsequently NF-kB was decreased in intestinal schistosomiasis 12 weeks after infection even though the parasite burden is still high [49]. TLR-related genes are generally decreased during the course of *Schistosoma* infection. TLR1, TLR3, TLR7, and TLR8 are strongly repressed, with the appearance of the eggs at week 8 after infection, and TLR3 shows most repression [50].

5. Helminth Immune Modulation through TLR

Inflammatory signal from TLR is a defensive degree of the host body to warrant elimination of harmful extortions posed by infectious agents as well as fastening the healing process. Conversely, the Th1 influenced inflammatory consequences orchestrated by TLRs also engage in destroying pathogenic infections but also can provoke decisive pathological effects [16]. Likewise, pathogen modified TLR signaling progresses to Th2 related response favorable for the pathogen that will lead disease progression [34]. Hence, a sufficient stability between pro- and anti-inflammatory immune responses is of massive significance to restore the normal physiological conditions of the host body during and after a pathogenic infection [40]. Wang et al. isolated immunomodulatory peptide called SJMHE1 from the HSP60 protein of *Schistosoma japonicum* and showed that small molecule peptide that has progressed during host-parasite interactions is of huge

significance in the search for novel anti-inflammatory agents and therapeutic goals for autoimmune diseases [51, 52].

TLRs trigger an intracellular signaling cascade through Toll/interleukin-1 receptor (TIR) and through the recruitment of adaptor molecules, such as TICAM-1, MyD88 [53], TRIF, and TRAM [54, 55]. These adaptor molecules act either independently or in combination, to induce transcription factors such as c-Jun-N-terminal kinases (JNK), mitogen-activated protein kinases (MAPK), p38, extracellular signal-regulated kinase (ERK), and NF-kB (nuclear factor kappa B), leading to the transcription of inflammatory and immunomodulatory genes including costimulatory molecules, cytokines, and chemokines [56–58]. Ongoing infections in the deficiency of certain TLR diverge adaptive responses, which aggravates the immunopathology of the host, which has been shown by various studies [35, 39]. More recently, the way TLR mediates interaction between these multicellular parasites and the host immune system has been well documented. During acute phase of helminth infections, DC promotes Th1 environment through the activation of TLR, which would match with induced Th1 responses [6, 38, 40]. Helminth antigen contains proteins, glycoproteins, and glycolipids. DC induces proinflammatory activation and maturation due to impassive behaviour on activation and the failure with helminth antigens. The calreticulin protein isolated from *Heligmosomoides polygyrus* can be able to provoke IL-4 secretion through triggering class A scavenger receptor [59]. Immunomodulatory activity by ES of different species is well characterized in the nematode *Heligmosomoides polygyrus* [60] and the trematode *Fasciola hepatica* [61]. *Acanthocheilonema viteae* ES-62 product contains glycoprotein, which triggers TLR4 in turn inducing Th2 type of immune response which in turn determines the phenotype of the APCs [62–64]. However, studies carried out on innate cells with *Schistosoma mansoni*'s soluble egg antigen (SEA) were not capable of eliciting such TLR response, instead dampening the release of proinflammatory cytokines with response to LPS [65, 66]. Several studies have shown that the inhibitory consequences of helminth-derived factors on TLR stimulated triggering as determined by proinflammatory cytokine secretion and expression of costimulatory molecules [67, 68].

Schistosome soluble egg antigen (SEA) and ES products freed by the egg stage of the parasite encompass effective Th2-inducing and immunomodulatory activity. SEA from *Schistosoma mansoni* was shown to be a tremendously strong inducer of Th2 responses, even in the absence of current infection or any supplement of adjuvant. *Schistosoma mansoni* also expresses glycans that have been shown to reveal immunomodulatory functions [69]. Schistosomal infection promotes the differentiation of DC and secretes IL10, thereby inducing Tregs, mediated through the downstream effect of TLR2 [64]. TLR2 is a receptor that plays an important role in filarial infection; the filarial endosymbiont *Wolbachia* is known to elicit immune responses through TLR2 and TLR4 and is known to be the major mediator of inflammatory responses in lymphatic filariasis and onchocerciasis [45, 46]. Studies carried out by [45] have revealed that, upon harboring *Wolbachia*, an endosymbiont within the filarial parasite, it can interact with the innate immune system through TLR2 and TLR4. Our own studies have shown that humans infected with filarial infection revealed a tardy response against APC and T cell specific TLR1, TLR2, TLR4, TLR9, and their ligands, expressing decreased proinflammatory cytokines [31, 32]. Thus, most of these infections impair the Th1 response, mainly through impairment of conventional DC maturation, and favor Th2 or regulatory immune response. Most of the helminth derivatives, including phosphorylcholine containing glycoprotein ES-62, induce anti-inflammatory or Th2 response in *Acanthocheilonema viteae* infection [62, 63, 70]. This nematode is of particular interest and it does not contain the endosymbiont bacteria *Wolbachia*. This relationship is essential to other filarial worms since its death through antibiotic treatment leads to worm sterility and death. The bacteria are also thought to mediate immune responses by triggering TLR2 and TLR4 [45]. Phospholipids from schistosomes and *Ascaris* worms also trigger TLR2 and the lysophosphatidyl serine could activate DCs to induce Th2 and IL-10-producing Treg. Several studies have demonstrated that ongoing infections in the absence of certain TLR deviates adaptive responses, which exacerbates the immunopathology of the host, reviewed in [71]. Initiation of alternatively activated macrophages (AAM) was expendable for the defending effect of *Litomosoides sigmodontis* infection on *E. coli*-provoked peritoneal sepsis, whereas TLR2-activation during the reprogramming of functional macrophages was crucial [72]. Fatty acid binding protein (FABP) plays an important role in parasite nutrition [61]. *Fasciola hepatica* fatty acid binding protein (Fh12) blocks induction of inflammatory mediators *in vitro* and *in vivo* and in doing so completely inhibits activation of TLR4 by LPS in a dose-dependent manner [73]. Nullification of the ES-62-mediated suppression of LPS leads to the production of IL-6, IL-12p70, and TNF-α by DCs. Thus, by exploiting this homeostatic regulatory mechanism, ES-62 can protect against abnormal inflammation, can support parasite survival, and disclose therapeutic potential in inflammatory disease [74].

Sustained infection or impairment of innate immune cells occurs due to degradation of antiparasitic molecules determined by the helminth parasites, which present inside the host [39]. Parasite survival is promoted in the host; by downregulation of an antigen specific T cell proliferation [75–77] ES-62 from rodent filarial nematode inhibit the activation of B and T cell [70]. Parasite derived molecules from *Schistosoma* are processed through TLR4 and MyD88 dependent pathway [78]. *Schistosoma* egg product primes DC to drive Th2 responses [39] and LFNP III stimulates IL-10 producing B1 cells in mice [79]. Rodent malarial secretory product induces immunomodulation by inhibition of B and T lymphocyte proliferation and inhibition of maturation of naïve DCs priming T cells and inhibition of IFN-γ, IL-12, and IL-17. Parasite derived lipids signal through TLR2 [80].

6. TLR Signaling

TLR signaling has been extensively investigated since 1999. The cytoplasmic domain of TLR, termed Toll/interleukin-1 receptor (TIR), is highly conserved and functions as binding site for downstream adaptor molecules. Signaling by TLRs involves a variety of adaptor proteins [81], the most common one being the myeloid differentiation marker 88 (MyD88), used by all TLRs, except TLR3 [82]. Downstream targets of MyD88 include nuclear factor kappa B (NF-κB), mitogen-activated protein kinases (MAPKs) (p38, JNK, and ERK1/2), KB kinase inhibitor (IKK), and interferon regulatory factors (IRF) [83]. After endocytosis into endosomes, both TLR3 and TLR4 induce IFN-β by downstream signaling mediated by the alternate adapter TRIF. TLR2 and TLR4 require use of TIRAP in addition to MyD88 for downstream signaling [84, 85]. MyD88−/− mice infected by *T. gondii* showed diminished IL-12 levels and Th1 cell responses.

TIR domain is a key molecular module of TLR mediated innate immune response pathways. All mammalian TLRs contain TIR domains in their C-terminal regions. Homo- or heterotypic dimerization of TIR domains is required to initiate downstream signaling. Similar to most of the microbes, helminth parasites evade host immune response by dampening TLR expression and downregulating the TLR mediated cell signaling [86], whereas helminth-derived molecules are capable of activating TLRs through a set of kinases, resulting in Th1 type of immune response. The nuclear factor kappa B (NF-kB) pathway activated by triggering TLRs, as a result of induction of inflammatory responses occurs. NF-kB pathway and interferon regulatory factor (IRF) pathway receive signals from activated Toll/IL-1R (TIR) domain and start signal through five different adaptor molecules. by binding with specific ligand and contact with the ligand, TLR recruits an adaptor protein either to TIR domain or with IL-1 receptor associated kinases (IRAKs) [87]. IRAKs have important role in the early stages of TLR signaling. MyD88 interacts with IRAK1 and subsequently recruits IRAK1 or IRAK2 [88].

MyD88 involves all TLR signaling pathways except TLR3. MyD88 binds with MyD88 adaptor-like (MAL) protein (MAL). MyD88 independent pathway is activated through binding of TLR3 to its adaptor molecule TIR-related adaptor protein inducing interferon (TRIF). TLR4 signals through TRIF binding to its adaptor TRIF-related adaptor molecule (TRAM) and also through MAL/MyD88 protein complex. It has been shown that TLR4 ligands like LPS are capable of strongly activating JNK, MAP-kinase, and ERK, whereas molecule like LNFPIII can phosphorylate only ERK. Besides, TLR4 can be activated through MyD88-independent pathway by interferons through TRAM/TRIF complex [58, 89, 90] NF-kB comprised p50, p65, p52, RelB, and c-Rel subunits [91]. After dimerization of the subunits translocation occurs in the nucleus and NF-kB binds to DNA. Ag receptors, apoptosis, and host defence genes are regulated by NF-kB inside the nucleus. During immune response against pathogens, sensory and effector functions of TLRs are involved in the production of proinflammatory cytokines which ultimately increases the function of APCs, which have potential for immediate response against particular pathogen [92]. Differential activation of MAPK p38 and extracellular-signal-regulated kinase (ERK) within DCs results in altered levels of DC maturation and cytokine production. Studies have proved that activation of p38 has pivotal role in the DC maturation and proinflammatory immune responses and ERK activation

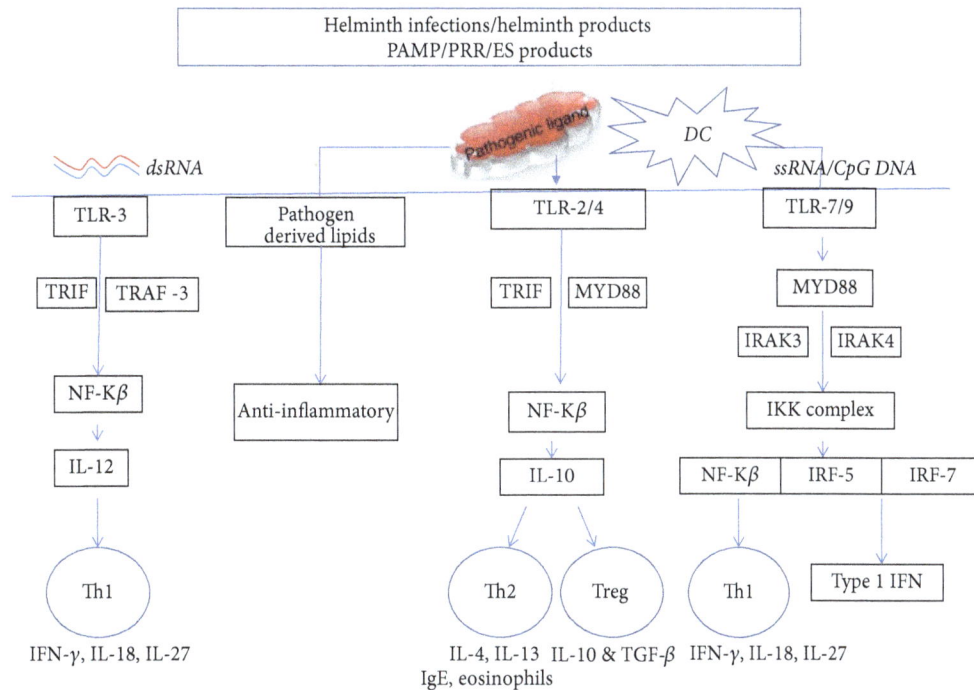

FIGURE 1: TLR interactions during helminth infections: depicting the involvement and induction of various TLRs during helminth infections and their byproducts. TLR pathways stimulation endorses specific Th environment.

has been much more needed for anti-inflammatory Th2 response [93, 94]. In SEA model, during infections, absence of TLR2 leads to enhanced the disease severity and Th1 and diminished Th2 responses [95]. DCs and B cells produce and activate IL-10 and TGF-β through MyD88 dependent pathway and the suppression of IL-12, IL-6, IL-1β, and TNF-α occurs by modulation of intracellular pathway of TLR2 by SEA [94]. *Acanthocheilonema viteae*, a rodent nematode product of ES-62 glycoprotein, inhibits both B and T cell activation and TLR4 via MyD88 dependent pathway [70].

Clinically, for asymptomatic filarial infections, the deleterious pathology can be evaded by possible mechanism by downregulation of TLR on APC and T cells [96]. Previous literature shows TLR signaling in repose to intracellular pathogens including parasitic protozoa [97]. *Wolbachia* extracts and *Wolbachia* surface protein (Wsp) can induce immune response through TLR2 and TLR4 [45, 46]. In mice model Hise et al. showed that Wsp induce an inflammatory response through TLR2 and TLR4 [98]. *Wolbachia* mediated inflammatory responses mediated by TLR2 and TLR6 dependent on MyD88 and TIRAP/MAL. For intracellular parasite *Leishmania donovani* infection, the activation of TLR4 is mediated by MyD88 [99].

TLR4 mediates accessibility to distinctive signaling pathway due to its cellular locations. Due to acidification of the endosome, this stimulates conformational changes that are the major properties to allow TLR4 differential use of adaptor proteins, involving distinct signaling pathways [100]. TLR8 recognizes viral ssRNA and endogenous RNA, such as microRNAs, resulting in the production of proinflammatory cytokines. Hence, localization sites of the receptors are crucial for the nucleic acid-sensing mode and downstream signaling

[101]. B cell differentiation and activation of TLR signaling in B cells are initiated through B cell receptor or CD40 ligation. B regulatory cells are characterized by enhanced IL-10 production and in contrast downregulation of inflammatory reaction triggered by priming of B cell by TLR ligands [102]. The schematic diagram (Figure 1) depicts the involvement and induction of various TLRs during helminth infections and the byproducts.

7. Helminths and Immunomodulatory Sphingolipids

The human immune system can interact with carbohydrate, (glyco)protein, and lipid products of pathogens. Numerous studies of helminths have shown that one of the major properties of eukaryotic lipids is an immunomodulatory effect. One mechanism is by induction of Treg cells; Treg induction by a lipid product produced by schistosomes has been demonstrated [80]. Treg induction by *Schistosoma mansoni* egg antigens involves TLR2 expression. Lipids may also induce Th2 responses. Th2 development can be induced by diacylated phosphatidylserine, a lipid fraction of schistosomes that induces DC maturation [103]; this effect required TLR2 [80]. Similar findings have been reported in the immune response to *Ascaris lumbricoides*. Glycans may also play a role in resistance to *Schistosoma japonicum* [104]. In addition to TLR2, TLR9 may promote host protective immune responses. For example, TLR9 expression decreased the antileishmanial response by lipophosphoglycan (LPG) and TLR2. *Leishmania major* parasite infected with macrophages showed increased levels of LPG leading to decreased levels of TLR9 in comparison with a *Leishmania major* parasite with decreased levels of

LPG. Study from Späth et al. demonstrated that *Leishmania* phosphoglycan$^{-/-}$ cells were unable to persist in activated macrophages but recalled the ability to endure indefinitely in the mammalian host without provoking disease in non-activated macrophages [105]. Activation of LPG helps in parasite survival in macrophages through TLR2 [106]. Thus, the induction of Tregs as well as other anti-inflammatory responses induced via interactions of a variety of helminth products with the innate immune system (APCs and iNKT cells) facilitates the survival of the helminth in the host and prevents inflammation.

8. Conclusion

TLRs provide a bridge between innate and acquired immunity. Moreover, TLRs not only are key players in the inflammatory process by promoting the production of inflammatory molecules, for example, cytokines and chemokines, but also function as regulatory (anti-inflammatory) contributors and appear to provide signals that are necessary for the resolution of excessive inflammation. In this review, we have explained how helminth-derived products, which provoke host responses, influence the immune system to prevent inflammatory diseases or immunopathology thus ensuring their survival in the host. One mechanism for the anti-inflammatory response induced by helminth-derived products is via their interaction with TLR 2/4 and TLR9. Sphingolipids and other lipids such as diacylated phosphatidylserine, glycans, and LPG lead to the induction of Th2 responses in helminth infection through stimulation of Tregs as well as other anti-inflammatory responses induced via interactions with the innate immune system enabling the survival of the helminth in the host and preventing inflammation. However, most of the factors that influence TLR induction of either proinflammatory or anti-inflammatory mediators are still to be elucidated. A further understanding of parasite derived TLR ligands can lead to innovative therapeutic and prophylactic strategies for parasitic infections.

Additional Points

A comprehensive search approach using keyword and subject headings was applied to PubMed. One hundred and six articles were included. In addition, search engine searches using Google and Google Scholar were performed under the query "TLR and Helminth, TLR and signaling pathways, Immune responses to helminth and TLR, Innate immune responses". The search was confined to articles written in English.

Acknowledgments

The authors would like to sincerely thank Dr. Subash Babu, Scientific Director, NIH-NIRT-ICER, and Dr. Anne Davidson, Investigator, Feinstein Institute for Medical Research, NY, USA, for their critical reading, comments, and proofreading. They would like to acknowledge the mentorship and support of late Dr. V. Kumaraswami, who has been instrumental in this review. Ramalingam Bethunaickan's work is supported by DBT Ramalingaswami Fellowship, Ministry of Science and Technology, India.

References

[1] P. Salgame, G. S. Yap, and W. C. Gause, "Effect of helminth-induced immunity on infections with microbial pathogens," *Nature Immunology*, vol. 14, no. 11, pp. 1118–1126, 2013.

[2] F. Mencl, M. Birkle, M. Blanda, and L. W. Gerson, "EMTs' knowledge regarding transmission of infectious disease," *Prehospital Emergency Care*, vol. 4, no. 1, pp. 57–61, 2000.

[3] S. Babu, S. Q. Bhat, N. P. Kumar et al., "Attenuation of toll-like receptor expression and function in latent tuberculosis by coexistent filarial infection with restoration following antifilarial chemotherapy," *PLOS Neglected Tropical Diseases*, vol. 3, no. 7, article no. e489, 2009.

[4] A. S. MacDonald, M. I. Araujo, and E. J. Pearce, "Immunology of parasitic helminth infections," *Infection and Immunity*, vol. 70, no. 2, pp. 427–433, 2002.

[5] R. M. Maizels, E. J. Pearce, D. Artis, M. Yazdanbakhsh, and T. A. Wynn, "Regulation of pathogenesis and immunity in helminth infections," *The Journal of Experimental Medicine*, vol. 206, no. 10, pp. 2059–2066, 2009.

[6] R. M. Maizels and M. Yazdanbakhsh, "Immune regulation by helminth parasites: cellular and molecular mechanisms," *Nature Reviews Immunology*, vol. 3, no. 9, pp. 733–744, 2003.

[7] G. M. Barton and R. Medzhitov, "Toll-like receptors and their ligands," in *Toll-Like Receptor Family Members and Their Ligands*, vol. 270 of *Current Topics in Microbiology and Immunology*, pp. 81–92, Springer Berlin Heidelberg, Berlin, Germany, 2002.

[8] A. L. Hart, H. O. Al-Hassi, R. J. Rigby et al., "Characteristics of intestinal dendritic cells in inflammatory bowel diseases," *Gastroenterology*, vol. 129, no. 1, pp. 50–65, 2005.

[9] A. Iwasaki and R. Medzhitov, "Toll-like receptor control of the adaptive immune responses," *Nature Immunology*, vol. 5, no. 10, pp. 987–995, 2004.

[10] M. S. Faria, F. C. G. Reis, and A. P. C. A. Lima, "Toll-like receptors in *Leishmania* infections: guardians or promoters?" *Journal of Parasitology Research*, vol. 2012, Article ID 930257, 12 pages, 2012.

[11] J. M. Blander and R. Medzhitov, "Regulation of phagosome maturation by signals from toll-like receptors," *Science*, vol. 304, no. 5673, pp. 1014–1018, 2004.

[12] S. Mariathasan and D. M. Monack, "Inflammasome adaptors and sensors: Intracellular regulators of infection and inflammation," *Nature Reviews Immunology*, vol. 7, no. 1, pp. 31–40, 2007.

[13] E. Lien and R. R. Ingalls, "Toll-like receptors," *Critical Care Medicine*, vol. 30, no. 1, pp. S1–S11, 2002.

[14] R. Medzhitov, P. Preston-Hurlburt, and C. A. Janeway Jr., "A human homologue of the Drosophila toll protein signals activation of adaptive immunity," *Nature*, vol. 388, no. 6640, pp. 394–397, 1997.

[15] K. Takeda and S. Akira, "Toll-like receptors," in *Current Protocols Immunology*, vol. 109, 14 edition, 2015.

[16] S. Akira, "Mammalian Toll-like receptors," *Current Opinion in Immunology*, vol. 15, no. 1, pp. 5–11, 2003.

[17] T. Kawai and S. Akira, "Pathogen recognition with Toll-like receptors," *Current Opinion in Immunology*, vol. 17, no. 4, pp. 338–344, 2005.

[18] L. H. Franco, A. K. Fleuri, N. C. Pellison et al., "Autophagy downstream of endosomal Toll-like receptor signaling in macrophages is a key mechanism for resistance to Leishmania major infection," *The Journal of Biological Chemistry*, vol. 292, no. 32, pp. 13087–13096, 2017.

[19] L. Yu, L. Wang, and S. Chen, "Endogenous toll-like receptor ligands and their biological significance," *Journal of Cellular and Molecular Medicine*, vol. 14, no. 11, pp. 2592–2603, 2010.

[20] S. M. Whitaker, M. Colmenares, K. G. Pestana, and D. McMahon-Pratt, "Leishmania pifanoi proteoglycolipid complex P8 induces macrophage cytokine production through Toll-like receptor 4," *Infection and Immunity*, vol. 76, no. 5, pp. 2149–2156, 2008.

[21] P. Kropf, M. A. Freudenberg, M. Modolell et al., "Toll-Like Receptor 4 Contributes to Efficient Control of Infection with the Protozoan Parasite Leishmania major," *Infection and Immunity*, vol. 72, no. 4, pp. 1920–1928, 2004.

[22] A. Krug, A. Towarowski, S. Britsch et al., "Toll-like receptor expression reveals CpG DNA as a unique microbial stimulus for plasmacytoid dendritic cells which synergizes with CD40 ligand to induce high amounts of IL-12," *European Journal of Immunology*, vol. 31, no. 10, pp. 3026–3037, 2001.

[23] H. Chi and R. A. Flavell, "Innate recognition of non-self nucleic acids," *Genome Biology*, vol. 9, no. 3, article no. 211, 2008.

[24] E. Latz, A. Verma, A. Visintin et al., "Ligand-induced conformational changes allosterically activate Toll-like receptor 9," *Nature Immunology*, vol. 8, no. 7, pp. 772–779, 2007.

[25] K. J. Ishii and S. Akira, "Innate immune recognition of, and regulation by, DNA," *Trends in Immunology*, vol. 27, no. 11, pp. 525–532, 2006.

[26] J. Liese, U. Schleicher, and C. Bogdan, "TLR9 signaling is essential for the innate NK cell response in murine cutaneous leishmaniasis," *European Journal of Immunology*, vol. 37, no. 12, pp. 3424–3434, 2007.

[27] F. H. A. Fakher, N. Rachinel, M. Klimczak, J. Louis, and N. Doyen, "TLR9-dependent activation of dendritic cells by DNA from Leishmania major favors TH1 cell development and the resolution of lesions," *The Journal of Immunology*, vol. 182, no. 3, pp. 1386–1396, 2009.

[28] B. Everts, R. Tussiwand, L. Dreesen et al., "Migratory CD103⁺ dendritic cells suppress helminth-driven type 2 immunity through constitutive expression of IL-12," *The Journal of Experimental Medicine*, vol. 213, no. 1, pp. 35–51, 2016.

[29] C. Maldonado, W. Trejo, A. Ramirez, and et al., "Lipophosphopeptidoglycan of Entamoeba histolytica induces an antiinflammatory innate immune response and downregulation of toll-like receptor 2 (TLR-2) gene expression in human monocytes," *Archives of Medical Research*, vol. 31, no. 4, pp. 71–73, 2000.

[30] R. T. Semnani, P. G. Venugopal, L. Mahapatra et al., "Induction of TRAIL- and TNF- -dependent apoptosis in human monocyte-derived dendritic cells by microfilariae of brugia malayi," *The Journal of Immunology*, vol. 181, no. 10, pp. 7081–7089, 2008.

[31] S. Babu, C. P. Blauvelt, V. Kumaraswami, and T. B. Nutman, "Diminished expression and function of TLR in lymphatic filariasis: a novel mechanism of immune dysregulation," *The Journal of Immunology*, vol. 175, no. 2, pp. 1170–1176, 2005.

[32] S. Babu, C. P. Blauvelt, V. Kumaraswami, and T. B. Nutman, "Cutting edge: diminished T cell TLR expression and function modulates the immune response in human filarial infection," *The Journal of Immunology*, vol. 176, no. 7, pp. 3885–3889, 2006.

[33] M. Fukata and M. T. Abreu, "Pathogen recognition receptors, cancer and inflammation in the gut," *Current Opinion in Pharmacology*, vol. 9, no. 6, pp. 680–687, 2009.

[34] K.-D. Lee, S.-M. Guk, and J.-Y. Chai, "Toll-like receptor 2 and Muc2 expression on human intestinal epithelial cells by gymnophalloides seoi adult antigen," *Journal of Parasitology*, vol. 96, no. 1, pp. 58–66, 2010.

[35] P. G. Venugopal, T. B. Nutman, and R. T. Semnani, "Activation and regulation of Toll-Like Receptors (TLRs) by helminth parasites," *Immunologic Research*, vol. 43, no. 1–3, pp. 252–263, 2009.

[36] S. J. Jenkins, D. Ruckerl, P. C. Cook et al., "Local macrophage proliferation, rather than recruitment from the blood, is a signature of TH2 inflammation," *Science*, vol. 332, no. 6035, pp. 1284–1288, 2011.

[37] R. M. Maizels, A. Balic, N. Gomez-Escobar, M. Nair, M. D. Taylor, and J. E. Allen, "Helminth parasites—Masters of regulation," *Immunological Reviews*, vol. 201, pp. 89–116, 2004.

[38] R. M. Anthony, L. I. Rutitzky, J. F. Urban, M. J. Stadecker, and W. C. Gause, "Protective immune mechanisms in helminth infection," *Nature Reviews Immunology*, vol. 7, no. 12, pp. 975–987, 2007.

[39] B. Everts, H. H. Smits, C. H. Hokke, and M. Yazdanbakhsh, "Helminths and dendritic cells: sensing and regulating via pattern recognition receptors, Th2 and Treg responses," *European Journal of Immunology*, vol. 40, no. 6, pp. 1525–1537, 2010.

[40] J. E. Allen and R. M. Maizels, "Diversity and dialogue in immunity to helminths," *Nature Reviews Immunology*, vol. 11, no. 6, pp. 375–388, 2011.

[41] L. Bosurgi, Y. G. Cao, M. Cabeza-Cabrerizo, and etal., "Macrophage function in tissue repair and remodeling requires IL-4 or IL-13 with apoptotic cells," *Science*, vol. 356, pp. 1072–1076, 2017.

[42] S. J. Van Dyken and R. M. Locksley, "Interleukin-4-and interleukin-13-mediated alternatively activated macrophages: Roles in homeostasis and disease," *Annual Review of Immunology*, vol. 31, pp. 317–343, 2013.

[43] A. C. La Flamme, K. Ruddenklau, and B. T. Bäckström, "Schistosomiasis decreases central nervous system inflammation and alters the progression of experimental autoimmune encephalomyelitis," *Infection and Immunity*, vol. 71, no. 9, pp. 4996–5004, 2003.

[44] D. Sewell, Z. Qing, E. Reinke et al., "Immunomodulation of experimental autoimmune encephalomyelitis by helminth ova immunization," *International Immunology*, vol. 15, no. 1, pp. 59–69, 2003.

[45] N. W. Brattig, C. Bazzocchi, C. J. Kirschning et al., "The major surface protein of Wolbachia endosymbionts in filarial nematodes elicits immune responses through TLR2 and TLR4," *The Journal of Immunology*, vol. 173, no. 1, pp. 437–445, 2004.

[46] M. J. Taylor, H. F. Cross, and K. Bilo, "Inflammatory responses induced by the filarial nematode Brugia malayi are mediated by lipopolysaccharide-like activity from endosymbiotic Wolbachia bacteria," *The Journal of Experimental Medicine*, vol. 191, no. 8, pp. 1429–1436, 2000.

[47] E. Moreau and A. Chauvin, "Immunity against helminths: interactions with the host and the intercurrent infections," *Journal of Biomedicine and Biotechnology*, vol. 2010, Article ID 428593, 9 pages, 2010.

[48] Y. Gao, M. Zhang, L. Chen, M. Hou, M. Ji, and G. Wu, "Deficiency in TLR2 but not in TLR4 impairs dendritic cells derived IL-10 responses to schistosome antigens," *Cellular Immunology*, vol. 272, no. 2, pp. 242–250, 2012.

[49] D. S. Ashour, Z. S. Shohieb, and N. I. Sarhan, "Upregulation of Toll-like receptor 2 and nuclear factor-kappa B expression in experimental colonic schistosomiasis," *Journal of Advanced Research*, vol. 6, no. 6, pp. 877–884, 2015.

[50] G. Cheng, R. Luo, C. Hu, J. Cao, and Y. Jin, "Deep sequencing-based identification of pathogen-specific microRNAs in the plasma of rabbits infected with Schistosoma japonicum," *Parasitology*, vol. 140, no. 14, pp. 1751–1761, 2013.

[51] X. Wang, J. Wang, Y. Liang et al., "Schistosoma japonicum HSP60-derived peptide SJMHE1 suppresses delayed-type hypersensitivity in a murine model," *Parasites & Vectors*, vol. 9, 2016.

[52] X. Wang, W. Wang, and P. Wang, "Long-term effectiveness of the integrated schistosomiasis control strategy with emphasis on infectious source control in China: a 10-year evaluation from 2005 to 2014," *Parasitology Research*, vol. 116, no. 2, pp. 521–528, 2017.

[53] L. K. Scalfone, H. J. Nel, L. F. Gagliardo et al., "Participation of MyD88 and interleukin-33 as innate drivers of Th2 immunity to Trichinella spiralis," *Infection and Immunity*, vol. 81, no. 4, pp. 1354–1363, 2013.

[54] T. Mustelin, "Restless T cells sniff and go," *Science*, vol. 313, no. 5795, pp. 1902-1903, 2006.

[55] K. Nagpal, T. S. Plantinga, C. M. Sirois et al., "Natural loss-of-function mutation of myeloid differentiation protein 88 disrupts its ability to form myddosomes," *The Journal of Biological Chemistry*, vol. 286, no. 13, pp. 11875–11882, 2011.

[56] E. F. Kenny and L. A. J. O'Neill, "Signalling adaptors used by Toll-like receptors: an update," *Cytokine*, vol. 43, no. 3, pp. 342–349, 2008.

[57] P. Manna, M. Ghosh, J. Ghosh, J. Das, and P. C. Sil, "Contribution of nano-copper particles to in vivo liver dysfunction and cellular damage: Role of IκBα/NF-κB, MAPKs and mitochondrial signal," *Nanotoxicology*, vol. 6, no. 1, pp. 1–21, 2012.

[58] L. A. J. O'Neill, K. A. Fitzgerald, and A. G. Bowie, "The Toll-IL-1 receptor adaptor family grows to five members," *Trends in Immunology*, vol. 24, no. 6, pp. 286–289, 2003.

[59] J. Rzepecka, S. Rausch, C. Klotz et al., "Calreticulin from the intestinal nematode Heligmosomoides polygyrus is a Th2-skewing protein and interacts with murine scavenger receptor-A," *Molecular Immunology*, vol. 46, no. 6, pp. 1109–1119, 2009.

[60] H. J. McSorley, J. P. Hewitson, and R. M. Maizels, "Immunomodulation by helminth parasites: defining mechanisms and mediators," *International Journal for Parasitology*, vol. 43, no. 3-4, pp. 301–310, 2013.

[61] M. W. Robinson, J. P. Dalton, B. A. O'Brien, and S. Donnelly, "Fasciola hepatica: The therapeutic potential of a worm secretome," *International Journal for Parasitology*, vol. 43, no. 3-4, pp. 283–291, 2013.

[62] M. Whelan, M. M. Harnett, K. M. Houston, V. Patel, W. Harnett, and K. P. Rigley, "A filarial nematode-secreted product signals dendritic cells to acquire a phenotype that drives development of Th2 cells," *The Journal of Immunology*, vol. 164, no. 12, pp. 6453–6460, 2000.

[63] H. S. Goodridge, S. McGuiness, K. M. Houston et al., "Phosphorylcholine mimics the effects of ES-62 on macrophages and dendritic cells," *Parasite Immunology*, vol. 29, no. 3, pp. 127–137, 2007.

[64] T. G. Moreels, R. J. Nieuwendijk, J. G. De Man et al., "Concurrent infection with Schistosoma mansoni attenuates inflammation induced changes in colonic morphology, cytokine levels, and smooth muscle contractility of trinitrobenzene sulphonic acid induced colitis in rats," *Gut*, vol. 53, no. 1, pp. 99–107, 2004.

[65] C. M. Kane, L. Cervi, J. Sun et al., "Helminth antigens modulate TLR-initiated dendritic cell activation," *The Journal of Immunology*, vol. 173, no. 12, pp. 7454–7461, 2004.

[66] M. Ritter, O. Gross, S. Kays et al., "Schistosoma mansoni triggers Dectin-2, which activates the Nlrp3 inflammasome and alters adaptive immune responses," *Proceedings of the National Acadamy of Sciences of the United States of America*, vol. 107, no. 47, pp. 20459–20464, 2010.

[67] K. Brännström, M. E. Sellin, P. Holmfeldt, M. Brattsand, and M. Gullberg, "The Schistosoma mansoni protein Sm16/SmSLP/SmSPO-1 assembles into a nine-subunit oligomer with potential To inhibit Toll-like receptor signaling," *Infection and Immunity*, vol. 77, no. 3, pp. 1144–1154, 2009.

[68] D. J. Dowling, C. M. Hamilton, S. Donnelly et al., "Major secretory antigens of the Helminth Fasciola hepatica activate a suppressive dendritic cell phenotype that attenuates Th17 cells but fails to activate Th2 immune responses," *Infection and Immunity*, vol. 78, no. 2, pp. 793–801, 2010.

[69] M. Okano, A. R. Satoskar, K. Nishizaki, M. Abe, and D. A. Harn Jr., "Induction of Th2 responses and IgE is largely due to carbohydrates functioning as adjuvants on Schistosoma mansoni egg antigens," *The Journal of Immunology*, vol. 163, no. 12, pp. 6712–6717, 1999.

[70] H. S. Goodridge, F. A. Marshall, K. J. Else et al., "Immunomodulation via novel use of TLR4 by the filarial nematode phosphorylcholine-containing secreted product, ES-62," *The Journal of Immunology*, vol. 174, no. 1, pp. 284–293, 2005.

[71] I. Ludwig-Portugall and L. E. Layland, "Tlrs, Treg, and B cells, an interplay of regulation during helminth infection," *Frontiers in Immunology*, vol. 3, article 8, 2012.

[72] F. Gondorf, A. Berbudi, B. C. Buerfent et al., "Chronic filarial infection provides protection against bacterial sepsis by functionally reprogramming macrophages," *PLoS Pathogens*, vol. 11, no. 1, Article ID e1004616, pp. 1–27, 2015.

[73] I. Martin, K. Cabán-Hernández, O. Figueroa-Santiago, and A. M. Espino, "Fasciola hepatica fatty acid binding protein inhibits TLR4 activation and suppresses the inflammatory cytokines induced by lipopolysaccharide in vitro and in vivo," *The Journal of Immunology*, vol. 194, no. 8, pp. 3924–3936, 2015.

[74] R. J. Eason, K. S. Bell, F. A. Marshall et al., "The helminth product, ES-62 modulates dendritic cell responses by inducing the selective autophagolysosomal degradation of TLR-transducers, as exemplified by PKCδ," *Scientific Reports*, vol. 6, Article ID 37276, 2016.

[75] E. Candolfi, C. A. Hunter, and J. S. Remington, "Mitogen- and antigen-specific proliferation of T cells in murine toxoplasmosis is inhibited by reactive nitrogen intermediates," *Infection and Immunity*, vol. 62, no. 5, pp. 1995–2001, 1994.

[76] W. J. Dai and B. Gottstein, "Nitric oxide-mediated immunosuppression following murine Echinococcus multilocularis infection," *The Journal of Immunology*, vol. 97, no. 1, pp. 107–116, 1999.

[77] K. W. Schleifer and J. M. Mansfield, "Suppressor macrophages in African trypanosomiasis inhibit T cell proliferative responses by nitric oxide and prostaglandins," *The Journal of Immunology*, vol. 151, no. 10, pp. 5492–5503, 1993.

[78] S. J. Jenkins, J. P. Hewitson, S. Ferret-Bernard, and A. P. Mountford, "Schistosome larvae stimulate macrophage cytokine production through TLR4-dependent and -independent pathways," *International Immunology*, vol. 17, no. 11, pp. 1409–1418, 2005.

[79] P. Velupillai, W. E. Secor, A. M. Horauf, and D. A. Harn, "B-1 Cell (CD5+B220+) outgrowth in murine schistosomiasis is genetically restricted and is largely due to activation by polylactosamine sugars," *The Journal of Immunology*, vol. 158, no. 1, pp. 338–344, 1997.

[80] D. Van der Kleij, E. Latz, J. F. H. M. Brouwers et al., "A novel host-parasite lipid cross-talk. Schistosomal lyso-phosphatidylserine activates toll-like receptor 2 and affects immune polarization," *The Journal of Biological Chemistry*, vol. 277, no. 50, pp. 48122–48129, 2002.

[81] T. M. Watters, E. F. Kenny, and L. A. J. O'Neill, "Structure, function and regulation of the Toll/IL-1 receptor adaptor proteins," *Immunology & Cell Biology*, vol. 85, no. 6, pp. 411–419, 2007.

[82] L. A. J. O'Neill and A. G. Bowie, "The family of five: TIR-domain-containing adaptors in Toll-like receptor signalling," *Nature Reviews Immunology*, vol. 7, no. 5, pp. 353–364, 2007.

[83] M. A. West, A. R. Prescott, M. C. Kui et al., "TLR ligand-induced podosome disassembly in dendritic cells is ADAM17 dependent," *The Journal of Cell Biology*, vol. 182, no. 5, pp. 993–1005, 2008.

[84] L. A. J. O'Neill, "The interleukin-1 receptor/Toll-like receptor superfamily: 10 Years of progress," *Immunological Reviews*, vol. 226, no. 1, pp. 10–18, 2008.

[85] B. A. Beutler, "TLRs and innate immunity," *Blood*, vol. 113, no. 7, pp. 1399–1407, 2009.

[86] J. I. Alvarez, "Inhibition of Toll like receptor immune responses by microbial pathogens," *Frontiers in Bioscience*, vol. 10, pp. 582–587, 2005.

[87] H. Wesche, W. J. Henzel, W. Shillinglaw, S. Li, and Z. Cao, "MyD88: an adapter that recruits IRAK to the IL-1 receptor complex," *Immunity*, vol. 7, no. 6, pp. 837–847, 1997.

[88] D. De Nardo, "Toll-like receptors: Activation, signalling and transcriptional modulation," *Cytokine*, vol. 74, no. 2, pp. 181–189, 2015.

[89] M. Yamamoto, S. Sato, H. Hemmi et al., "Role of adaptor TRIF in the MyD88-independent toll-like receptor signaling pathway," *Science*, vol. 301, no. 5633, pp. 640–643, 2003.

[90] K. A. Fitzgerald, D. C. Rowe, B. J. Barnes et al., "LPS-TLR4 signaling to IRF-3/7 and NF-κB involves the toll adapters TRAM and TRIF," *The Journal of Experimental Medicine*, vol. 198, no. 7, pp. 1043–1055, 2003.

[91] S. Delhalle, R. Blasius, M. Dicato, and M. Diederich, "A beginner's guide to NF-κB signaling pathways," *Annals of the New York Academy of Sciences*, vol. 1030, pp. 1–13, 2004.

[92] R. T. Semnani, H. Sabzevari, R. Iyer, and T. B. Nutman, "Filarial antigens impair the function of human dendritic cells during differentiation," *Infection and Immunity*, vol. 69, no. 9, pp. 5813–5822, 2001.

[93] S. Agrawal, A. Agrawal, B. Doughty et al., "Cutting edge: different Toll-like receptor agonists instruct dendritic cells to induce distinct Th responses via differential modulation of extracellular signal-regulated kinase-mitogen-activated protein kinase and c-Fos," *The Journal of Immunology*, vol. 171, no. 10, pp. 4984–4989, 2003.

[94] J. Correale and M. Farez, "Helminth antigens modulate immune responses in cells from multiple sclerosis patients through TLR2-dependent mechanisms," *The Journal of Immunology*, vol. 183, no. 9, pp. 5999–6012, 2009.

[95] L. E. Layland, R. Rad, H. Wagner, and C. U. Prazeres da Costa, "Immunopathology in schistosomiasis is controlled by antigen-specific regulatory T cells primed in the presence of TLR2," *European Journal of Immunology*, vol. 37, no. 8, pp. 2174–2184, 2007.

[96] R. T. Semnani and T. B. Nutman, "Toward an understanding of the interaction between filarial parasites and host antigen-presenting cells," *Immunological Reviews*, vol. 201, pp. 127–138, 2004.

[97] G. Zhang and S. Ghosh, "Negative regulation of toll-like receptor-mediated signaling by Tollip," *The Journal of Biological Chemistry*, vol. 277, no. 9, pp. 7059–7065, 2002.

[98] A. G. Hise, K. Daehnel, I. Gillette-Ferguson et al., "Innate immune responses to endosymbiotic Wolbachia bacteria in Brugia malayi and Onchocerca volvulus are dependent on TLR2, TLR6, MyD88, and Mal, but not TLR4, TRIF, or TRAM," *The Journal of Immunology*, vol. 178, no. 2, pp. 1068–1076, 2007.

[99] J. Paul, K. Naskar, S. Chowdhury, N. Alam, T. Chakraborti, and T. De, "TLR4-mediated activation of MyD88 signaling induces protective immune response and IL-10 down-regulation in Leishmania donovani infection," *Indian Journal of Biochemistry and Biophysics*, vol. 51, no. 6, pp. 531–541, 2014.

[100] M. Gangloff, C. J. Arnot, M. Lewis, and N. J. Gay, "Functional insights from the crystal structure of the N-terminal domain of the prototypical toll receptor," *Structure*, vol. 21, no. 1, pp. 143–153, 2013.

[101] T. Imanishi, C. Ishihara, M. E. L. S. G. A. Badr et al., "Nucleic acid sensing by T cells initiates Th2 cell differentiation," *Nature Communications*, vol. 5, 2014.

[102] L. E. P. M. Van Der Vlugt, S. Haeberlein, W. De Graaf, T. E. D. Martha, and H. H. Smits, "Toll-like receptor ligation for the induction of regulatory B cells," *Methods in Molecular Biology*, vol. 1190, pp. 127–141, 2014.

[103] E. Panther, M. Idzko, S. Corinti et al., "The influence of lysophosphatidic acid on the functions of human dendritic cells," *The Journal of Immunology*, vol. 169, no. 8, pp. 4129–4135, 2002.

[104] Y. Y. M. Yang, X. H. Li, K. Brzezicka et al., "Specific anti-glycan antibodies are sustained during and after parasite clearance in Schistosoma japonicum-infected rhesus macaques," *PLOS Neglected Tropical Diseases*, vol. 11, no. 2, Article ID e0005339, 2017.

[105] G. F. Späth, L. A. Garraway, S. J. Turco, and S. M. Beverley, "The role(s) of lipophosphoglycan (LPG) in the establishment of Leishmania major infections in mammalian hosts," *Proceedings of the National Acadamy of Sciences of the United States of America*, vol. 100, no. 16, pp. 9536–9541, 2003.

[106] S. Srivastava, S. P. Pandey, M. K. Jha, H. S. Chandel, and B. Saha, "Leishmania expressed lipophosphoglycan interacts with Toll-like receptor (TLR)-2 to decrease TLR-9 expression and reduce anti-leishmanial responses," *Clinical & Experimental Immunology*, vol. 172, no. 3, pp. 403–409, 2013.

15

Checklists of Parasites of Farm Fishes of Babylon Province, Iraq

Furhan T. Mhaisen[1] and Abdul-Razzak L. Al-Rubaie[2]

[1]*Tegnervägen 6B, 641 36 Katrineholm, Sweden*
[2]*Department of Biological Control Technology, Al-Musaib Technical College, Al-Furat Al-Awsat Technical University, Al-Musaib, Iraq*

Correspondence should be addressed to Furhan T. Mhaisen; mhaisenft@yahoo.co.uk

Academic Editor: José F. Silveira

Literature reviews of all references concerning the parasitic fauna of fishes in fish farms of Babylon province, middle of Iraq, showed that a total of 92 valid parasite species are so far known from the common carp (*Cyprinus carpio*), the grass carp (*Ctenopharyngodon idella*), and the silver carp (*Hypophthalmichthys molitrix*) as well as from three freshwater fish species (*Carassius auratus*, *Liza abu*, and *Heteropneustes fossilis*) which were found in some fish farms of the same province. The parasitic fauna included one mastigophoran, three apicomplexans, 13 ciliophorans, five myxozoans, five trematodes, 45 monogeneans, five cestodes, three nematodes, two acanthocephalans, nine arthropods, and one mollusc. The common carp was found to harbour 81 species of parasites, the grass carp 30 species, the silver carp 28 species, *L. abu* 13 species, *C. auratus* one species, and *H. fossilis* one species. A host-parasite list for each fish species was also provided.

1. Introduction

Although fish farming in Iraq started in 1955 with a small pond in Al-Zaafaraniya, south of Baghdad city [1], an advance was achieved in fish farming industry in Iraq during the seventies and early eighties of the last century when many fish farms were established especially in the middle of Iraq [2]. However, such achievement was hindered due to consequences of the war situations during 1980–1988 and 1991 as well as the economic sanction imposed by the UN against Iraq on August 6, 1990. During the last few years, a great advance was achieved in fish farming in general and fish cages in particular due to the increasing demand on fish protein as well as the increasing investment in fish-culture industry in most provinces of Iraq. According to the statistics, a total of 441 working fish farms are scattered in Iraq [3]. Of these farms, a total of 72 working fish farms are situated in Babylon province alone with a water area of 44.5% of the total water area of fish farms in Iraq.

Under extensive fish culture and inadequate administrative and control measures, fish farms are vulnerable to great hazards due to the infection with parasites and other disease agents [2, 4, 5]. Many parasite species can easily spread among fishes suffering from crowd and bad managements, especially those parasites with direct life cycles [6].

In connection with the parasites of cultured fishes of Babylon province, Mhaisen et al. [7] surveyed the literature on the parasitic fauna of fishes of Al-Furat Fish Farm (previously known as Babylon Fish Farm), which is the biggest fish farm in Babylon province, and showed that the parasitic fauna of fishes of that farm included 60 valid parasite species (10 protozoans, three myxozoans, one trematode, 29 monogeneans, five cestodes, three nematodes, two acanthocephalans, six crustaceans, and one mollusc larva). In addition, some investigations on other fish farms in Babylon province were done. Such data are scattered in different local journals, unpublished theses, and few other sources. Therefore, the present paper was aimed at gathering data from the literature concerning all fish farms of Babylon province and providing a list of parasite species according to their major groups as well as a host-parasite list for cultured fishes of these farms and some other fish species found in such farms. Such parasite list will help owners of fish farms and fish veterinarians to know what sort of parasites are found in their fish farms, which will help them later in taking appropriate measures for their control.

2. Sources and Methods

A total of 50 references (33 published articles, 12 unpublished theses, two unpublished reports, one book, one conference abstract, and one review article) dealing with the parasites of farm fishes of Babylon province were used to prepare the present paper. Data from such references was gathered to provide host-parasite and parasite-host lists. The systematic account of these parasites is based on some electronic sites [8–12] as well as some taxonomic references [13–16].

The index-catalogue of parasites and disease agents of fishes of Iraq [17] was used to indicate the total number of fish hosts harbouring each parasite species in the whole waters of Iraq.

3. Parasitological Investigations Achieved on Fish Farms of Babylon Province

Few fish farms in Babylon province were surveyed for some parasitic infections. Fishes from Al-Furat Fish Farm received the greatest attention in this respect. So far 25 chronologically arranged references [18–42] were concerned with the parasitic fauna of Al-Furat Fish Farm. Only seven references [43–49] were concerned with the parasitic fauna of Al-Shark Al-Awsat Fish Farm. The literature concerned with fish parasites of other farms in Babylon province included those from Al-Latifiya Fish Farm [6, 50–52]; Al-Bajaa Fish Farm [26]; Abdul-Razzak Al-Janabi Fish Farm; Fawzi Al-Janabi Fish Farm and Ali Al-Hayali Fish Farm [53]; three fish farms at Al-Iskandariya district: Abdul-Hadi Al-Matloob Fish Farm, Hussain Al-Gaiem Fish Farm, and Maki Chinak Fish Farm [54]; Technical Institute of Al-Musaib Fish Farm [55]; and Al-Manahil (Al-Bilad) Fish Farm at Al-Iskandariya district [56]. In addition, surveys were done from some unnamed fish ponds such as those at Al-Mahaweel district [57], Al-Musaib district [58], Al-Iskandariya district [59], and Sadat Al-Hindiya district [60] as well as some other unnamed farms in the province [61–65].

4. Results and Discussion

Surveying the literature concerning the parasites so far recorded from fish farms of Babylon province showed the presence of 92 parasite species. These parasites included one mastigophoran, three apicomplexans, 13 ciliophorans, five myxozoans, five trematodes, 45 monogeneans, five cestodes, three nematodes, two acanthocephalans, nine arthropods, and one mollusc. The common carp was found to harbour 81 species of parasites, the grass carp 30 species, the silver carp 28 species, *L. abu* 13 species, *C. auratus* one species, and *H. fossilis* one species. The layout and names of the major taxonomic groups (phyla and classes) followed a checklist of an FAO Fisheries Technical Paper [66]. These major groups represent the concerned phyla of the parasites, but due to the great numbers of parasite species of the phylum Platyhelminthes, its three classes (Trematoda, Monogenea, and Cestoda) were applied in addition to their phylum.

5. Major Groups of Parasitic Fauna: Parasite-Host List

The parasite-host list is arranged in the major groups (phyla or classes) of parasitic fauna according to Kirjušina and Vismanis [66]. For each major group, a list of species together with their hosts and concerned references is given. To economize space, names of fish farms are not given here as they can be easily detected from the previous subtitle "Parasitological Investigations Achieved on Fish Farms of Babylon Province." Also the systematic account of all major groups is given down to the specific name of all parasites. For each parasite species, all records in farm fishes in Babylon province are given together with the first record of each concerned parasite in Iraq as well as the present number of all hosts so far known in Iraq for each concerned species based on the index-catalogue of parasites and disease agents of fishes of Iraq [17].

5.1. Phylum Mastigophora. The phylum Mastigophora is represented in farm fishes of Babylon province with only one parasite species of the genus *Ichthyobodo*. The systematic account of this parasite, followed by parasite-host list, is given here.

Phylum Mastigophora

　Class Kinetoplastidea

　　Order Kinetoplastida

　　　Family Bodonidae

　　　　Ichthyobodo necator (Henneguy, 1884) Pinto, 1928

Ichthyobodo necator (Henneguy, 1884) Pinto, 1928, was erroneously reported as *Costia necatrix* from the skin and gills of *C. carpio* [27]. The first record of *C. necatrix* in Iraq was from body surface of *H. fossilis* from Al-Ashar Canal at Basrah [67]. Seven fish host species are so far known for this parasite (as *C. necatrix*) in Iraq [17].

5.2. Phylum Apicomplexa. The phylum Apicomplexa, which is known as phylum Myzozoa according to WoRMS [12], is represented in farm fishes of Babylon province with three species; two of them belonged to the genus *Eimeria* and one unspecified species to the genus *Haemogregarina*.

Phylum Apicomplexa

　Class Sporozoa

　　Order Eucoccidiorida

　　　Family Eimeriidae

　　　　Eimeria dogieli (Dogiel, 1948) Pellerdy, 1963

　　　　Eimeria mylopharyngodoni Chen, 1956

　　　Family Haemogregarinidae

　　　　Haemogregarina sp.

Eimeria dogieli (Dogiel, 1948) Pellerdy, 1963, was recorded from the intestine of *C. carpio* [27]. So far, this is the only record of *E. dogieli* from fishes of Iraq [17].

Eimeria mylopharyngodoni Chen, 1956, was recorded from *C. carpio* [49]. The specific name was misspelled as *mylopharyngodon* and no authority, site of infection, parasite description, and illustration were given for this parasite by Hussain et al. [49]. So this record is considered as questionable especially if we take in consideration that Hussain et al. [49] examined *C. carpio* externally while *E. mylopharyngodoni* is known to infect intestine, kidneys, and liver of fishes [68]. This is the only record of *E. mylopharyngodoni* from fishes of Iraq [17].

Haemogregarina sp. was found on gills of *C. carpio* [49]. In Iraq, three species of *Haemogregarina* were so far recorded from blood of three fish species in Basrah province only [17]. So we think that the record of *Haemogregarina* sp. from gills of *C. carpio* by Hussain et al. [49] with neither description nor a good illustration is considered as questionable.

5.3. Phylum Ciliophora. The phylum Ciliophora is represented in farm fishes of Babylon province with 13 species, three of which belonged to the genera *Chilodonella*, *Ichthyophthirius*, and *Tripartiella*, five to the genus *Apiosoma*, and four to the genus *Trichodina* in addition to unspecified species of the genus *Trichodina*.

Phylum Ciliophora

    Class Kinetophragminophorea

        Order Cyrtophorida

            Family Chilodonellidae

                Chilodonella cyprini (Moroff, 1902) Strand, 1928

    Class Oligohymenophorea

        Order Hymenostomatida

            Family Ichthyophthiriidae

                Ichthyophthirius multifiliis Fouquet, 1876

        Order Petrichida

            Family Epistylididae

                Apiosoma amoebae (Grenfell, 1887) Lom, 1966
                Apiosoma cylindriformis (Chen, 1955)
                Apiosoma minuta (Chen, 1961) Lom, 1966
                Apiosoma piscicola Blanchard, 1885
                Apiosoma poteriformis (Timofeev, 1962) Lom, 1966

        Order Mobilida

            Family Trichodinidae

                Trichodina cottidarum Dogiel, 1948
                Trichodina domerguei (Wallengren, 1897)

Trichodina gracilis Polyanskii, 1955
Trichodina nigra Lom, 1960
Trichodina sp.
Tripartiella amurensis (Chan, 1961)

Chilodonella cyprini (Moroff, 1902) Strand, 1928, was recorded from skin, buccal cavity, and gills of *C. idella* [27, 29, 43, 51], skin, fins, buccal cavity, and gills of *C. carpio* [27, 30, 43, 51, 53], and skin and gills of *H. molitrix* [27, 43, 51]. The first record of *C. cyprini* in Iraq was from skin, buccal cavity, and gills of *Mystus pelusius* from Tigris River at Baghdad [69]. So far 13 fish host species are known for this parasite in Iraq [17].

Ichthyophthirius multifiliis Fouquet, 1876, was recorded from skin and gills of *C. idella* [27, 29, 43, 51, 56], skin, fins, and gills of *C. carpio* [25, 27, 29, 30, 35, 43, 45, 47, 49, 51, 53–56, 59, 60, 65], and skin and gills of *H. molitrix* [27, 29, 45, 47, 51]. The first record of *I. multifiliis* in Iraq was from the skin and gills of *Chelon subviridis* (reported as *Mugil dussumieri*) from Tigris River near Baghdad [70]. So far this parasite has 35 fish host species in Iraq [17].

Apiosoma amoebae (Grenfell, 1887) Lom, 1966, was reported from skin, buccal cavity, and gills of *C. idella* [20, 24, 27], skin and buccal cavity of *C. carpio* [27, 55], and gills and buccal cavity of *H. molitrix* [20, 27]. It is appropriate to mention here that *A. amoebae* was reported as *Glossatella amoebae* [27, 55]. The first record of *A. amoebae* in Iraq was from the skin, buccal cavity, and gills of *C. idella* from Babylon Fish Farm [20]. So far this parasite has five fish host species in Iraq [17].

Apiosoma cylindriformis (Chen, 1955) was reported from gills of *C. idella* [20, 24, 27], buccal cavity and gills of *C. carpio* [27], and gills of *H. molitrix* [20, 21, 27]. Al-Zubaidy [27] reported *A. cylindriformis* as *Glossatella cylindriformis*. The first record of *A. cylindriformis* in Iraq was from gills of *C. idella* and *H. molitrix* from Babylon Fish Farm [20]. So far, *A. cylindriformis* has seven fish host species in Iraq [17].

Apiosoma minuta (Chen, 1961) Lom, 1966, was recorded from skin of *C. carpio* [45, 47]. This was its first record in Iraq. So far, *A. minuta* has two fish host species in Iraq as it was recently recorded from *Luciobarbus xanthopterus* by Al-Salmany [71].

Apiosoma piscicola Blanchard, 1885, was recorded from skin, buccal cavity, and gills of *C. idella* [27, 51], skin, buccal cavity, and gills of *C. carpio* [27, 29, 51, 55], and buccal cavity and gills of *H. molitrix* [20, 21, 27, 29, 51]. It is appropriate to mention here that *A. piscicola* was reported as *Glossatella piscicola* [27]. *A. piscicola* was recorded for the first time in Iraq from skin, buccal cavity, and gills of *C. idella*, *C. carpio* and *H. molitrix* from Al-Suwaira and Al-Latifiya Fish Farms [51]. So far, this parasite has ten fish host species in Iraq [17].

Apiosoma poteriformis (Timofeev, 1962) Lom, 1966, was recorded from skin and gills of *C. idella* [20, 24, 27] and buccal cavity and gills of *C. carpio* [27]. It is appropriate to mention here that *A. poteriformis* was reported as *Glossatella poteriformis* by Al-Zubaidy [27]. *A. poteriformis* was recorded for the first time in Iraq from gills of *C. idella* from Babylon Fish Farm [20]. So far, this parasite has three fish host species in Iraq [17].

Trichodina cottidarum Dogiel, 1948, was recorded from skin and gills of *C. carpio* [45, 47, 49, 54, 59, 60] and skin and gills of *H. molitrix* [45, 47]. *T. cottidarum* was recorded for the first time in Iraq from gills of *C. carpio* from a manmade lake at Baghdad city [72]. So far, this parasite has 13 fish host species in Iraq [17].

Trichodina domerguei (Wallengren, 1897) was recorded from skin, fins, buccal cavity, and gills of *C. idella* [24, 27, 29, 43, 51, 56], skin, fins, buccal cavity, and gills of *C. carpio* [25, 27, 29, 30, 35, 43, 51, 53, 55, 56, 59], skin, buccal cavity, and gills of *H. molitrix* [20, 27, 51], and skin and gills of *L. abu* [23, 56]. The first record of *T. domerguei* in Iraq was from skin, fins, and gills of eight freshwater fish species from Tigris River, Al-Tharthar Lake, and fish markets in Baghdad city [73]. So far, *T. domerguei* has 39 host species in Iraq [17] and, therefore, it is the most prevalent ciliate species among fishes of Iraq.

Trichodina gracilis Polyanskii, 1955, was recorded from skin of *C. carpio* [45, 47]. This was its first record in Iraq. So far, it has three fish host species in Iraq [17].

Trichodina nigra Lom, 1960, was recorded from skin and gills of the three carp species: *C. idella* [29], *C. carpio* [27, 29, 35, 49, 55, 59, 60], and *H. molitrix* [27, 29]. The first record of *T. nigra* in Iraq was from skin and gills of *C. carpio* and gills of *H. molitrix* from Babylon Fish Farm [27]. So far, *T. nigra* has nine host species in Iraq [17].

Unidentified specimen of *Trichodina* was recorded from *L. abu* [50] with no mention to site of infection. In addition to 24 recognized *Trichodina* species so far recorded from fishes of Iraq, some unidentified species of *Trichodina* were so far recorded from six fish species [17].

Tripartiella amurensis (Chan, 1961) was recorded from skin of *C. carpio* [45, 47]. This was its first record in Iraq. No more records are so far known for *T. amurensis* in Iraq [17].

5.4. Phylum Myxozoa. The phylum Myxozoa is represented in farm fishes of Babylon province with five species: one species belonged to the genus *Myxobilatus* and four species to the genus *Myxobolus*.

Phylum Myxozoa

 Class Myxosporea

  Order Bivalvulida

   Family Sphaerosporidae
    Myxobilatus legeri (Cépède, 1905)
   Family Myxobolidae
    Myxobolus dogieli Bykhovskaya-Pavlovskaya & Bykhovski, 1940
    Myxobolus muelleri Bütschli, 1882
    Myxobolus oviformis Thélohan, 1892
    Myxobolus pfeifferi Thélohan, 1895

Myxobilatus legeri (Cépède, 1905), erroneously reported as *Myxobllatus legerl*, with no given authority, description, and illustration, was recorded from skin and gills of *C. carpio* [54]. This was the only record of *M. legeri* in Iraq [17].

Myxobolus dogieli Bykhovskaya-Pavlovskaya & Bykhovski, 1940, was recorded from kidneys and gallbladder of *L. abu* [56]. The first record of *M. dogieli* in Iraq was mainly from the external surface of heart, liver, and ovaries of *L. abu* from Tigris River at Baiji town [74]. So far, *M. dogieli* has nine host species in Iraq [17].

Myxobolus muelleri Bütschli, 1882, was recorded from intestine and liver of *C. carpio* [27] with the specific name spelled as *mülleri*. The first record of *M. muelleri* in Iraq was from gills of *Luciobarbus xanthopterus*, reported as *B. xanthopterus* [70]. So far, *M. muelleri* has eight host species in Iraq [17].

Myxobolus oviformis Thélohan, 1892, was recorded from skin, intestine, and kidneys of *C. carpio* [27, 54, 59]. The first report of *M. oviformis* in Iraq was from gill arches and heart of four fish species [70]. So far, *M. oviformis* has 20 fish host species in Iraq [17].

Myxobolus pfeifferi Thélohan, 1895, was recorded from gills, intestine, liver, kidneys, and gallbladder of *C. idella* [29], gills, gallbladder, intestine, kidneys, and liver of *C. carpio* [25, 27, 29, 53, 59], gills, liver, and intestine of *H. molitrix* [27, 29], and gills, intestinal wall, and gonads of *L. abu* [23]. The first report of *M. pfeifferi* in Iraq was from gills of *Acanthobrama marmid* from Tigris River at Mosul city [75]. So far, *M. pfeifferi* is the prevalent myxozoan among fishes of Iraq as it has 35 fish host species [17].

5.5. Phylum Platyhelminthes: Class Trematoda. The class Trematoda of the phylum Platyhelminthes is represented in farm fishes of Babylon province with five species; two species belonged to the genera *Apharyngostrigea* and *Ascocotyle* and three species to the genus *Diplostomum*.

Phylum Platyhelminthes

 Class Trematoda

  Order Diplostomida

   Family Strigeidae
    Apharyngostrigea cornu (Zeder, 1800)
   Family Diplostomidae
    Diplostomum indistinctum (Guberlet, 1923) Hughes, 1929
    Diplostomum paraspathaceum Schigin, 1965
    Diplostomum spathaceum (Rudolphi, 1819) Olsson, 1876

  Order Plagiorchiida

   Family Heterophyidae
    Ascocotyle coleostoma (Looss, 1896) Looss, 1899

Apharyngostrigea cornu (Zeder, 1800) was recorded as metacercaria from mesentery, coelom and liver of *C. carpio* [52]. This was its first report in Iraq. No more records are so far known on the occurrence of *A. cornu* from fishes of Iraq [17]. The adult worm of this parasite was detected from the intestine of the purple heron *Ardea purpurea* in Bahr Al-Najaf Depression [76].

Ascocotyle coleostoma (Looss, 1896) Looss, 1899, was recorded as metacercaria from gills of *H. fossilis* [56]. This parasite was reported for the first time in Iraq from gills of *H. fossilis* and *L. abu* from Diyala River [77]. *A. coleostoma* has so far 34 fish host species in Iraq [17]. The adult worm of *A. coleostoma* was detected from the grey heron *A. cinerea* in Babylon (now Al-Furat) Fish Farm [78].

Diplostomum indistinctum (Guberlet, 1923) Hughes, 1929, was recorded as metacercaria from eyes of *H. molitrix* [43]. The first occurrence of metacercariae of *D. indistinctum* was from eyes of *Luciobarbus esocinus*, reported as *B. esocinus* from fish market in Mosul city [79]. No more records are so far known on the occurrence of this parasite from fishes of Iraq [17].

Diplostomum paraspathaceum Schigin, 1965, was recorded as metacercaria from eyes of both *C. idella* [43] and *C. carpio* [43]. This was its first record in Iraq. No more hosts are so far known for this parasite in Iraq [17].

Diplostomum spathaceum (Rudolphi, 1819) Olsson, 1876, was recorded as metacercaria from eyes of the three carp species: *C. idella* [43, 56], *C. carpio* [29, 43], and *H. molitrix* [43, 56]. The first occurrence of metacercariae of *D. spathaceum* was from eyes of *C. luteus*, reported as *B. luteus*, *Cyprinion macrostomum*, and *C. carpio* from Dokan Lake [80]. Thirty-four hosts are so far known for this parasite in fishes of Iraq [17]. Metacercariae of *Diplostomum* spp. are responsible for worm cataract which causes fish blindness [81]. Adult worms of this parasite were found in the intestine of some fish-eating birds such as the silver gull *Larus argentatus* in Bahr Al-Najaf Depression [76].

5.6. Phylum Platyhelminthes: Class Monogenea. The class Monogenea of the phylum Platyhelminthes is represented in farm fishes of Babylon province with 45 species: 12 species of the genus *Gyrodactylus*, 26 species of *Dactylogyrus*, and one species of each of the genera *Pseudacolpenteron*, *Diplozoon*, *Eudiplozoon*, *Paradiplozoon*, and *Microcotyle* in addition to some unidentified species of *Dactylogyrus* and *Diplozoon*. It is appropriate to mention here that this group is considered as Monogenea by some electronic sites [9–12] but as Monogenoidea in some references [14, 66].

Phylum Platyhelminthes

Class Monogenea

Order Gyrodactylidea

Family Gyrodactylidae

Gyrodactylus baicalensis Bogolepova, 1950
Gyrodactylus ctenopharngodontis Ling in Gusev, 1952
Gyrodactylus elegans von Nordmann, 1832
Gyrodactylus kherulensis Ergens, 1974
Gyrodactylus macracanthus Hukuda, 1940

Gyrodactylus malmbergi Ergens, 1961
Gyrodactylus markevitschi Kulakovskaya, 1952
Gyrodactylus medius Kathariner, 1895
Gyrodactylus menschikowi Gvosdev, 1950
Gyrodactylus salaris Malmberg, 1957
Gyrodactylus sprostonae Ling, 1962
Gyrodactylus vicinus Bychowsky, 1957

Order Dactylogyridea

Family Dactylogyridae

Dactylogyrus achmerowi Gusev, 1955
Dactylogyrus amurensis Akhmerov, 1952
Dactylogyrus anchoratus (Dujardin, 1845) Wagener, 1857
Dactylogyrus arcuatus Yamaguti, 1942
Dactylogyrus barbioides Gusev, Ali, Abdul-Ameer, Amin & Molnár, 1993
Dactylogyrus cornu Linstow, 1878
Dactylogyrus crassus Kulwiec, 1927
Dactylogyrus ctenopharyngodonis Achmerow, 1952
Dactylogyrus dogieli Gusev, 1953
Dactylogyrus ergensi Molnár, 1964
Dactylogyrus extensus Mueller & Van Cleave, 1932
Dactylogyrus gobii Gvosdev, 1950
Dactylogyrus hypophthalmichthys Akhmerov, 1952
Dactylogyrus inexpectatus Izjumova, in Gusev, 1955
Dactylogyrus jamansajensis Osmanov, 1958
Dactylogyrus lamellatus Akhmerov, 1952
Dactylogyrus latituba Gusev, 1955
Dactylogyrus lopuchinae Jukhimenko, 1981
Dactylogyrus minutus Kulwiec, 1927
Dactylogyrus navicularis A. Gusev, 1955
Dactylogyrus phoxini Malevitskaia, 1949
Dactylogyrus propinquus Bychowsky, 1931
Dactylogyrus sahuensis Ling in Chen et al., 1973
Dactylogyrus simplex Bychowsky, 1936
Dactylogyrus skrjabini Akhmerov, 1954
Dactylogyrus vastator Nybelin, 1924
Dactylogyrus spp.
Pseudacolpenteron pavlovskii Bychowsky & Gussev, 1955

Order Mazocraeidea

Family Diplozoidae

Diplozoon paradoxum Nordmann, 1832

Diplozoon sp.

Eudiplozoon nipponicum (Goto, 1891)

Paradiplozoon barbi (Reichenbach-Klinke, 1951)

Family Microcotylidae

Microcotyle donavini van Beneden & Hesse, 1863

Gyrodactylus baicalensis Bogolepova, 1950, was recorded from skin, fins, and gills of *C. carpio* [25, 27, 29, 52]. The first report of *G. baicalensis* in Iraq was from skin, buccal cavity, and gills of *C. carpio* from Al-Suwaira and Al-Latifiya Fish Farms [52]. So far, *G. baicalensis* has eight fish host species in Iraq [17].

Gyrodactylus ctenopharngodontis Ling in Gusev, 1952, was recorded from skin, fins, buccal cavity, and gills of *C. idella* [24, 27]. The first report of *G. ctenopharngodontis* in Iraq was from gills of *C. idella* from Babylon Fish Farm [24]. No more hosts are so far recorded for this parasite in Iraq [17].

Gyrodactylus elegans von Nordmann, 1832, was recorded from skin, fins, buccal cavity, and gills of both *C. idella* [27, 29, 52] and *C. carpio* [25–27, 29, 30, 32, 33, 43, 50, 52, 53, 55, 56, 60] as well as from gills of *H. molitrix* [29] and from *L. abu* [50] with no mention to site of infection. The first report of *G. elegans* in Iraq was from *C. carpio* from Al-Zaafaraniya Fish Farm and *L. abu* from Al-Latifiya Fish Farm [50]. So far, *G. elegans* has 23 fish host species in Iraq [17].

Gyrodactylus kherulensis Ergens, 1974, was recorded from skin and gills of *C. idella* [27] and skin, fins, and gills of *C. carpio* [18, 27]. The first report of *G. kherulensis* in Iraq was from gills of *C. carpio* from Babylon Fish Farm [18]. So far, *G. kherulensis* has four fish host species in Iraq [17].

Gyrodactylus macracanthus Hukuda, 1940 (reported as *G. paralatus* Gusev, 1955), was recorded from skin and gills of *C. carpio* [27] and skin, fins, and buccal cavity of *H. molitrix* [27]. This was its first report in Iraq. Later on, it was reported from both hosts [82] as *G. paralatus* also. According to Gussev [83] and Pugachev et al. [14], *G. paralatus* is a synonym of *G. macracanthus*. No more hosts are so far known for *G. macracanthus* or its synonym *G. paralatus* from fishes of Iraq [17].

Gyrodactylus malmbergi Ergens, 1961, was recorded from skin, fins, and gills of *C. carpio* [27] and from skin and gills of *H. molitrix* [27]. This was its first report in Iraq and no more hosts are so far known for *G. malmbergi* from fishes of Iraq [17].

Gyrodactylus markevitschi Kulakovskaya, 1952, was recorded from skin and gills of *C. carpio* [27, 45, 47, 60]. The first report of this parasite in Iraq was from gills of *Capoeta trutta* (reported as *Varicorhinus trutta*) from Tigris River at Baiji town [74]. So far, *G. markevitschi* has six host species in Iraq [17].

Gyrodactylus medius Kathariner, 1895, was recorded from skin and fins of *C. carpio* [27]. This was its first record in Iraq. The year of authority of this parasite was erroneously given as 1893 instead of 1895 by Al-Zubaidy [27] as well as by three other references according to Mhaisen and Abdul-Ameer [15]. Also the authorship of this parasite was given as

Katheriner instead of Kathariner according to MonoDB [10]. Now, this parasite has three host species in Iraq [17].

Gyrodactylus menschikowi Gvosdev, 1950, was recorded from skin and gills of *C. carpio* [60]. The first report of this parasite in Iraq was from gills and skin of *C. carpio* and skin, fins, and gills of *L. abu* both from Hilla River [84]. So far, no more hosts for this parasite are known in Iraq [17].

Gyrodactylus salaris Malmberg, 1957, was recorded from skin and gills of *C. carpio* [27, 29]. The first report of this parasite in Iraq was from gills and skin of *C. carpio* from Al-Furat Fish Farm [27]. The year of authority of this parasite was reported as 1956 instead of 1957 by Al-Zubaidy [27] and Al-Jadoaa [29]. No more hosts for this parasite are so far known in Iraq [17].

Gyrodactylus sprostonae Ling, 1962, was recorded from skin and fins of *C. carpio* [27]. This was its first report in Iraq. Now, this parasite has seven host species in Iraq [17].

Gyrodactylus vicinus Bychowsky, 1957, was recorded from skin, fins, and gills of *C. carpio* [27, 35]. The first report of this parasite in Iraq was from skin, fins, and gills of *C. carpio* from Al-Furat Fish Farm [27]. Now, it has three host species in Iraq [17].

Finally, the unidentified *Gyrodactylus* species reported from *C. carpio* [34, 38] were the same 11 species which had been recorded in Al-Zubaidy [27]. In Iraq, so far 15 fish host species were reported for some unspecified *Gyrodactylus* species [17].

Dactylogyrus achmerowi Gusev, 1955, was recorded from gills of *C. carpio* [19, 27, 29, 30, 32, 33, 36, 54, 55, 60]. The first report of *D. achmerowi* in Iraq was from gills of *C. carpio* from Al-Wahda Fish Hatchery at Al-Suwaira and Babylon Fish Farm [19]. Now, it has 11 host species in Iraq [17].

Dactylogyrus amurensis Akhmerov, 1952, was recorded from gills of *C. carpio* [55]. This was its first report in Iraq. However, neither description and measurements nor illustration was given. So far, this parasite has two host species in Iraq [17].

Dactylogyrus anchoratus (Dujardin, 1845) Wagener, 1857, was recorded from gills of *C. carpio* [45, 47]. The first report and description of *D. anchoratus* in Iraq were from gills of *C. carpio* from Tigris River at Al-Zaafaraniya [85, 86]. Now, it has seven fish host species in Iraq [17].

Dactylogyrus arcuatus Yamaguti, 1942, was recorded from gills of *C. idella* [29] and skin, buccal cavity, and gills of *C. carpio* [25, 27, 35, 36, 49, 52, 55, 59, 60]. The first report of *D. arcuatus* in Iraq was from skin, buccal cavity, and gills of *C. carpio* from Al-Suwaira and Al-Latifiya Fish Farms [52]. Now, it has seven fish host species in Iraq [17].

Dactylogyrus barbioides Gusev, Ali, Abdul-Ameer, Amin & Molnár, 1993, was recorded from gills of *C. carpio* [60]. *D. barbioides* was described as a new species from gills of *Barbus grypus* from Tigris River near Baiji town [87]. Now, it has three fish host species in Iraq [17].

Dactylogyrus cornu Linstow, 1878, was recorded from gills of *C. carpio* [27, 59]. *D. cornu* was recorded for the first time in Iraq from gills of five fish species from Diyala River [88]. So far, it has 13 fish host species in Iraq [17].

Dactylogyrus crassus Kulwiec, 1927, was recorded from gills of *C. carpio* [45, 47, 55, 60]. *D. crassus* was recorded for the first time in Iraq from gills of *C. carpio* from Al-Shark Al-Awsat Fish Farm [45]. It is appropriate to mention here that the specific name *crassus* was misspelled as *carassus* by Al-Rubaie et al. [55]. No more hosts are so far known for this parasite in Iraq [17].

Dactylogyrus ctenopharyngodonis Achmerow, 1952, was recorded from gills of *C. idella* from Al-Shark Al-Awsat Fish Farm [43]. This is the only report on the occurrence of *D. ctenopharyngodonis* in fishes of Iraq [17].

Dactylogyrus dogieli Gusev, 1953, was recorded from gills of *C. carpio* in fish cages and an earthen pond in Sadat Al-Hindiya [60]. The first report of *D. dogieli* was from five fish species from Euphrates River at Al-Musaib city [89] and its full description and illustration were published later by Al-Sa'adi et al. [90]. Six host species are so far known for *D. dogieli* in Iraq [17].

Dactylogyrus ergensi Molnár, 1964, was recorded from gills of *C. carpio* from Al-Furat Fish Farm [27]. This was its first report in Iraq. No more hosts were reported for *D. ergensi* in Iraq [17].

Dactylogyrus extensus Mueller & Van Cleave, 1932, was recorded from gills of *C. idella* [29], buccal cavity and gills of *C. carpio* [25, 27, 29, 30, 32, 33, 36, 43, 45, 47, 52, 54, 55, 59, 60], and gills of *H. molitrix* [29, 45, 47]. The first report of *D. extensus* in Iraq was from the buccal cavity and gills of *C. carpio* from Al-Suwaira and Al-Latifiya Fish Farms [52]. *D. solidus* which was also recorded from the same host by Salih et al. [52] as well as by Mhaisen & Abul-Eis [25] and Al-Rubaie et al. [55] is considered as a synonym of *D. extensus* according to Gibson et al. [13]. *D. extensus* and its synonym *D. solidus* have so far 17 fish host species in Iraq [17].

Dactylogyrus gobii Gvosdev, 1950, was recorded from skin and gills of *C. carpio* [45, 47, 60]. *D. gobii* was recorded for the first time in Iraq from gills of *C. carpio* from Al-Shark Al-Awsat Fish Farm [45]. Now, it has three host species in Iraq [17].

Dactylogyrus hypophthalmichthys Akhmerov, 1952, was recorded from skin, buccal cavity, and gills of *H. molitrix* [21, 27, 43, 45, 47, 52, 56]. It is reliable to state here that *D. hypophthalmichthys* was reported as *Neodactylogyrus hypophthalmichthys* by Asmar et al. [56]. The first report of *D. hypophthalmichthys* in Iraq was from the buccal cavity and gills of *H. molitrix* from Al-Suwaira and Al-Latifiya Fish Farms [52]. *H. molitrix* is the only host so far known for *D. hypophthalmichthys* in Iraq [17].

Dactylogyrus inexpectatus Izjumova, in Gusev, 1955, was recorded from skin and gills of *C. idella* [27, 29, 52], gills of *C. carpio* [27, 29], and gills of *H. molitrix* [29]. The first report of *D. inexpectatus* in Iraq was from skin and gills of *C. idella* from Al-Suwaira and Al-Latifiya Fish Farms [52]. Now, it has five host species in Iraq [17].

Dactylogyrus jamansajensis Osmanov, 1958, was recorded from gills of *C. carpio* [45, 47]. The first report of *D. jamansajensis* in Iraq was from gills of *C. luteus* from manmade lakes, north of Baghdad [91]. Now, it has five host species in Iraq [17].

Dactylogyrus lamellatus Akhmerov, 1952, was recorded from skin, fins, buccal cavity, and gills of *C. idella* [24, 27, 43, 52, 56] and gills of *C. carpio* [27]. The first report of *D.*

lamellatus in Iraq was from the skin, buccal cavity, and gills of *C. idella* from Al-Suwaira and Al-Latifiya Fish Farms [52]. Now, it has three host species in Iraq [17].

Dactylogyrus latituba Gusev, 1955, was recorded from gills of both *C. idella* [27] and *C. carpio* [25, 27] as well as from the buccal cavity and gills of *H. molitrix* [27]. The first report of *D. latituba* in Iraq was from gills of *C. luteus* from manmade lakes, north of Baghdad [91]. Now, it has four host species in Iraq [17].

Dactylogyrus lopuchinae Jukhimenko, 1981, was recorded from gills of *C. carpio* [45, 47, 54]. *D. lopuchinae* was recorded for the first time in Iraq from gills of *C. carpio* from Al-Shark Al-Awsat Fish Farm [45]. No more hosts are so far known for this parasite in Iraq [17].

Dactylogyrus minutus Kulwiec, 1927, was recorded from skin, fins, and gills of *C. carpio* [27, 30, 32, 33, 45, 47, 55, 56, 59, 60]. The first report on this parasite in Iraq was from gills of *C. carpio* from Tigris River at Al-Zaafaraniya, south of Baghdad, and Al-Qadisia Dam Lake [85], while its description and illustration were given later by Mhaisen et al. [86]. So far, *D. minutus* has 12 fish species in Iraq [17].

Dactylogyrus navicularis A. Gusev, 1955, was recorded from fins, buccal cavity, and gills of *C. carpio* [27, 35, 60]. The first report of *D. navicularis* in Iraq was from the buccal cavity, fins, and gills of *C. carpio* from Al-Furat Fish Farm [27]. *C. carpio* is the only host so far known for *D. navicularis* in Iraq [17].

Dactylogyrus phoxini Malevitskaia, 1949, was recorded from skin and gills of *C. carpio* [45, 47, 60]. The first record of *D. phoxini* in Iraq was in June 1995 from gills of *C. carpio* from Tigris River at Al-Zaafaraniya but the report was published later by Balasem et al. [92]. This parasite has so far only two host species in Iraq [17].

Dactylogyrus propinquus Bychowsky, 1931, was recorded from gills of *C. carpio* [27, 35, 49]. The first report of *D. propinquus* in Iraq was from gills of *C. carpio* from Al-Furat Fish Farm [27]. *C. carpio* is the only host so far known for *D. propinquus* in Iraq [17].

Dactylogyrus sahuensis Ling in Chen et al., 1973, was recorded from fins and gills of *C. carpio* [27]. The first report of *D. sahuensis* in Iraq was from fins and gills of *C. carpio* from Al-Furat Fish Farm [27]. *C. carpio* is the only host so far known for *D. sahuensis* in Iraq [17].

Dactylogyrus simplex Bychowsky, 1936, was recorded from *C. carpio* [49] with no mention to site of infection. The first report of *D. simplex* in Iraq was from gills of *C. carpio* from the new fish farm of the Fish Research Center at Al-Zaafaraniya [93]. *D. simplex* has so far three host species in Iraq [17].

Dactylogyrus skrjabini Akhmerov, 1954, was recorded from buccal cavity and gills of *C. carpio* [27, 45, 47] and buccal cavity and gills of *H. molitrix* [21, 27, 52]. The first report of *D. skrjabini* in Iraq was from buccal cavity and gills of *H. molitrix* from Al-Suwaira and Al-Latifiya Fish Farms [52]. Now, *D. skrjabini* has six host species in Iraq [17].

Dactylogyrus vastator Nybelin, 1924, was recorded from gills of *C. idella* [27] and skin and gills of *C. carpio* [26, 27, 30, 32, 33, 35, 36, 43, 49, 52, 53, 56, 59, 60, 64]. The first report of *D. vastator* from Iraq was from skin and gills of *C. macrostomum* from Tigris River at Baghdad [94]. So far, *D. vastator* was reported from 33 fish host species from north, middle, and south of Iraq [17].

Unidentified *Dactylogyrus* species were recorded from skin and gills of *C. idella* [56] and from skin, buccal cavity, and gills of *C. carpio* [34, 38, 56]. Some of these specimens were larval stages [56], while the unidentified *Dactylogyrus* species of Al-Zubaidy et al. [34, 38] were the same 15 species which were recorded in Al-Zubaidy [27]. In Iraq, so far nine fish host species were reported for some unspecified *Dactylogyrus* species [17].

Pseudacolpenteron pavlovskii Bychowsky & Gussev, 1955, was recorded from fins and gills of *C. carpio* [25, 27, 45, 47] and gills of *H. molitrix* [27]. The first report of *P. pavlovskii* from Iraq was from skin and gills of *C. carpio* from Babylon Fish Farm [25]. This species has so far only two host species in Iraq [17].

Diplozoon paradoxum Nordmann, 1832, was recorded from gills of *C. carpio* [60]. This parasite was reported for the first time in Iraq from gills of *Carasobarbus luteus*, reported as *Barbus luteus*, from Al-Husainia creek, Karbala province [95]. Now, it has five fish hosts in Iraq [17].

Unidentified *Diplozoon* species were recorded from gills of *C. carpio* [43]. Some other unidentified *Diplozoon* species occurred as larvae in 12 fish host species in Iraq [17].

Eudiplozoon nipponicum (Goto, 1891) was recorded from gills of *C. carpio* [60]. This parasite was recorded for the first time in Iraq from gills of *C. carpio* from a manmade lake in Baghdad [96] as *Diplozoon nipponicum* but then it was reported by its valid name *E. nipponicum* by all subsequent researchers. So far, three host species are known for *E. nipponicum* in Iraq [17].

Paradiplozoon barbi (Reichenbach-Klinke, 1951) was recorded from gills of *C. carpio* [27] as *Diplozoon barbi*. This parasite was reported for the first time in Iraq from gills of *Chondrostoma nasus*, *C. regium*, and *C. carpio* from Tigris River at Baghdad [97] as *Diplozoon barbi*. Also, all the subsequent records in the Iraqi literature, except the checklists of Mhaisen and Abdul-Ameer [16], referred to this parasite as *D. barbi*. According to Khotenovsky [98], *D. barbi* is a synonym of *P. barbi*. Eight host species are so far known for this parasite in Iraq [17].

Microcotyle donavini van Beneden & Hesse, 1863, was recorded from gills of *L. abu* from Babylon Fish Farm [22]. This was its first report in Iraq. Ten host species are so far known for *M. donavini* in Iraq [17].

5.7. Phylum Platyhelminthes: Class Cestoda. The class Cestoda of the phylum Platyhelminthes is represented in farm fishes of Babylon province with five species: one species of each of the genera *Bothriocephalus*, *Ligula*, and *Neogryporhynchus* as well as two species of *Proteocephalus*.

Phylum Platyhelminthes

Class Cestoda

Order Bothriocephalidea

Family Bothriocephalidae

Bothriocephalus acheilognathi Yamaguti, 1934

Order Diphyllobothriidea

Family Diphyllobothriidae

Ligula intestinalis (L., 1758) Bloch, 1782

Order Proteocephalidea

Family Proteocephalidae

Proteocephalus osculatus (Goeze, 1782) Nybelin, 1942
Proteocephalus torulosus (Batsch, 1786) Nufer, 1905

Order Cyclophyllidea

Family Dilepididae

Neogryporhynchus cheilancristrotus (Wedl, 1855) Baer & Bona, 1960

Bothriocephalus acheilognathi Yamaguti, 1934, was recorded from the intestine of both *C. idella* [24, 56] and *C. carpio* [25–27, 52, 53, 62]. It is appropriate to mention here that this worm was reported by its synonym *B. opsariichthydis* by Salih et al. [52] and Al-Zubaidy [27]. The first report of *B. acheilognathi* in Iraq was from the intestine of *C. carpio* from different fish farms near Baghdad [99]. Two other species of *Bothriocephalus*, *B. gowkongensis* Yeh, 1955, and *B. opsariichthydis* Yamaguti, 1934, were also reported from Iraq [17]. According to Molnár [100], both these two species are considered as synonyms of *B. acheilognathi*. At the present time, *B. acheilognathi* and both of its above-named synonyms has so far a total of 21 host species in Iraq [17].

Ligula intestinalis (L., 1758) Bloch, 1782, was recorded from the body cavity of both *C. idella* [24, 27, 29] and *C. carpio* [27]. *L. intestinalis* was reported for the first time in Iraq as a plerocercoid from the body cavity of *Leuciscus vorax* (reported as *A. vorax*) from Shatt Al-Arab River [101]. So far, this species has 13 fish host species in Iraq [17]. In Iraq, the adult stage of *L. intestinalis* was reported from the intestine of the moorhen *Gallinula chloropus chloropus* from around Baghdad [102].

Proteocephalus osculatus (Goeze, 1782) Nybelin, 1942, was recorded from the intestine of *C. carpio* [27]. The first report of this parasite in Iraq was from the intestine of *L. vorax* (reported as *A. vorax*) from Al-Tharthar Lake [103]. So far, this species has eight fish host species in Iraq [17].

Proteocephalus torulosus (Batsch, 1786) Nufer, 1905, was recorded from intestine of *C. carpio* [27]. The first report of this parasite in Iraq was from the intestine of *C. carpio* from a fish farm near Baghdad city [99]. So far, this species has two fish host species in Iraq [17].

Neogryporhynchus cheilancristrotus (Wedl, 1855) Baer & Bona, 1960, was recorded from the intestine of *C. carpio* by Al-Zubaidy [27] as *Gryporhynchus cheilancristrotus*. The first report of *N. cheilancristrotus* in Iraq was from the intestine of *L. abu* from Diyala River [104]. So far, this species has four fish host species in Iraq [17].

5.8. Phylum Nematoda. The phylum Nematoda is represented in farm fishes of Babylon province with three species:
unidentified larval species of the genus *Contracaecum* as
well as one species of each of the genera *Cucullanus* and
Rhabdochona.

Phylum Nematoda

    Class Secernentea

        Order Ascaridida

            Family Anisakidae

                Contracaecum spp.

            Family Cucullanidae

                Cucullanus cyprini Yamaguti, 1941

        Order Spirurida

            Family Rhabdochonidae

                Rhabdochona hellichi (Srámek, 1901)

Unidentified larval species of *Contracaecum* was recorded
from the intestinal wall, body cavity, liver, spleen, heart, and
gonads of *C. carpio* [27, 29] and intestinal wall of *L. abu*
[22, 37, 42]. The first report of *Contracaecum* spp. larvae
in Iraq was from the body cavity and different viscera of
10 fish species from different inland waters of Iraq [70].
Contracaecum spp. larvae have so far 40 fish host species in
Iraq [17]. Adult worms of *Contracaecum* spp. were detected
from six species of aquatic birds in Iraq, *Egretta alba*, *E.
garzetta*, *Ardeola ralloides*, *Botaurus stellaris*, *Ardea purpurea*,
and *Ceryle rudis*, from Bahr Al-Najaf Depression [76].

Cucullanus cyprini Yamaguti, 1941, was recorded from
intestine of *C. carpio* [27]. The first report of this parasite
in Iraq was from the intestine of *Alburnus caeruleus* and
Luciobarbus xanthopterus (reported as *B. xanthopterus*) from
Al-Tharthar Lake [103]. So far, this species has 15 fish host
species in Iraq [17].

Rhabdochona hellichi (Srámek, 1901), erroneously
reported as *R. bellichi*, was recorded from intestine of the
three carp species: *C. idella* [29], *C. carpio* [29], and *H.
molitrix* [29]. Ali et al. [105] reported this parasite (also
erroneously as *R. bellichi*) from the intestine and coelom of
L. xanthopterus (reported as *B. xanthopterus*), *H. fossilis*, and
Mystus pelusius (reported as *M. halepensis*). Eight fish species
are so far known for this parasite in Iraq [17].

5.9. Phylum Acanthocephala. The phylum Acanthocephala is
represented in farm fishes of Babylon province with two valid
species of the genus *Neoechinorhynchus*.

Phylum Acanthocephala

    Class Eoacanthocephala

        Order Neoechinorhynchida

            Family Neoechinorhynchidae

                Neoechinorhynchus iraqensis Amin,
                Al-Sady, Mhaisen & Bassat, 2001

                Neoechinorhynchus rutili (Müller,
                1780) Hamann, 1892

Neoechinorhynchus iraqensis Amin, Al-Sady, Mhaisen &
Bassat, 2001, was recorded from intestine of both *C. carpio*
[27] and *L. abu* [22, 50]. It is appropriate to mention here that
this species was reported as *N. agilis* from *C. carpio* and *L. abu*
by Al-Zubaidy [27] and Ali et al. [22], respectively. *N. agilis*
is a misidentification of *N. iraqensis* [106]. The first report of
N. iraqensis was as *species de novo* from the intestine of *L. abu*
from the Euphrates River at Al-Fallujah region [107], while its
description was given later by Amin et al. [108]. *N. iraqensis*
and the misidentified *N. agilis* have so far 24 fish host species
in Iraq [17].

Neoechinorhynchus rutili (Müller, 1780) Hamann, 1892,
was recorded from the intestine of the three carp species:
C. idella [29], *C. carpio* [27, 29], and *H. molitrix* [29]. *N.
rutili* was firstly recorded by Herzog [70] from *L. xanthopterus*
(reported as *B. xanthopterus*) from Tigris and Diyala rivers
near Baghdad and from *L. abu* (reported as *Mugil abu*) from
Citscher Oasis near Al-Fallujah. *N. rutili* has so far 16 fish host
species in Iraq [17].

5.10. Phylum Arthropoda: Subphylum Crustacea. The subphylum Crustacea of the phylum Arthropoda is represented in
farm fishes of Babylon province with nine species: one species
of each of the genera *Argulus*, *Dermoergasilus*, *Paraergasilus*,
Lamproglena, and *Lernaea* and three species of *Ergasilus* in
addition to unidentified species of *Ergasilus*.

Phylum Arthropoda

    Subphylum Crustacea

        Class Maxillopoda

            Order Arguloida

                Family Argulidae
                Argulus foliaceus (L., 1758)

            Order Cyclopoida

                Family Ergasilidae
                Dermoergasilus varicoleus Ho, Jayara-
                jan & Radhakrishnan, 1992
                Ergasilus barbi Rahemo, 1982
                Ergasilus mosulensis Rahemo, 1982
                Ergasilus sieboldi von Nordmann, 1832
                Ergasilus sp.
                Paraergasilus inflatus Ho, Khamees &
                Mhaisen, 1996
                Family Lernaeidae
                Lamproglena pulchella von Nordmann,
                1832
                Lernaea cyprinacea L., 1758

Argulus foliaceus (L., 1758) was recorded from gills of
the three carp species: *C. idella* [27], *C. carpio* [25, 27], and
H. molitrix [27] as well as from fins of *C. auratus* [58]. *A.
foliaceus* was reported for the first time in Iraq [70] from skin
of *C. carpio* from Al-Zaafaraniya Fish-Culture Station and

C. luteus (reported as *B. luteus*) from Al-Habbaniyah Lake [70]. *A. foliaceus* has so far 16 fish host species in Iraq [17].

Dermoergasilus varicoleus Ho, Jayarajan & Radhakrishnan, 1992, was recorded from gills of *L. abu* [31, 40]. This crustacean was reported for the first time in Iraq from gills of *L. abu* from Shatt Al-Arab River [109]. *D. varicoleus* has so far nine fish host species in Iraq [17].

Ergasilus barbi Rahemo, 1982, was recorded from gills of *L. abu* [46]. This crustacean was firstly detected from gills of *B. grypus* from Tigris River at Mosul city by Fattohy [75] and its full description as a new species was achieved by Rahemo [110]. *E. barbi* has so far 13 fish host species in Iraq [17].

Ergasilus mosulensis Rahemo, 1982, was recorded from gills of the three carp species, *C. idella* [27], *C. carpio* [27], and *H. molitrix* [27] as well as gills of *L. abu* [46]. This crustacean was firstly detected from gills of *L. abu* from Tigris River at Mosul city by Fattohy [75] and its full description as a new species was achieved by Rahemo [110]. *E. mosulensis* has so far 23 fish host species in Iraq [17].

Ergasilus sieboldi von Nordmann, 1832, was recorded from gills, buccal cavity, and skin of the three carp species: *C. idella* [27, 29], *C. carpio* [25, 27, 29, 51, 55], and *H. molitrix* [27, 29, 56]. The first report of *E. sieboldi* in Iraq was from gills of *L. vorax* (reported as *A. vorax*) from Al-Habbaniyah Lake [70]. *E. sieboldi* has so far 26 fish host species in Iraq [17].

Ergasilus sp. was recorded from *L. abu* [50] with no mention to site of infection. In addition to 11 species of *Ergasilus* so far recorded from fishes of Iraq, some specimens of unidentified *Ergasilus* species were also reported from 12 fish species in Iraq [17].

Paraergasilus inflatus Ho, Khamees & Mhaisen, 1996, was recorded from gills of *H. molitrix* [27]. This crustacean was reported as a new species from gill rakers of *L. abu* from Shatt Al-Arab River [111]. *D. varicoleus* has so far seven fish host species in Iraq [17].

Lamproglena pulchella von Nordmann, 1832, was recorded from gills of *C. carpio* [27]. The first report of *L. pulchella* in Iraq was from gills of *Chondrostoma regium* and *Capoeta trutta* (reported as *Varicorhinus trutta*) from Tigris River at Mosul city [112]. *L. pulchella* has so far 19 fish host species in Iraq [17].

Lernaea cyprinacea L., 1758, was recorded from skin, fins, and gills of the three carp species, *C. idella* [24, 27–29, 43, 51], *C. carpio* [6, 25–31, 39, 41, 43, 44, 48, 51, 53–55, 57, 61, 63], and *H. molitrix* [21, 27–29, 43], as well as from gills of *L. abu* [23]. The first report of the anchor worm *L. cyprinacea* in Iraq was from skin, fins, buccal cavity, pharyngeal cavity, gills, and anus of seven freshwater fish species from Al-Zaafaraniya Fish-Culture Station [113]. *L. cyprinacea* is the commonest crustacean among fishes of Iraq as it has so far 30 fish host species in Iraq [17].

5.11. Phylum Mollusca. The phylum Mollusca is represented in farm fishes of Babylon province with only one parasite species of the genus *Unio*. The systematic account of this parasite, followed by parasite-host list, is given here.

Phylum Mollusca

Class Bivalvia

Order Unionoida

Family Unionidae

Unio pictorum (Linnaeus, 1758)

Unio pictorum (Linnaeus, 1758) was recorded from gills of both *C. carpio* [27, 29] and *H. molitrix* [27]. The first report of the glochidial larvae of *U. pictorum* in Iraq was from gills of eight freshwater fish species from Diyala River [104]. It is appropriate to mention here that the authority of *U. pictorum* was erroneously stated as Zhadin, 1938, in all the Iraqi literature except Al-Salmany [71]. *U. pictorum* has so far 24 fish host species in Iraq [17].

6. Host-Parasite List

The following host-parasite list for fish parasites in fish farms of Babylon province is compiled. For each host, the scientific names of all recorded parasites are alphabetically enlisted under their major parasitic groups. To economize space, references of previous records for each parasite species are not given here. These can be obtained from the account of each concerned parasite species in the part of major groups of parasitic fauna within the results and discussion part.

6.1. Cyprinus carpio Linnaeus, 1758

Mastigophora: *Ichthyobodo necator*.

Apicomplexa: *Eimeria dogieli, E. mylopharyngodoni*, and *Haemogregarina* sp.

Ciliophora: *Apiosoma amoebae, A. cylindriformis, A. minuta, A. piscicola, A. poteriformis, Chilodonella cyprini, Ichthyophthirius multifiliis, Trichodina cottidarum, T. domerguei, T. gracilis, T. nigra*, and *Tripartiella amurensis*.

Myxozoa: *Myxobilatus legeri, Myxobolus muelleri, M. oviformis*, and *M. pfeifferi*.

Trematoda: *Apharyngostrigea cornu, Diplostomum paraspathaceum*, and *D. spathaceum*.

Monogenea: *Dactylogyrus achmerowi, D. amurensis, D. anchoratus, D. arcuatus, D. barbioides, D. cornu, D. crassus, D. dogieli, D. ergensi, D. extensus, D. gobii, D. inexpectatus, D. jamansajensis, D. lamellatus, D. latituba, D. lopuchinae, D. minutus, D. navicularis, D. phoxini, D. propinquus, D. sahuensis, D. simplex, D. skrjabini, D. vastator, Dactylogyrus spp., Diplozoon paradoxum, Diplozoon sp., Eudiplozoon nipponicum, Gyrodactylus baicalensis, G. elegans, G. kherulensis, G. macracanthus, G. malmbergi, G. markewitschi, G. medius, G. menschikowi, G. salaris, G. sprostonae, G. vicinus, Paradiplozoon barbi*, and *Pseudacolpenteron pavlovskii*.

Cestoda: *Bothriocephalus acheilognathi, Ligula intestinalis, Neogryporhynchus cheilancristrotus, Proteocephalus osculatus,* and *P. torulosus.*

Nematoda: *Contracaecum* sp., *Cucullanus cyprini,* and *Rhabdochona hellichi.*

Acanthocephala: *Neoechinorhynchus iraqensis* and *N. rutili.*

Crustacea: *Argulus foliaceus, Ergasilus mosulensis, E. sieboldi, Lamproglena pulchella,* and *Lernaea cyprinacea.*

Mollusca: *Unio pictorum.*

6.2. Ctenopharyngodon idella (Valenciennes, 1844)

Ciliophora: *Apiosoma amoebae, A. cylindriformis, A. piscicola, A. poteriformis, Chilodonella cyprini, Ichthyophthirius multifiliis, Trichodina domerguei,* and *T. nigra.*

Myxozoa: *Myxobolus pfeifferi.*

Trematoda: *Diplostomum paraspathaceum* and *D. spathaceum.*

Monogenea: *Dactylogyrus arcuatus, D. ctenopharyngodonis, D. extensus, D. inexpectatus, D. lamellatus, D. latituba, D. vastator, Dactylogyrus* sp., *Gyrodactylus ctenopharyngodontis, G. elegans,* and *G. kherulensis.*

Cestoda: *Bothriocephalus acheilognathi* and *Ligula intestinalis.*

Nematoda: *Rhabdochona hellichi.*

Acanthocephala: *Neoechinorhynchus rutili.*

Crustacea: *Argulus foliaceus, Ergasilus mosulensis, E. sieboldi,* and *Lernaea cyprinacea.*

6.3. Hypophthalmichthys molitrix (Valenciennes, 1844)

Ciliophora: *Apiosoma amoebae, A. cylindriformis, A. piscicola, Chilodonella cyprini, Ichthyophthirius multifiliis, Trichodina cottidarum, T. domerguei,* and *T. nigra.*

Myxozoa: *Myxobolus pfeifferi.*

Trematoda: *Diplostomum indistinctum* and *D. spathaceum.*

Monogenea: *Dactylogyrus extensus, D. hypophthalmichthys, D. inexpectatus, D. latituba, D. skrjabini, Gyrodactylus elegans, G. macracanthus, G. malmbergi,* and *Pseudacolpenteron pavlovskii.*

Nematoda: *Rhabdochona hellichi.*

Acanthocephala: *Neoechinorhynchus rutili.*

Crustacea: *Argulus foliaceus, Ergasilus mosulensis, E. sieboldi, Lernaea cyprinacea,* and *Paraergasilus inflatus.*

Mollusca: *Unio pictorum.*

6.4. Carassius auratus (Linnaeus, 1758)

Crustacea: *Argulus foliaceus.*

6.5. Heteropneustes fossilis (Bloch, 1794)

Trematoda: *Ascocotyle coleostoma.*

6.6. Liza abu (Heckel, 1843)

Ciliophora: *Trichodina domerguei* and *Trichodina* sp.

Myxozoa: *Myxobolus dogieli* and *M. pfeifferi.*

Monogenea: *Gyrodactylus elegans* and *Microcotyle donavini.*

Nematoda: *Contracaecum* sp.

Acanthocephala: *Neoechinorhynchus iraqensis.*

Crustacea: *Dermoergasilus varicoleus, Ergasilus barbi, E. mosulensis, Ergasilus* sp., and *Lernaea cyprinacea.*

Competing Interests

The authors declare that there are not any competing interests related to the publication of this paper.

References

[1] M. I. Al-Hamed, "Carp culture in Iraq," *Iraqi Journal of Agriculture Research*, vol. 1, no. 2, pp. 14–23, 1960.

[2] F. T. Mhaisen, "A review on the parasites and diseases in fishes of ponds and farms of Iraq," *Iraqi Journal of Veterinary Science*, vol. 6, no. 2, pp. 20–28, 1993.

[3] Ministry of Agriculture, *Statistical Data on Fish Farms in Different Provinces of Iraq up to 31 December 2014*, Ministry of Agriculture, Baghdad, Iraq, 2014.

[4] F. T. Mhaisen, "The role of wild fishes in fish farms of Iraq from parasitological and pathological points of view," *Iraqi Journal of Veterinary Medicine*, vol. 17, pp. 126–136, 1993.

[5] F. T. Mhaisen, "Natural enemies of farm fishes with special emphasis on fish farms of Iraq," *Al-Tharwa Al-Samakia (Fisheries)*, vol. 14, pp. 92–98, 1996.

[6] F. T. Mhaisen, *Diseases and Parasites of Fishes*, Basrah University Press, 1983.

[7] F. T. Mhaisen, K. S. Al-Niaeem, and A. B. Al-Zubaidy, "Literature review on fish parasites of Al-Furat Fish Farm, Babylon province, Iraq," *Iraqi Journal of Aquaculture*, vol. 9, no. 1, pp. 95–122, 2012.

[8] EOL Encyclopedia of Life On-Line Database, 2015, http://www.eol.org.

[9] ITIS Integrated Taxonomic Information System On-Line Database, 2015, http://www.itis.gov.

[10] MonoDB Web-Host for the Monogenea, 2015, http://www.monodb.org/.

[11] PESI Pan-European Species Dictionaries Infrastructure, 2015, http://www.eu-nomen.eu/portal/webservices.php.

[12] WoRMS World Register of Marine Species, 2015, http://www.marinespecies.org.

[13] D. I. Gibson, T. A. Timofeeva, and P. I. Gerasev, "A catalogue of the nominal species of the monogenean genus *Dactylogyrus* Diesing, 1850 and their host genera," *Systematic Parasitology*, vol. 35, no. 1, pp. 3–48, 1996.

[14] O. N. Pugachev, P. I. Gerasev, A. V. Gussev, R. Ergens, and I. Khotenowsky, Eds., *Guide to Monogenoidea of Freshwater Fish of Palaearctic and Amur Regions*, Edizioni Ledizioni Ledi, Milano, Italy, 2009.

[15] F. T. Mhaisen and K. N. Abdul-Ameer, "Checklists of *Gyrodactylus* species (Monogenea) from fishes of Iraq," *Basrah Journal of Agriculture Science*, vol. 26, no. 1, pp. 8–25, 2013.

[16] F. T. Mhaisen and K. N. Abdul-Ameer, "Checklists of diplozoid species (Monogenea) from fishes of Iraq," *Bulletin of Iraq Natural History Museum*, vol. 13, no. 2, pp. 95–111, 2014.

[17] F. T. Mhaisen, "Index-catalogue of parasites and disease agents of fishes of Iraq," 2015.

[18] N. M. Ali, F. T. Mhaisen, E. S. Abul-Eis, and L. S. Kadim, "First occurrence of the monogenetic trematode *Gyrodactylus kherulensis* Ergens, 1974 in Iraq on the gills of the common carp *Cyprinus carpio*," *Journal of Biological Science Research*, vol. 19, no. 3, pp. 659–664, 1988.

[19] F. T. Mhaisen, N. M. Ali, E. S. Abul-Eis, and L. S. Kadim, "First record of *Dactylogyrus achmerowi* Gussev, 1955 with an identification key for the dactylogyrids of fishes of Iraq," *Journal of Biological Science Research*, vol. 19, pp. 887–900, 1988.

[20] N. M. Ali, F. T. Mhaisen, and E. S. Abul-Eis, "Three stalked ciliates (Scyphidia: Peritrichia) new to the parasitic fauna of the fishes of Iraq," in *Proceedings of the Fifth Scientific Research Council*, vol. 5, pp. 218–224, 1989.

[21] N. M. Ali, F. T. Mhaisen, E. S. Abul-Eis, and L. S. Kadim, "Parasites of the silver carp *Hypophthalmichthys molitrix* from Babylon Fish Farm, Hilla, Iraq," *Rivista di Idrobiologia*, vol. 28, no. 1-2, pp. 151–154, 1989.

[22] N. M. Ali, F. T. Mhaisen, E. S. Abul-Eis, and L. S. Kadim, "Helminth parasites of the mugilid fish *Liza abu* (Heckel) inhabiting Babylon Fish Farm, Hilla, Iraq," in *Proceedings of the Fifth Scientific Research Council*, vol. 5, pp. 225–233, 1989.

[23] F. T. Mhaisen, N. M. Ali, E. S. Abul-Eis, and L. S. Kadim, "Protozoan and crustacean parasites of the mugilid fish *Liza abu* (Heckel) inhabiting Babylon Fish Farm, Hilla, Iraq," *Journal of Biological Science Research*, vol. 20, no. 3, pp. 517–525, 1989.

[24] F. T. Mhaisen, N. M. Ali, E. S. Abul-Eis, and L. S. Kadim, "Parasitological investigation on the grass carp (*Ctenopharyngodon idella*) of Babylon Fish Farm, Hilla, Iraq," *Iraqi Journal of Biological Sciences*, vol. 10, no. 1, pp. 89–96, 1990.

[25] F. T. Mhaisen and E. S. Abul-Eis, "Parasites of the common carp (*Cyprinus carpio*) in Babylon Fish Farm, Hilla, Iraq, cited in F. T. Mhaisen, N. R. Khamees and S. A. M. Al-Daraji, Parasites and disease agents of carps in Iraq: a check-list," *Basrah Journal of Agriculture Science*, vol. 4, no. 1, pp. 133–139, 1991.

[26] F. T. Mhaisen, A. N. Balasem, G. H. Al-Khateeb, S. M. J. Al-Shaikh, and J. M. Al-Jawda, "Survey of parasites of farm fishes in three provinces in mid Iraq," *Al-Tharwa Al-Samakia*, vol. 13, pp. 84–87, 1993.

[27] A. B. Al-Zubaidy, *Studies on the parasitic fauna of carps in Al-Furat Fish Farm, Babylon province, Iraq [Ph.D. thesis]*, University of Babylon, 1998.

[28] F. H. A. Al-Dulaimi, *Distribution of infection with Lernaea cyprinacea L. in carps and its control by using some plant extracts [M.S. thesis]*, University of Babylon, 2002.

[29] N. A. A. Al-Jadoaa, *The parasitic infections and pathological changes of some local and cultured fishes from Al-Qadisiya and Babylon provinces [Ph.D. thesis]*, Al-Qadisiya University, 2002.

[30] N. A. Al-Zamily, *Efficiency of some plant extracts in eradication of monogenetic trematodes parasitizing skin and gills of the common carp (Cyprinus carpio L.) [M.S. thesis]*, University of Baghdad, 2002.

[31] R. A. Kadim, *Investigation of crustacean parasites on some species of fishes in Al-Furat Fish Farm and Al-Mahaweel Drainage Collector in Babylon province [M.S. thesis]*, University of Babylon, 2003.

[32] F. T. Mhaisen, M. T. Al-Kaisey, and N. A. Al-Zamily, "Effect of the water extracts of red pepper and colocynth in treating the common carp infected with skin and gill trematodes," *Iraqi Journal of Science and Technology*, vol. 2, no. 2, pp. 5–14, 2005.

[33] N. A. Al-Zamily, F. F. Mhaisen, and M. T. Al-Kaisey, "Efficiency of the water extract of pomegranate fruit shells in treating the common carp infected with skin and gill trematodes," *Ibn Al-Haitham Journal for Pure and Applied Science*, vol. 19, no. 4, pp. 1–12, 2006.

[34] A. B. Al-Zubaidy, F. S. Al-Zubadi, and F. T. Mhaisen, "Treatment of the common carp (*Cyprinus carpio*) infected with monogenetic trematodes by using some plant extracts," *Journal of Um Salama for Science*, vol. 4, no. 1, pp. 22–27, 2007.

[35] S. M. A. Al-Haider, *Studying the biotic and abiotic factors affecting in survival rate of larvae and fingerling rearing ponds at Al-Furat Company-Babilon [M.S. thesis]*, Foundation of Technical Education, Baghdad, Iraq, 2008.

[36] G. B. Al-Oumashi, "A study on carp infection dynamics by worm (*Dactylogyrus*) in Al-Forat Fish Farm," *Al-Qadisiya Journal of Pure Science*, vol. 13, no. 2, pp. 8–13, 2008.

[37] A. B. Al-Zubaidy, "Prevalence and densities of *Contracaecum* sp. larvae in *Liza abu* (Heckel, 1843) from different Iraqi water bodies," *Journal of King Abdulaziz University, Marine Science*, vol. 20, pp. 3–17, 2009.

[38] A. B. Al-Zubaidy, F. T. Mhaisen, and F. S. Al-Zubaidi, "Study on the parasitic fauna of carps in Al-Furat Fish Farm, Iraq. 1: treatment of the common carp (*Cyprinus carpio*) infected with monogenetic trematodes by using some plant extracts," *Journal of King Abdulaziz University, Marine Science*, vol. 20, pp. 155–170, 2009.

[39] R. A. Kadhim, "Resistance of common carp fishes *Cyprinus carpio* (L.) to reinfection by anchor worm *Lernaea cyprinacea* (L.)," *Al-Qadisiyah Journal of Science*, vol. 14, no. 3, pp. 49–58, 2009.

[40] R. A. Kadim and A. B. Al-Zubaidy, "Treatment of *Liza abu* infected with the parasitic crustacean *Dermoergasilus varicoleus* by using boiling water extracts of *Artemisia herba-alba* (L.) and *Petroselinum crispum* (Mill.)," *Journal of Babylon University, Pure Science*, vol. 17, no. 2, pp. 518–528, 2009.

[41] R. A. Kadim and A. B. Al-Zubaidy, "The pathological effects of the anchor worm *Lernaea cyprinacea* L. on the common carp *Cyprinus carpio* L.," *Journal of Babylon University, Pure Science*, vol. 17, no. 3, pp. 929–935, 2009.

[42] A. B. Al-Zubaidy, "Studies on the food, size and endoparasites of *Liza abu* from river and farm in Babylon, middle of Iraq," In press.

[43] S. K. Muhammed, *An external and eye parasite survey for carp fishes in Al-Eskandaryia region (Babylon) [M.S. thesis]*, University of Baghdad, 2000.

[44] A. N. Yaseen, *Using of some crude plant extracts in treating the common carp, Cyprinus carpio infected with the anchor worm Lernaea cyprinacea [M.S. thesis]*, University of Baghdad, 2000.

[45] H. T. Hussain, *Ectoparasitic infection of the common carp and silver carp fingerlings stocked under winter in Al-Shark Al-Awsat Fish Farm, Babylon province [M.S. thesis]*, Foundation of Technical Education, Baghdad, Iraq, 2005.

[46] F. H. A. Al-Dulaimi, A. K. A. Al-Hamiary, and A. H. A. Al-Khafaji, "Infection of *Liza abu* with parasitic crustaceans in waters of a fish farm in Al-Eskandaria district, Babylon province," *Journal of Babylon University, Pure Science*, vol. 12, no. 3, pp. 749–758, 2006.

[47] H. T. Hussain, F. T. Mhaisen, and A. L. Al-Rubaie, "Ectoparasitic infection of the common carp and silver carp fingerlings stocked during winter in Al-Shark Al-Awsat Fish Farm, Babylon province," *Journal of Babylon University, Pure Science*, vol. 14, no. 3, pp. 204–219, 2007.

[48] A. N. Yaseen, F. T. Mhaisen, and M. T. Al-Kaisey, "Effect of aqueous and alcoholic extracts of leaves of henna *Lawsonia inermis* in treating the common carp *Cyprinus carpio* L. infected with the anchor worm, *Lernaea cyprinacea*," *Iraqi Journal of Agriculture*, vol. 14, no. 5, pp. 150–156, 2009.

[49] H. T. Hussain, E. H. Howaidi, T. S. Naif, K. R. Abd, and M. H. Takheal, "Study the relationship between the length of the fish and the incidence external parasites of common carp *Cyprinus carpio* in the Al-Shark Al-Awsat Fish Farm in Babylon Province," *Journal of Al-Qadisiyah for Pure Science*, vol. 18, no. 1, pp. 1–6, 2013.

[50] M. D. Ali and F. Shaaban, "Some species of parasites of freshwater fish raised in ponds and in Tigris-Al-Tharthar canal region," in *Proceedings of the 7th Scientific Conference of the Iraqi Veterinary Medical Association*, pp. 44–46, Mosul, Iraq, October 1984.

[51] N. M. Ali, N. E. Salih, and K. N. Abdul-Ameer, "Protozoa and Crustacea infesting three species of carp raised in ponds in Iraq," *Journal of Biological Science Research*, vol. 19, no. 2, pp. 387–394, 1988.

[52] N. E. Salih, N. M. Ali, and K. N. Abdul-Ameer, "Helminthic fauna of three species of carp raised in ponds in Iraq," *Journal of Biological Science Research*, vol. 19, no. 2, pp. 369–386, 1988.

[53] F. T. Mhaisen, A. N. Balasem, G. H. Al-Khateeb, S. M. J. Al-Shaikh, J. M. Al-Jawda, and S. M. Haiawi, "Survey of parasites of three fish farms at Al-Latifiya, south Baghdad," *Marina Mesopotamica*, vol. 8, no. 2, pp. 218–224, 1993.

[54] H. T. Hussain, E. H. Hwaidi, H. H. Elewi, and H. M. AbidAli, "Survey of ectoparasitic infections on the common carp *Cyprinus carpio* in three fish farms at Al-Eskandriya, Babylon province," *Scientific Journal of University of Kerbala*, vol. 9, no. 1, pp. 126–131, 2011.

[55] A. L. Al-Rubaie, H. T. Hussain, and K. N. Abdul-Ameer, "The external parasites of the common carp (*Cyprinus carpio*) in Technical Institute of Al-Mussayab Fish Farm," *Journal of Babylon University, Science*, vol. 14, no. 3, pp. 46–50, 2007.

[56] K. R. Asmar, A. N. Balasem, J. M. Al-Jawda, A. J. Hummadi, and T. K. Adday, "Study of the parasitic infections in some fish farms," Report 7050-PO 152-2001, 2001.

[57] A. A. E. Al-Jubory, *Influence of fertilizing the ponds of Cyprinus carpio on the infection with Lernaea cyprinacea [Higher Dipl. thesis]*, Foundation of Technical Education, Baghdad, Iraq, 2009.

[58] F. H. A. Al-Dulaimi, "Infection with a fish louse *Argulus foliaceus* L. in a goldfish (*Carassius auratus*) at earthern pond and aquarium fish in Babylon province, Iraq," *Journal of Babylon University, Pure Science*, vol. 18, no. 2, pp. 468–473, 2010.

[59] H. T. Hussain, T. S. Naief, and E. H. Hwaidi, "Comparative study of external parasitic infection in the common carp (*Cyprinus carpio*) bred in monoculture and polyculture," *Scientific Journal of University of Kerbala*, vol. 9, no. 4, pp. 64–71, 2011.

[60] N. T. M. Al-Taei, *Study of some of the environmental aspects for a group of the external parasitic animals for common carp Cyprinus carpio L. in cages and pond at Al-Saddah/Babylon province [M. Tech. Thesis]*, Foundation of Technical Education, Baghdad, Iraq, 2013.

[61] H. M. J. Al-Sardee, *Effect of quicklime on common carp infested with Lernaea cyprinacea L. [M.S. thesis]*, University of Baghdad, 1992.

[62] S. M. J. Al-Shaikh, "Study of some important and common fish diseases in Iraq and methods of their control and prevention," in *Proceedings of the Joint Symposium of the Arab Union of Fish Producers and Arab Union Councils of Scientific Research*, pp. 133–137, Baghdad, Iraq, December 1993.

[63] S. M. J. Al-Shaikh and H. M. J. Al-Sardee, "Gross pathological and histological changes in infection with the anchor worm and its treatment with quicklime," *Al-Tharwa Al-Samakia (Fisheries)*, vol. 13, pp. 95–97, 1993.

[64] S. M. J. Al-Shaikh, "Pathogenesis and treatment of Monogenea (*Dactylogyrus*) in fishes," *Veterinarian*, vol. 9, no. 3, pp. 213–217, 1999.

[65] A. L. Al-Rubaie, "Efficiency of propolis on ciliate protozoan parasite *Ichthyophthirius multifiliis* of *Cyprinus carpio*," *Euphrates Journal of Agriculture Science*, vol. 1, no. 3, pp. 54–60, 2009.

[66] M. Kirjušina and K. Vismanis, "Checklist of the parasites of fishes of Latvia," FAO Fisheries Technical Report 369/3, FAO, 2007.

[67] M. N. Bhatti, "A note on the occurrence of costiasis disease in the stinging catfish, *Heteropneustes fossilis* (Bloch) from Basrah waters," *Arab Gulf*, vol. 11, no. 1, p. 216, 1979.

[68] D. W. Duszynski, L. Couch, and S. J. Upton, *Coccidia (Eimeriidae) of Cypriniformes (Cyprinids)*, 2015, http://www.k-state.edu/parasitology/worldcoccidia/FISHBIB.

[69] N. M. Ali, N. E. Salih, and K. N. Abdul-Ameer, "Parasitic fauna of some freshwater fishes from Tigris River, Baghdad, Iraq. I: Protozoa," *Journal of Biological Science Research*, vol. 18, no. 2, pp. 11–17, 1987.

[70] P. H. Herzog, "Untersuchungen über die parasiten der süßwasserfische des Irak," *Archiv für Fischereiwissenschaften*, vol. 20, no. 2-3, pp. 132–147, 1969.

[71] S. O. K. Al-Salmany, *Parasitic infection of some fish species from Euphrates River at Al-Qaim city, Anbar province [M.S. thesis]*, University of Tikrit, 2015.

[72] K. N. Abdul-Ameer, "The first record of the ciliated protozoan *Trichodina cottidarum* in Iraq on the gills of the common carp *Cyprinus carpio*," *Ibn Al-Haitham Journal for Pure and Applied Science*, vol. 17, no. 3, pp. 1–6, 2004.

[73] M. Shamsuddin, I. A. Nader, and M. J. Al-Azzawi, "Parasites of common fishes from Iraq with special reference to larval form of *Contracaecum* (Nematoda: Heterocheilidae)," *Bulletin of the Biological Research Centre, Baghdad*, vol. 5, pp. 66–78, 1971.

[74] K. N. Abdul-Ameer, *Study of the parasites of freshwater fishes from Tigris River in Salah Al-Dien province, Iraq [M.S. thesis]*, University of Baghdad, 1989.

[75] Z. I. Fattohy, *Studies on the parasites of certain teleostean fishes from the River Tigris, Mosul, Iraq [M.S. thesis]*, University of Mosul, 1975.

[76] H. M. H. Al-Awadi, F. T. Mhaisen, and F. F. Al-Joborae, "Helminth parasitic fauna of aquatic birds in Bahr Al-Najaf depression, mid Iraq," *Bulletin of Iraq Natural History Museum*, vol. 11, no. 2, pp. 7–15, 2010.

[77] N. M. Ali, A. R. Al-Jafery, and K. N. Abdul-Ameer, "New records of three digenetic trematodes on some freshwater fishes from Diyala River, Iraq," in *Proceedings of the Fourth Scientific Conference, Scientific Research Council*, vol. 5, pp. 10–19, 1986.

[78] F. T. Mhaisen and E. S. Abul-Eis, "Parasitic helminths of eight species of aquatic birds in Babylon Fish Farm, Hilla, Iraq," *Zoology in the Middle East*, vol. 7, pp. 115–120, 1992.

[79] T. I. Al-Alousi, S. M. J. Al-Shaikh, and N. R. Abdul-Rahman, "Incidence of metacercariae of *Diplostomum* Nordmann, 1832 in Iraqi freshwater fish," *Journal of Veterinary Parasitology*, vol. 2, no. 1, p. 75, 1988.

[80] S. M. A. Abdullah, *Survey of the parasites of fishes of Dokan Lake [M.S. thesis]*, University of Salahaddin, 1990.

[81] F. T. Mhaisen, "Worm cataract in freshwater fishes of Iraq," *Ibn Al-Haitham Journal for Pure and Applied Science*, vol. 17, no. 3, pp. 25–33, 2004.

[82] S. M. A. Abdullah, "Parasitic fauna of some freshwater fishes from Darbandikhan Lake, north of Iraq," *Journal of Dohuk University*, vol. 8, no. 1, pp. 29–35, 2005.

[83] A. V. Gussev, "Parasitic metazoans: class Monogenea," in *Key to the Parasites of Freshwater Fish*, O. N. Bauer, Ed., vol. 2 of *Fauna of the USSR*, pp. 1–424, Nauka, St. Petersburg, Russia, 1985.

[84] A. B. Al-Zubaidy, "First record of three monogenic parasites species from Iraqian freshwater fishes," *Journal of King Abdulaziz University, Marine Sciences*, vol. 18, no. 1, pp. 83–94, 2008.

[85] F. T. Mhaisen, A. N. Balasem, G. H. Al-Khateeb, and K. R. Asmar, "Recording of five monogenetic trematodes for the first time from fishes of Iraq," in *Proceedings of the Abstracts of the 14th Scientific Conference of the Iraqi Biological Society*, Najaf, Iraq, March 1997.

[86] F. T. Mhaisen, A. N. Balasem, G. H. Al-Khateeb, and K. R. Asmar, "Recording of five monogenetic trematodes for the first time from fishes of Iraq," *Bulletin of Iraq Natural History Museum*, vol. 10, no. 1, pp. 31–38, 2003.

[87] A. V. Gussev, N. M. Ali, K. N. Abdul-Ameer, S. M. Amin, and K. Molnár, "New and known species of *Dactylogyrus* Diesing, 1850 (Monogenea, Dactylogyridae) from cyprinid fishes of the River Tigris, Iraq," *Systematic Parasitology*, vol. 25, no. 3, pp. 229–237, 1993.

[88] N. M. Ali, A. R. Al-Jafery, and K. N. Abdul-Ameer, "New records of three monogenetic trematodes on some freshwater fishes from Diyala River, Iraq," *Journal of Biological Science Research*, vol. 17, no. 2, pp. 253–266, 1986.

[89] B. A.-H. E. Al-Sa'adi, *The parasitic fauna of fishes of Euphrates River: applied study in Al-Musaib city [M. Tech. thesis]*, Foundation of Technical Education, Baghdad, Iraq, 2007.

[90] B. A. Al-Sa'adi, F. T. Mhaisen, and A.-R. L. Al-Rubaie, "The first record of two monogeneans: *Dactylogyrus dogieli* Gussev, 1953 and *Octomacrum europaeum* Roman & Bykhovskii, 1956 from fishes of Iraq," *Basrah Journal of Agriculture Science*, vol. 26, no. 1, pp. 1–7, 2013.

[91] N. M. Ali, E. S. Abul-Eis, and K. N. Abdul-Ameer, "On the occurrence of fish parasites raised in manmade lakes," in *Proceedings of the 6th Conference of the European Ichthyologists*, Budapest, Hungary, August 1988.

[92] A. N. Balasem, F. T. Mhaisen, K. R. Asmar, J. M. Al-Jawda, and T. K. Adday, "Record of two species of the monogenetic

[92cont] trematodes genus *Dactylogyrus* for the first time in Iraq on gills of the cyprinid fish *Alburnus caeruleus*," *Bulletin of Iraq Natural History Museum*, vol. 10, no. 4, pp. 11–16, 2009.

[93] A. A. Sadek, *Ectoparasites of the common carp (Cyprinus carpio L.) fingerlings intensively stocked during autumn and winter [M.S. thesis]*, University of Baghdad, 1999.

[94] N. M. Ali, N. E. Salih, and K. N. Abdul-Ameer, "Parasitic fauna of some freshwater fishes from Tigris River, Baghdad, Iraq. II: Trematoda," *Journal of Biological Science Research*, vol. 18, no. 2, pp. 19–27, 1987.

[95] A. A. J. J. Al-Saadi, *Ecology and taxonomy of parasites of some fishes and biology of Liza abu from Al-Husainia creek in Karbala province [Ph.D. thesis]*, University of Baghdad, 2007.

[96] F. S. Al-Nasiri, "First occurrence of the monogenetic trematode *Diplozoon nipponicum* Goto, 1891 in Iraq from common carp *Cyprinus carpio* (Pisces)," *Iraqi Journal of Agriculture*, vol. 8, no. 6, pp. 95–99, 2003.

[97] A.-R. A.-M. Rasheed, "First record of *Diplozoon barbi* Reichenbach-Klinke, 1951 from some freshwater fishes from Tigris River, Baghdad, Iraq," *Zanco*, vol. 2, no. 3, pp. 5–15, 1989.

[98] I. A. Khotenovsky, *Monogenea: Suborder Octomacrinea Khotenovsky*, vol. 132 of *Fauna of the USSR*, Nauka, St. Petersburg, Russia, 1985.

[99] K. A. Khalifa, "Occurrence of parasitic infections in Iraqian fish ponds," in *Proceedings of the Abstracts of 2nd Scientific Conference*, p. 333, Arab Biological Union, Fés, Morocco, March 1982.

[100] K. Molnár, "On the synonyms of *Bothriocephalus acheilognathi* Yamaguti," *Parasitologica Hungarica*, vol. 10, pp. 61–62, 1977.

[101] Z. I. Al-Hasani, "Occurrence of two known helminthic parasites in two vertebrate hosts collected from Basrah, Iraq," *Dirasat*, vol. 12, no. 7, p. 25, 1985.

[102] M. K. Mohammad, A. A. Al-Moussawi, and M. K. Jasim, "The parasitic fauna of the moorhen *Gallinula chloropus chloropus* L. in the Middle of Iraq," *Bulletin of Iraq Natural History Museum*, vol. 9, no. 4, pp. 41–49, 2002.

[103] A. A. J. J. Al-Saadi, *A survey of alimentary canal helminths of some species of fishes from Tharthar Lake [M.S. thesis]*, University of Baghdad, 1986.

[104] N. M. Ali, A. R. Al-Jafery, and K. N. Abdul-Ameer, "Parasitic fauna of freshwater fishes in Diyala River, Iraq," *Journal of Biological Science Research*, vol. 18, no. 1, pp. 163–181, 1987.

[105] N. M. Ali, N. E. Salih, and K. N. Abdul-Ameer, "Parasitic fauna of some freshwater fishes from Tigris River, Baghdad, Iraq. IV: Nematoda," *Journal of Biological Science Research*, vol. 18, no. 3, pp. 35–45, 1987.

[106] F. T. Mhaisen, "Literature review and check lists of acanthocephalans of fishes of Iraq," *Al-Mustansiriya Journal of Science*, vol. 13, no. 1, pp. 13–25, 2002.

[107] R. S. Al-Sady, *Description of a new species of Acanthocephala (Neoechinorhynchus iraqensis) and some ecological aspects of its infection to the mugilid fish Liza abu from Al-Faluja region, Al-Anbar province with observations on the experimental infection [M.S. thesis]*, University of Baghdad, 2000.

[108] O. M. Amin, R. S. S. Al-Sady, F. T. Mhaisen, and S. F. Bassat, "*Neoechinorhynchus iraqensis* sp. n. (Acanthocephala: Neoechinorhynchidae) from the freshwater mullet, *Liza abu* (Heckel), in Iraq," *Comparative Parasitology*, vol. 68, no. 1, pp. 108–111, 2001.

[109] N. R. Khamees and F. T. Mhaisen, "Two copepod crustaceans as additional species to the parasitic fauna of fishes of Iraq," *Basrah Journal of Science*, vol. 13, no. 1, pp. 49–56, 1995.

[110] Z. I. F. Rahemo, "Two new species of *Ergasilus* (Copepoda: Cyclopoida) from the gills of two Iraqi freshwater fishes," *Bulletin of Basrah Natural History Museum*, vol. 5, pp. 39–59, 1982.

[111] J.-S. Ho, N. R. Khamees, and F. T. Mhaisen, "Ergasilid copepods (Poecilostomatoida) parasitic on the mullet *Liza abu* in Iraq, with the description of a new species of *Paraergasilus* Markevich, 1937," *Systematic Parasitology*, vol. 33, no. 2, pp. 79–87, 1996.

[112] Z. I. F. Rahemo, "Recording of two new hosts of *Lamproglena pulchella* Nordmann, 1832 (Crustacea: Decapoda) in Iraq," *Iraqi Journal of Biological Sciences*, vol. 5, no. 1, pp. 82–83, 1977.

[113] M. I. Al-Hamed and L. Hermiz, "Experiments on the control of anchor worm (*Lernaea cyprinacea*)," *Aquaculture*, vol. 2, no. 1, pp. 45–51, 1973.

16

Molecular Cloning and Characterization of *Babesia orientalis* Rhoptry Neck 2 *Bo*RON2 Protein

Ngabu Malobi,[1] Lan He,[1,2] Long Yu,[1] Pei He,[1] Junwei He,[1]
Yali Sun,[1] Yuan Huang,[1] and Junlong Zhao[1,2]

[1]State Key Laboratory of Agricultural Microbiology, College of Veterinary Medicine,
 Huazhong Agricultural University, Wuhan, Hubei 430070, China
[2]Key Laboratory of Animal Epidemical Disease and Infectious Zoonoses, Ministry of Agriculture,
 Huazhong Agricultural University, Wuhan, China

Correspondence should be addressed to Lan He; helan@mail.hzau.edu.cn and Junlong Zhao; zhaojunlong@mail.hzau.edu.cn

Academic Editor: José F. Silveira

Babesiosis caused by *Babesia orientalis* is one of the most prevalent infections of water buffalo transmitted by *Rhipicephalus haemaphysaloides* causing a parasitic and hemolytic disease. The organelles proteins localized in apical membrane especially rhoptries neck and microneme protein form a complex called moving junction important during invasion process of parasites belonging to apicomplexan group, including *Babesia* species. A truncated fragment coding a 936 bps fragment was cloned in pMD-19T and subcloned into pET32 (a)+ expression vector, expressed in *E. coli* BL21. Purified recombinant *Bo*RON2 was used to produce polyclonal antibody against *Bo*RON2. Here, we identified the full sequence of gene encoding the rhoptry neck 2 protein that we named *Bo*RON2 which is 4035 bp in full-length open reading frame without introns, encoding a polypeptide of 1345 amino acids. Western blot of r*Bo*RON2 probed with buffalo positive serum analysis revealed a band of around 150 kDa in parasite lysates, suggesting an active involvement during invasion process. These findings most likely are constructive in perspective of ongoing research focused particularly on water buffalo babesiosis prevention and therapeutics and globally provide new information for genes comparative analysis.

1. Introduction

Protozoan parasites are a significant cause of mortality in human and animals worldwide. They belong to the phylum Apicomplexa, specifically characterized by the presence of complex specialized organelles at their apical end, and are responsible of important diseases, such as malaria which causes more than 1 million deaths worldwide, toxoplasmosis, cryptosporidiosis, coccidiosis, and babesiosis [1]. Babesiosis caused by *Babesia* genus causes a huge economic loss to the livestock industry worldwide [2, 3]. The life cycle of these organisms is very complex, involving different stages. In *Plasmodium* parasites causing malaria, for example, the first stage of infection starts by sporozoites injection into bloodstream by female *Anopheles* (infecting hepatics cells in mammals and salivary glands) followed by entry in liver cells and division named preerythrocytic stage, followed by the release of merozoites into bloodstream which initiate the asexual parasite multiplication stage, and these merozoites' differentiation leads to male and female gametocytes that invade mosquito midgut cells in which the gametocytes fusion produces a zygote that develops into a motile ookinete that penetrates the midgut wall and forms oocysts [4]. The oocyst enlarged over time and burst leading to sporozoites release which migrate to the mosquito salivary gland where they become susceptible of next infection through blood meal [5].

The parasite invasion in host cell is a crucial step in apicomplexan biology infection process, as their extracellular life is limited to a short period of time. This process is achieved in less than 10 seconds and is powered by the glideosome, a macromolecular complex consisting of adhesive proteins and an actomyosin system anchored in the inner membrane complex of the parasite [6]. This invasion involves sequential secretion of the contents of two secretory organelles from the apical complex: the micronemes and rhoptries [7]. Commonly the proteins localized in rhoptries and micronemes mediate interaction between host receptor and parasite during invasion [8]. From the apical complex, the parasite moves through this ring-like structure which is referred to as the moving junction (MJ) and whose function is to generate the parasitophorous membrane from the invaginated host plasma membrane [9]. The importance of rhoptries necks proteins in *P. falciparum* and in *T. gondii* as model for apicomplexan parasite [10] has raised concern in identifying the functional characterization of these proteins in other genera, such as *Neospora, Theileria, Babesia, Cryptosporidium, and other Plasmodium* species. The current model suggests that rhoptry neck protein 2 (RON2) attaches the moving junction to the host cell membrane via its predicted transmembrane domains serving as a receptor for apical membrane antigen 1 (AMA-1), which anchors RONs proteins of junction complex to the parasite surface [10], proving the high importance of rhoptry neck 2 protein function in apicomplexan parasites group.

Previous studies on *T. gondii* identified four rhoptries neck proteins (RON2, RON4, RON5, and RON8) found to be exported from apical rhoptries forming a complex at the host parasite interface [10–12]; however in *Plasmodium* only the 3 orthologues RON2, RON4, and RON5 found in *Toxoplasma* have been identified [13, 14]. The micronemal protein AMA-1 plays an important role during the tight junction formation with rhoptries proteins and anchors the merozoite to host surface cell leading to the invasion [15]; additionally, although the invasion inhibitory of AMA-1/RON2 interaction has been proved to be crucial for *T. gondii* and *P. falciparum* from preventing entry in their respective host cell, the intraspecies interaction is conserved [16].

Among *Babesia* species, recently, the proteins RON2 of *B. microti* and *B. divergens* were reported to be also homologues to other RON2 proteins of apicomplexan and the purified IgG from *B. divergens* RON2 antibodies was shown to be able to inhibit the invasion over 36 hours up to 44% [17]. However, in *B. orientalis*, AMA-1 protein was identified and has showed to have a similar and relative conservation of domain three (DIII) and the presence of four cysteine motifs among *Babesia* species [18]; thus the identification of *B. orientalis* RON2 protein and molecular cloning to evaluate the antigenicity of this protein are going to make a new step in order to understand this parasite infection and biochemical regulation mechanism.

2. Materials and Method

2.1. Merozoite Preparation. *B. orientalis* collected from water buffalo was cultured in our laboratory as described [19] and kept in nitrogen liquid [20]. The merozoite antigen of *B. orientalis* was prepared from blood infected by this parasite according to the modification saponin lysis method [21].

Erythrocytes from *B. orientalis* have been washed 3 times by using phosphate buffered saline (PBS) followed by suspension in 9 mL of red blood cells (RBC) lysis buffer (Tiangen, China) and incubation at 37°C for 5 min. After solubilization, the antigen was clarified by centrifugation at 10,000 ×g for 1 h, resuspended in 1 mL PBS, and then stored at −20°C for further use.

2.2. Genomic DNA and cDNA Preparation. For DNA and RNA extraction, we collected blood sample from experimentally infected water buffalo with 3% parasitemia into EDTA tubes (Qingdao Pharmacypro Co., Ltd.). Genomic DNA (gDNA) was isolated by using QIAmp DNA Mini Kit™ (Qiagen Hilden, Germany) following the manufacturer protocol. The collected gDNA was stored at −20°C for further use. Total RNA was extracted from 250 μL of red blood cells by using TRIzol RNA extraction kit (Invitrogen, USA); RNase inhibitor (RNase OUT recombinant ribonuclease inhibitor Invitrogen, USA) was added; then the cDNA was synthesized by reverse transcription using FastQant RT Kit (Tiangen Biotech (Beijing) Co., Ltd.).

2.3. Identification, Amplification, and Sequencing of BoRON2. The degenerate primer was designed through the consensus sequence found using ClustalW [22] from alignment based upon similarity of related species published RON2 of *B. bovis* (XM_001608765.1), *B. bigemina* (KU696964.1), *B. divergens* (GU 198499.2), and *B. microti* (XM_01279409.1) nucleotides sequences from NCBI. The amplified PCR product supposed to represent the partial *B. orientalis* RON2 was separated in 2% agarose gel ethidium bromide and cleaned using QIAquick gel extraction kit (Qiagen, China) and cloned into pMD-19T cloning vector (Takara Biotechnology, China) for sequencing. The partial amplified sequence was identified in a cDNA library contig in the complete genome sequence of *B. orientalis* (unpublished), and then the complete open reading frame was identified by overlapping contigs. By using nBLAST search, the sequence has been confirmed as RON2 regarding the high similarity with its orthologues. The complete sequence was obtained by another PCR amplification using a specific pair of primers (Table 2) and the resulting product was sequenced for confirmation.

2.4. Sequence Analysis. The obtained *BoRON2* sequence was analyzed by using bioinformatic tools. SP.4.1 software was used to predict the presence of signal peptide [23]. The transmembrane domain was predicted with TMHMM v.2.0 server [24] and low complexity regions were analyzed by using SMART [25]. Alpha helix and beta sheet were predicted by using Psi Pred server [26]. The presence of repeat was assessed by the sequence tandem repeat using the modeling software XSTREAM [27]. Mega 6 was implemented for phylogenetic tree construction [28]. The tertiary dimensional structure was predicted by using I-Tasser standalone package [29]. For identification of probable anchor location motif scan has been used through Expasy server (http://prosite.expasy.org/). ClustalW was used for alignment and manual correction was applied by using Bioedit [22, 30].

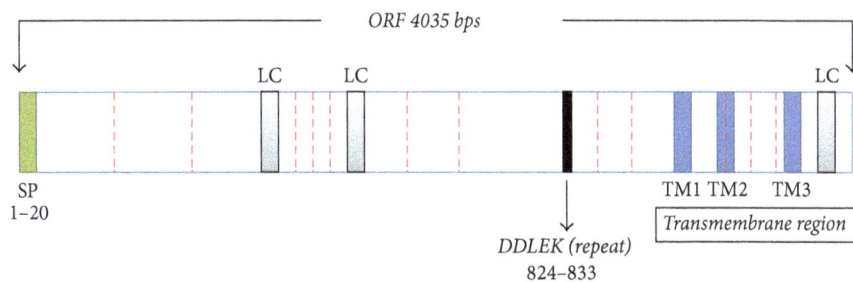

FIGURE 1: Schematic representation of *Babesia orientalis* RON2 gene with an open reading frame of 4035 bps length, signal peptide is shown is green (1–20), the three transmembrane (TM) domains are shown in blue, in grey are the predicted low complexity (LC) regions, and in red are the twelve cysteine residues contained in that gene. The repeat motif is shown in black between the amino acid regions 824 and 833.

2.5. Recombinant Expression and Antibody Production.

The gDNA fragment coding a C-terminal region which represents $BoRON2_{914-1226}$ peptide was amplified to produce a polyclonal antibody by using the expression vector p ET-32(a)+ to generate a recombinant His tag RON2 protein within *SalI* and *BamhI* restrictions sites enzymes underlined with a specific primer (Table 2). After successful sequencing the resulting plasmid was transformed into bacteria *E. coli* BL21 strain (Transgen, China). The recombinant protein was expressed after inducing 16 hours at 18°C with IPTG and visualized through SDS-PAGE. The protein expressed as inclusions bodies was solubilized in 8 M urea and eluted at 150 mM via His tag protein pure Ni-NTA resin charged column following the manufacturer's instruction (Transgen, China). BCA protein assay kit (Beyotime Institute of Biotechnology, China) was used to determine the protein concentration. To generate antirecombinant *BoRON2*-C serum, 2 rabbits were immunized subcutaneously with 500 μg of r*BoRON2*-C emulsified in Freund's complete adjuvant (Sigma, USA) first time at day 0 and then emulsified in Freund's incomplete adjuvant (FIA) the next 4 times, respectively, at days 20, 35, 45, and 60 followed by antisera collection.

2.6. Western Blot Analysis.

The reactivity of *BoRON2*-C protein for immune response was analyzed by western blot with *B. orientalis* erythrocytes infected serum and His tag antibody used as positive control and normal serum as negative control. The recombinant *BoRON2*-C was subjected under reducing condition to 12% SDS-PAGE analysis and transferred to a membrane nitrocellulose (Millipore USA) and then blocking buffer for fixation (1% BSA in TBS with 5% tween) followed by 2 h incubation moderately by shaking at room temperature and probed with secondary antibody Ig rabbit anti-goat. To identify the native *BoRON2*-C in merozoite, rabbit antisera collected from r*BoRON2*-C immunized rabbit was reacted with parasite lysates subjected to SDS-PAGE 12% followed by transfer electroblotting onto nitrocellulose membrane and then probed with rabbit polyclonal antiserum at 1:500 dilutions overnight at 4°C in TBST, after 5 times of washing with TBST, the secondary conjugate rabbit anti-goat antibody was added as final incubation.

3. Result

3.1. Molecular Characterization and Bioinformatic Analysis.

The obtained gDNA of *Babesia orientalis* was amplified and the complete RON2 gene sequence was identified by using specific primer as indicated on Table 2. *BoRON2* gene is 4035 bp nucleotides in length, encoding for 1345 amino acid residues with one copy of imperfect repeat sequence one time repeated in positions 824–833 (DDLEK–DDSEK). The amino acid sequence contains 12 cysteine residues as shown in Figure 1 and was predicted 150 kDa by using a computer based molecular weight calculator [31]. The phylogenetic tree shows that the more close species based on alignment of *BoRON2* protein as query is *B. bovis* (T2Bo strain) RON2 protein belonging into the same clade (Figure 2). They share 72% similarity, followed by *B. bigemina* with 70% of similarity. The amplification of cDNA has shown an amplicon of similar size with gDNA on gel electrophoresis (Figure 3(a)). Finally, the sequencing of cDNA has confirmed the intronless of this gene. The phylogenetic tree of *BoRON2* protein sequence with homologues was performed to identify the distance between them. The neighbor joining tree shows values of each branch with 0.1 as the length of branch (Figure 2). The protein has 7 conserved cysteine residues among aligned apicomplexan orthologues RON2, and 4 of them are located in C-terminal region alternating with transmembrane domains (Figure 5). In fact, many molecular relevant malaria antigens, actively functional, contain cysteine motifs in their structure forming disulfides bridges and/or occasioning three-dimensional structures fulfilling their biological function [32]. In *P. falciparum* the protein rhoptry neck 6 *Pf*RON6 is conserved among apicomplexans except for rodent malaria orthologue, and rich cysteine region was shown to be involved in parasite survival function [33]. The molecular scanning has revealed a putative phosphotidylinositol specific lipase domain at amino acid positions 1026 to 1086 probably involved in membrane trafficking during parasite invasion and a putative anchoring site during infective process. *BoRON2* protein has a putative signal peptide within its 20 first amino acids characterized by the presence of three transmembrane domains all located toward the C-terminal region between residues 1110 and 1251 and three low complexity regions. The relative conservation in C-terminal fragment covering

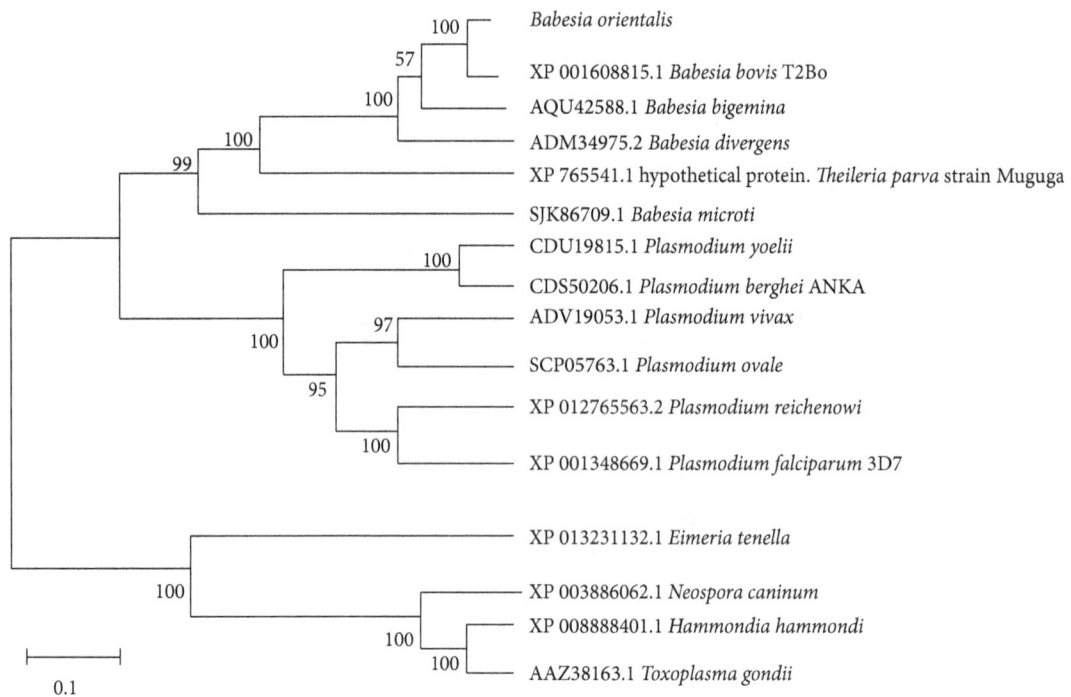

FIGURE 2: *Relationship of BoRON2 with other apicomplexan related RON2 proteins*. The neighbor joining tree was constructed using protein sequence of *Babesia orientalis* RON2 with its homologues found in NCBI, and the respective accessions numbers are indicated. The scale bar shows the length of branch; every 0.1 nucleotides' difference per 100 is represented. *B. orientalis* RON2 is in the same clade with *B. bovis*, *B. bigemina*, and *B. divergens*, more close to *B. bovis*.

the region of amino acid comprising between 1018 and 1295 (Figure 5) is most likely due to the presence in that region of active site for *B. orientalis* RON2 interaction and important domains during the invasion process. Neighbor joining phylogenetic tree was built from the protein sequence of *B. orientalis* RON2 submitted on GenBank with closely related apicomplexan parasites species as indicated with respective accession number (Table 1) which was obtained by using protein search on NCBI. The tertiary structure sequence is predominantly dominated by the presence of alpha helix (Figure 6) without coiled coils motifs structure in contrary to some reported RON2 orthologues such as *P. falciparum* and *P. vivax* [34], which may explain one of the host specificity reasons.

3.2. BoRON2 Immunoblotting Identification. The recombinant BoRON2 antibody from amplification of 936 bp (Figure 3(b)) located between the amino acid regions 914 and 1226 was expressed as His Thioredoxin tag fusion native protein. A strong band of 53 kDa was detected when probing the recombinant BoRON2 with positive *B. orientalis* serum of water buffalo (Figure 4(a)). The collected sera from rabbits immunized with BoRON2-C antigen have been used for *B. orientalis* RON2 protein identification in parasite lysates through immunoblot assay. As indicated on Figure 4(b) a band at ~150 kDa in parasite lysates was observed consistent with the computational predicted size, but no band was observed with the preimmune serum (PI).

TABLE 1: Apicomplexan species used for alignment and phylogenetic tree.

Organism	Accession number
Babesia bovis	XM_001608815.1
Babesia bigemina	AQU42588.1
Babesia microti	SJK86709.1
Babesia divergens	ADM34975.2
Theileria parva	XP_765541.1
Neospora caninum	XP_003886062.1
Toxoplasma gondii	AAZ38163.1
Eimeria tenella	XP_013231132.1
Plasmodium vivax	ADV19053.1
Plasmodium falciparum	XP_001348669.1
Plasmodium reichenowi	XP_012765563.2
Babesia orientalis	To be submitted
Hammondia hammondi	XP_008888401.1
Plasmodium ovale	SCP05763.1
Plasmodium berghei	CDS50206.1
Plasmodium yoelii	CDU19815.1

4. Discussion

Apicomplexan phylum contains obligate intracellular parasites characterized by a particular process during host cell invasion involving the secretory organelles essentially rhoptries and micronemes proteins [35]. They form a distinctive

TABLE 2: Set of primers used for amplification in this work.

Primer	Sequences (5'-3')	Remarks
F0	TGGATGMAATGATTTCTGGACCC	Degenerate
R0	ACGCACGTTTCCACTTGTTGT	primer
F1	ATGTTTGCGGTTACCCTGGCAACGATCACACTAGTG	ORF
R1	TCAATTAAATACAGTGTATGAGAAGTTGTCGTCAGC	
Fa	CGGGATCCATGGCGTTCCAAAGAGCTGCTA	Antibody
Ra	GCGTCGACTCATTGTACACCTACTGAGAGTGC	production

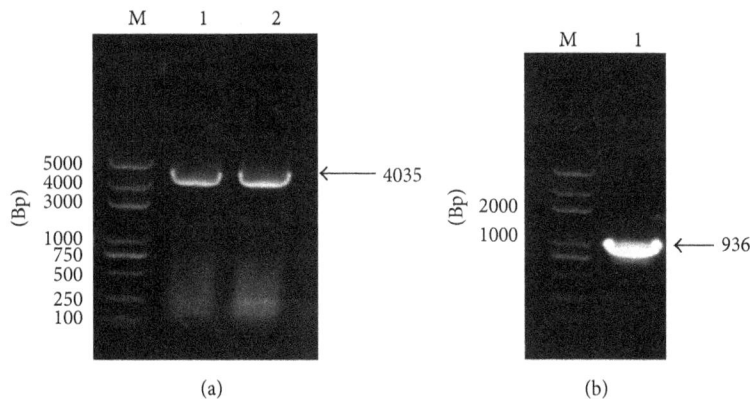

(a)   (b)

FIGURE 3: *PCR amplification of rhoptry neck 2 gene of Babesia orientalis from gDNA and cDNA and amplification of fragment used for polyclonal antibody production.* (a) M: nucleotide marker (15000 bps), line 1: amplification of *Bo*RON2 genomic DNA open reading frame showing a specific band of 4035 bp, and line 2: amplification of cDNA showing to have the same size with genomic DNA. (b) Line 1, the amplicon of product used for polyclonal antibody.

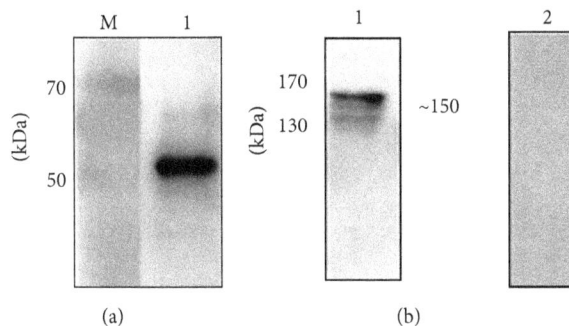

(a)   (b)

FIGURE 4: *Western blot of recombinant protein BoRON2-C.* (a) A specific band of 53 kDa was detected in reaction of recombinant *Bo*RON2 with anti *B. orientalis* water buffalo serum on line 1 and line M the prestained protein marker. (b) Immunoblot for detection of *Bo*RON2 protein in parasite lysates. Line 1, a unique band of around 150 kDa was detected in parasite lysate probed with polyclonal *Bo*RON2 antibody; line 2: no band detected with preimmune serum in parasite lysates.

structure called moving junction, in the space between host plasma membrane and parasite [33]. Most of these proteins of moving junction identified in *T. gondii* have been found to be well conserved among apicomplexan parasites, with clear homologues identified in other species such as *Plasmodium* [36]. The rhoptries neck protein 2 has been shown to be conserved among different reported homologous

species [37]. Numerous studies previously have reported both in *Plasmodium* and in *Toxoplasma* the direct interaction between AMA-1 and RON2 [38], and the occurrence of RON4 in homologues *Plasmodium* sp. and *Babesia* sp. has suggested the conservation of moving junction components across apicomplexan phylum [12].

In this work, we reported the RON2 protein of *B. orientalis*, which has been shown to be conserved to reported homologues, signifying the importance of this protein in biology of this parasite in infection stages. To our knowledge, this study is the first work to report molecular identification and characterization of *B. orientalis* RON2. The recombinant polyclonal antibody *Bo*RON2 raised was able to detect a unique band at ~150 kDa in parasite lysates, which was consistent with the predicted size. As reported previously, the generated polyclonal antibody RON2 of *B. microti* was able to detect a ~170 kDa predominant band consistent to the expected size and a product at ~55 kDa with no clear identity, but *B. divergens* antibodies identified also a consistent dominant band at ~170 kDa and 2 secondary bands at ~130 kDa and ~60 kDa [17]. Furthermore, in study on *P. falciparum* RON2 protein, sera in schizont stage have reacted with a band larger than 250 kDa corresponding to the predicted size and an additional band of 80 kDa without specific identity, but probably a Pf RON2 processing product [39]. However *P. vivax* RON2 polyclonal antibody has detected 2 bands (~220 kDa and ~185 kDa) different to the expected size (240 kDa) suggesting not only a proteolytic process but

```
Species                                                                                                                                               Position
Babesia_orientalis    - - S T Q E N L G I P S I N P L Y T R M A P D E R K V E F Q Q G M C G Q H C G A I W R A L L A F T M N A L R S P A S I K   1052
Babesia_bovis         - - S T Q E N L G I P G I N P L Y T R M A P D E R K L E F Q Q G M C G Q H C G A I W R A L L A F T M N A L R S P A S I K   1074
Babesia_microti       - E T I G G N P G L P E I S I R Y P H M S I E E R K I E F Q H S Q C A D H C I S I W R S L I A F T L N T L N N P A A I K   1199
Babesia_divergens     - - G A E E N M G I P S I N P L Y V R M S P D E R K V E F Q Q G M C G Q H C G A I W R A L L A F T M N T V R S P A S I K   1059
Babesia_bigemina      - - G T Q E N F G I P S V N P L Y T R M T P D E R K V E F Q Q G M C G Q H C G A I W R A L L A F T M N T M R S P A S I K   1062
Theileria_parva       - - D V P L N Y G M P T I S N L Y G R M S V D E R R I E F Q Q S M C S E H C G A I W K A I L A F T I S T L R N P G S I K   505
Neospora_canin        A A A T A R N Q G F L S L H Y D Y A H L S E A D R R K E F Q Q S M C M E H C E A I W K L I M A F V M P N L Q N P K K L K   1170
Eimeria_tenella       A A A A R R S A G F L S V H P E L A A L G P A A R E A E F Q N S M C M D H C E A L W T L V T S L V F S A L Q N P R K W R   1136
Toxoplasma_gondii     A A A T A R N Q G F L S I H Y D Y A N L P E E E R K K E F Q R S M C M E Q C E A L W K L V M A F V M P N L Q N P K K L K   1073
Plasmodium_falcipa    - - - V A R N P G M M A I N P K Y A E L S H E N R L R E L Q N S M C A D H C S S V W K V I S S F A L H H L K N P D S L H   1893
Plasmodium_vivax      - - - V A R N P G M V A I N P A Y A Q L N N E E R M K E L Q N S M C A D H C S A L W K T I S T F A L Q H L K N P E S L H   1907
Plasmodium_reichen    - - - V A R N P G M M A I N P K Y A E L S H E N R L R E L Q N S M C A D H C S S V W K V I S S F A L H H L K N P D S L H   1856

Babesia_orientalis    T F E K T L R Q G T S L K D M D K P E F V N S L R F I L H G D A M L H M Y D S M L P K K M K R E L R A I K Y G K A F Y F   1112
Babesia_bovis         T F E N T L R Q G T S L K D M E K P E F V N S L R F I L K G D A M L H M Y D S M L P R K M K R E L R A I K Y G K A F Y F   1134
Babesia_microti       Q F E K S L S S N S S L N D M S K P E Y I N S F K Y I L K G D S V L H M Y D N M L P R K V K R E I K A L K Y G K A F Y F   1259
Babesia_divergens     T F E K T L S Q N K S L K D M D S P E F V N S M R F I L K G D A M L H M Y D S M L P K K M K H E L R A I K Y G K A F Y F   1119
Babesia_bigemina      T F E K S L S Q N N T L Q D M E K P E F V N S M R F I L K G D A M L H M Y D S M L P K K M K N E L R A I K Y G K A F Y F   1122
Theileria_parva       S Y E K N L S S T T S L S E L N S P N Y V N N V R F I L K G D A W T G F Y D T L L P K S M K K E L E V M E F G K S F Y I   565
Neospora_canin        G Y E K D F S D A K E I E R L N S P H H V N A F R F G L S - - V Q I D F F D N M L D K T S K K N L K A M K Y G A S T W F   1228
Eimeria_tenella       D Y E K E V G G A A A A A A L A D P R R V N S F R L G L S - - V Q T D F F E N V L D K K S K R N I Q K M K F G G G S W F   1194
Toxoplasma_gondii     G Y E K D F S G A K E I E K L N S P H H V N A F R F S L S - - V Q I D F F D N M L D K T S K K N L K A M K Y G A S T W F   1131
Plasmodium_falcipa    T Y E S K F S K N S F G N K I D D K D F V H N F K M I L G G D A V L H Y F D N L L P K T M K K D L K A M K Y G V S L T S   1953
Plasmodium_vivax      S Y E S K F S K N S F G N K I D D Q N F V N N F K M I L G G D A V L H Y F D V L L P K S M K K E L K A M K N G V S L S S   1967
Plasmodium_reichen    T Y E S K F S K N S F G N K I D D K D F V H N F K M I L G G D A V L H Y F D N L L P K T M K K D L K A M K Y G V S L T S   1916

Babesia_orientalis    A N V M K M A S T L L G I I G F R Y T S N M L R I Q A P Y F G N M I V R W D R E R E K S R S K A I F S Y L S I G T M A T   1172
Babesia_bovis         A N I M K M A S T L L G M I G F R Y T S N M L R I Q A P Y F G N M I V R W D R E R E K S R S K A I F S Y L S I G T M A T   1194
Babesia_microti       A S I M K V A S M L F G G M G Y P Y V S R M L S I Q A P Y L G N F V V N W Q Q K R K S S R I G E I S G Y I G L G S I A S   1319
Babesia_divergens     A N V M K M A S K L L K I M G Y T Y T S N M L R I Q A P Y F G N F I V Q W D R E R E K S K S K A I F S Y L S I G T M A T   1179
Babesia_bigemina      A N I M K V A S N V L G I L G Y R Y T S N M L R I Q A P Y F G N F V V Q W D R E R E K S R S K A I F A Y L S I G T M A A   1182
Theileria_parva       A N I L K L A S V L M N R M G Y V T A T T M K V Q A P Y F G N F T T E W M K E R K K N R T K I L F S A L A L G T M A T   625
Neospora_canin        S Y A M K F A G Q V N S E M G N P N L G T A L Y M Q A P Y Y G D Y I R K W M E E R R A A R K Q A I I G V L T L G M M G L   1288
Eimeria_tenella       A Y A L L A A R L H R G L G H S D L A T F F S F Q A P Y L G H F V L Q W Q Q Q R R E A R R K A L L S M L S L G F F F A   1254
Toxoplasma_gondii     T Y A M K L A G Q V N S E M G N P N L G T A L Y M Q A A Y Y G N V I R K W M E E R R K S R K Q A I I G V L T L G M M G L   1191
Plasmodium_falcipa    A Y S L K L T K I I F S Q M Q L P Y L S Q M F Y M Q A P Y F G H F I G K W Q K K R Q Q S R L K E I M S F M T L G S L S A   2013
Plasmodium_vivax      A F S L K L T K I I F S E L Q L P Y L S Q M F Y T Q A P Y F G H F I G K W Q K E R E K S R M K E I L G F M T L G T L S A   2027
Plasmodium_reichen    A Y S L K L T K I I F S Q M Q L P Y L S Q M F Y M Q A P Y F G H F I G K W Q K K R Q Q S R L K E I M S F M T L G S L S A   1976

Babesia_orientalis    Y S I M Q C A D I A Q H A A D V G V G P A Q S C F I M V K P P - - S L H C V L Q P A E A L M K S A L S V G V Q D T L A V   1230
Babesia_bovis         Y S I L Q C A D I A Q H A A D V G V G P A E S C F I M V K P P - - A L H C V L K P V E T L M K S A L T I G V Q D T L A V   1252
Babesia_microti       H A I L S G M D I A Q H A A D V G I G P P E T C W F V P R P R P G R K L C I A E P I K S I A T T A T Q T A V Q D V F S V   1379
Babesia_divergens     Y S I M Q C A E I T Q H A F D V G S G P V E S C F M L I K P P - - Q M H C V L Q P A Q A I A K S A L T I G L Q D A L S V   1237
Babesia_bigemina      Y S V M Q C A E I T Q H A A D V G Q G P V E S C F M L I K P P - - R M H C L L Q P A E T I V K T A L S I G V Q D A L S V   1240
Theileria_parva       Y T V L E C M D I A Q H A V D M G H P P V E T C W Y L V K P P - - S M H C A I E P I S N L A I S A S V A I R D V F S S   683
Neospora_canin        Y S L L S V T D I V Q H M E D I G G A P P A S C V T N E I L G - - - V T C A P Q A I A K A T T S A A Q V A T Q D F L K V   1345
Eimeria_tenella       Y T F I S V S D I T Q H L N D S G L G P A V E C L E N L V V G - - - P V C P A A V V A P A V R S A A A A A A A D V F K V   1311
Toxoplasma_gondii     Y A L L N V A D I V Q H M E D I G G A P P V S C V T N E I L G - - - V T C A P Q A I A K A T T S A A R V A T Q D F L K V   1248
Plasmodium_falcipa    Y T L F S A M D I T Q Q A K D I G A G P V A S C F T T R M S P P - Q Q I C L N S V V N T A L S T S T Q S A M K C V F S V   2072
Plasmodium_vivax      Y T I L S A M D I S Q H A T D I G M G P A T S C Y T S T I P P P - K Q V C I Q Q A V K A T L T S S T Q A C M K S V F S V   2086
Plasmodium_reichen    Y T L F S A M D I T Q Q A K D I G T G P V A S C F T T R M S P P - Q Q I C L N S V V N T A L S T S T Q S A M K C V F S V   2035

Babesia_orientalis    T V L G L I G P Y F F L P M A A Y A S W N I L K H H F K - L H R L D I A L S G T F K R M W S K I S S L S V T K K L T G W   1289
Babesia_bovis         T V L G L I G P Y F F L P M M A Y A S W N I L K H H F K V L H R L D I A L S G T F K R M W S K I S S L S V T K K L T S W   1312
Babesia_microti       G L M A S I G P Y F I L P M A G L A A W S I L K S Q F K I L D R L Q T A F T G L F S K F F A T I A N G S G I K K I K R W   1439
Babesia_divergens     T V L A V I G P Y F F L P M A V H A S W Q I L K H H F K V I H R L D I A A S A A F S K M W S K I S S S S I A K R L S K W   1297
Babesia_bigemina      T V M A V I G P Y F F L P M A A Y A S W Q I L K H H F K V I H R L D I A L S N T F K R M W S K I S A S S I G K K L T A W   1300
Theileria_parva       T I L A L T G P Y M M I P M G I Y A G W T L L K R Q F K I L H R L D M A I S S V F S R L W R R V N T K D M I R S I A N L   743
Neospora_canin        G L F A G L A P Y L M L P M A I V S V W N I L K S E I K I L L Q F E M A V K H M F S R L K R W L A A P - - - - - F K N W   1400
Eimeria_tenella       G L F G L L T P Y L V W P M A A A A W Q L L R S E F K V L L Q F E M S L K S L F S R F S S W V R R P - - - - - F A R W   1366
Toxoplasma_gondii     G L F A G M A P Y L M L P M A V V S V W N I L K S E I K V L L Q F E M A L K H T F T R L K R W L A A P - - - - - F K N W   1303
Plasmodium_falcipa    G L F A S I G P Y L F A P M A G L A V W N I L K S E F K V L Q R I D M A L K N V F K N M W N K F L S L K G I S K L R G I   2132
Plasmodium_vivax      G L F A S I G P Y L F A P M A G L A V W N V L K S E F K V L Q R V D M A L K S V F K N M W R K F L S I K G I R K L K Y I   2146
Plasmodium_reichen    G L F A S I G P Y L F A P M A G L A V W N I L K S E F K V L E R I D M A L K N V F K N M W N K F L S L K G I S K L R G I   2095
```

FIGURE 5: *Amino acid alignment of Babesia orientalis RON2 and other orthologues.* The alignment showing only C-terminal the most conserved region (flanking the region of AA sequence comprising between 1018 and 1295) was generated by using ClustalW. The conserved residues sequences are drawn in black boxes and the four conserved cysteine in C-terminal region in red line among the all proteins RON2 sequences aligned and the three predicted transmembrane regions of *Bo*RON2 are underlined in blue, the 2 arrows indicating the 5′ and 3′ direction of sequence.

also a particular behavior during SDS-PAGE running [34]; curiously the same behavior for *T. gondii* RON2 characterized by an abnormal migration was observed, although the polyclonal antibody *T. gondii* RON2 has detected a band less than the predicted size (~150 kDa) cleaved to generate a 120 kDa protein [10].

The phylogenetic and BLAST analysis indicate the closeness of taxonomical relationship with other apicomplexan rhoptries proteins RON2, confirming, like for many reported *B. orientalis* genes such as AMA-1, Hsp20, Cox1, and Cob genes, that *B. bovis* is phylogenetically the closest among all related apicomplexan parasites with 72% sequence similarity;

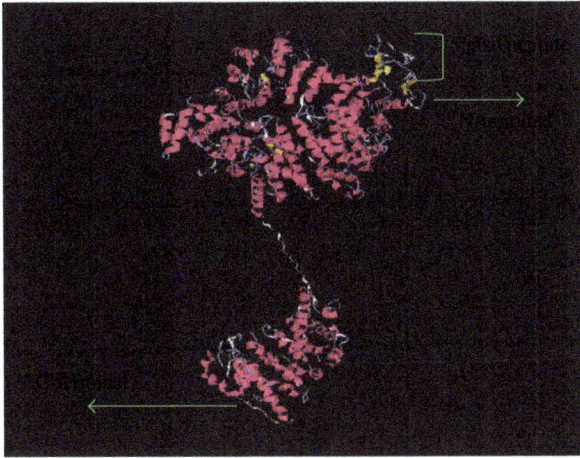

FIGURE 6: *Babesia orientalis RON2 tertiary prediction structure.* The predicted model with the highest C-score of *B. orientalis* RON2 protein.

additionally the apicoplast genome is more similar to *Babesia bovis* apicoplast genome structure [40]. The analysis of cDNA and gDNA has revealed that there are no introns. It has been also reported for AMA-1 nucleotide sequence of *B. orientalis* the absence of introns [17], supposing likely that the splicing process involving the apical complex proteins in this genus is a probable inexistent event during protein translation for invasion establishment process. The signal peptide is followed by N-terminal region relatively not conserved harboring 2 low complexity domains needed to be analyzed for functional identification. We noticed the presence of 12 cysteine residues; three between the two hydrophobic regions close to N-terminal region are conserved. The conserved C-terminal region contains 4 conserved cysteine residues, supposing the metabolic importance of that region in this genus; in fact most of apicomplexan parasites RON2, the cysteine-rich globular domain in merozoite surface, or apical proteins are involved in crucial interaction between ligand and surfaces receptors and also during invasion between host and parasite [41–43]. Additionally in *P. falciparum* merozoite and *T. gondii* tachyzoite the C-terminal recombinant protein RON2 was able to block the host cell invasion [15]. The relative conservation of *Babesia orientalis* C-terminal region presumes most likely the active site for interaction with AMA-1 facilitating this parasite invasion.

5. Conclusion

The underlying mechanism of buffalo babesiosis caused by *B. orientalis* in molecular level passes inevitably through RON2 protein identification. Herein we reported *B. orientalis* RON2 which has been identified to be conserved among related species, with a probable C-terminal physiologically and relatively more active. The abundance of cysteine residue is most likely relevant to its function during invasion, which unlikely makes this protein hard for solubilization in active form during experiment for the expression in bacteria. The polyclonal serum has been able to detect the native protein in parasite lysates suggesting *Bo*RON2 as a good potential candidate for vaccine and diagnosis of disease. Therefore, the limitation of doing more experiments for biochemistry and physiology purposes is due to the complication of culturing in vitro *B. orientalis* parasite.

Authors' Contributions

Particularly Junlong Zhao and Lan He contributed to experimental design, data analysis, and editing; Long Yu, Pei He, Junwei He, Yali Sun, and Yuan Huang participated in experiment and laboratory work. All authors have approved the manuscript.

Acknowledgments

This study was supported by the National Basic Science Research Program (973 Program) of China (Grant no. 2015CB150302), the Fundamental Research Funds for the Central Universities (2662015PY006), and Huazhong Agricultural University Scientific & Technological Self-Innovation Foundation (Program no. 2662014BQ020).

References

[1] L. Liu, H. Johnson, S. Cousens et al., "Global, regional and national causes of child mortality: an update systematic analysis for 2010 with time trends since 2000," *The Lancet*, vol. 379, no. 9832, pp. 2151–2161, 2012.

[2] W. C. Brown and G. H. Palmer, "Designing blood-stage vaccines against Babesia bovis and B. bigemina," *Parasitology Today*, vol. 15, no. 7, pp. 275–281, 1999.

[3] D. T. Dewaal, "Global importance of piroplasmosis," *The Journal of Protozoology*, vol. 10, pp. 106–127, 2000.

[4] T. Bousema, L. Okell, I. Felger, and C. Drakeley, "Asymptomatic malaria infections: Detectability, transmissibility and public health relevance," *Nature Reviews Microbiology*, vol. 12, no. 12, pp. 833–840, 2014.

[5] W. J. R. Stone, M. Eldering, G.-J. Van Gemert et al., "The relevance and applicability of oocyst prevalence as a read-out for mosquito feeding assays," *Scientific Reports*, vol. 3, article 3418, 2013.

[6] C. Opitz and D. Soldati, "A dynamic complex powering gliding motion and host cell invasion by Toxoplasma gondii," *Molecular Microbiology*, vol. 45, no. 3, pp. 597–604, 2002.

[7] V. B. Carruthers and L. D. Sibley, "Sequential protein secretion front three distinct organelles of Toxoplasma gondii accompanies invasion of human fibroblasts," *European Journal of Cell Biology*, vol. 73, no. 2, pp. 114–123, 1997.

[8] D. Gaur and C. E. Chitnis, "Molecular interactions and signaling mechanisms during erythrocyte invasion by malaria parasites," *Current Opinion in Microbiology*, vol. 14, no. 4, pp. 422–428, 2011.

[9] E. Suss-Toby, J. Zimmerberg, and G. E. Ward, "Toxoplasma invasion: The parasitophorous vacuole is formed from host cell plasma membrane and pinches off via a fission pore," *Journal of Cell Biology*, vol. 93, no. 16, pp. 8413–8418, 1996.

[10] K. W. Straub, S. J. Cheng, C. S. Sohn, and P. J. Bradley, "Novel components of the Apicomplexan moving junction reveal conserved and coccidia-restricted elements," *Cellular Microbiology*, vol. 11, no. 4, pp. 590–603, 2009.

[11] D. L. Alexander, J. Mital, G. E. Ward, P. Bradley, and J. C. Boothroyd, "Identification of the moving junction complex of Toxoplasma gondii: a collaboration between distinct secretory organelles.," *PLoS Pathogens*, vol. 1, no. 2, p. e17, 2005.

[12] M. Lebrun, A. Michelin, H. El Hajj et al., "The rhoptry neck protein RON4 relocalizes at the moving junction during Toxoplasma gondii invasion," *Cellular Microbiology*, vol. 7, no. 12, pp. 1823–1833, 2005.

[13] D. L. Alexander, S. Arastu-Kapur, J.-F. Dubremetz, and J. C. Boothroyd, "Plasmodium falciparum AMA1 binds a rhoptry neck protein homologous to TgRON4, a component of the moving junction in Toxoplasma gondii," *Eukaryotic Cell*, vol. 5, no. 7, pp. 1169–1173, 2006.

[14] C. R. Collins, C. Withers-Martinez, F. Hackett, and M. J. Blackman, "An inhibitory antibody blocks interactions between components of the malarial invasion machinery," *PLoS Pathogens*, vol. 5, no. 1, article e1000273, 2009.

[15] D. Richard, C. A. MacRaild, D. T. Riglar et al., "Interaction between *Plasmodium falciparum* apical membrane antigen 1 and the rhoptry neck protein complex defines a key step in the erythrocyte invasion process of malaria parasites," *The Journal of Biological Chemistry*, vol. 285, no. 19, pp. 14815–14822, 2010.

[16] M. Lamarque, S. Besteiro, J. Papoin et al., "The RON2-AMA1 interaction is a critical step in moving junction-dependent invasion by apicomplexan parasites," *PLoS Pathogens*, vol. 7, no. 2, article e1001276, 2011.

[17] R. L. Ord, M. Rodriguez, J. R. Cursino-Santos et al., "Identification and characterization of the rhoptry neck protein 2 in Babesia divergens and B. microti," *Infection and Immunity*, vol. 84, no. 5, pp. 1574–1584, 2016.

[18] L. He, L. Fan, J. Hu et al., "Characterisation of a Babesia orientalis apical membrane antigen, and comparison of its orthologues among selected apicomplexans," *Ticks and Tick-borne Diseases*, vol. 6, no. 3, pp. 290–296, 2015.

[19] L. He, H.-H. Feng, Q.-L. Zhang et al., "Development and evaluation of real-time PCR assay for the detection of babesia orientalis in water buffalo (Bubalus bubalis, Linnaeus, 1758)," *Journal of Parasitology*, vol. 97, no. 6, pp. 1166–1169, 2011.

[20] Z. L. Liu, L. H. Ma, B. A. Yao, and J. L. Zhao, "Test of infected cattle by bite *Rhipicephalus haemaphysaloides* and injection with parasitized blood of buffalo with *Babesia*," *Acta Zootechnica Sin*, vol. 26, no. 5, pp. 468–472, 1995.

[21] P. A. Conrad, K. Iams, W. C. Brown, B. Sohanpal, and O. K. ole-MoiYoi, "DNA probes detect genomic diversity in Theileria parva stocks," *Molecular and Biochemical Parasitology*, vol. 25, no. 3, pp. 213–226, 1987.

[22] M. A. Larkin, G. Blackshields, N. P. Brown et al., "Clustal W and clustal X version 2.0," *Bioinformatics*, vol. 23, no. 21, pp. 2947-2948, 2007.

[23] T. N. Petersen, S. Brunak, G. Von Heijne, and H. Nielsen, "Discriminating signal peptides from transmembrane regions: signal P 4.0," *Nature Methods*, vol. 8, no. 10, pp. 785-786, 2011.

[24] A. Krogh, B. Larsson, G. Von Heijne, and E. L. L. Sonnhammer, "Predicting transmembrane protein topology with a hidden Markov model: application to complete genomes," *Journal of Molecular Biology*, vol. 305, no. 3, pp. 567–580, 2001.

[25] I. Letunic, T. Doerks, and P. Bork, "SMART 7: recent updates to the protein domain annotation resource," *Nucleic Acids Research*, vol. 40, no. 1, pp. D302–D305, 2012.

[26] D. W. A. Buchan, F. Minneci, T. C. O. Nugent, K. Bryson, and D. T. Jones, "Scalable web services for the PSIPRED protein analysis workbench," *Nucleic Acids Research*, vol. 41, pp. W340–W348, 2013.

[27] A. M. Newman and J. B. Cooper, "A practical algorithm for identification and architecture modeling of tandem repeats in protein sequences," *BMC Bioinformatics*, vol. 8, article 382, 2007.

[28] K. Tamura, G. Stecher, D. Peterson, A. Filipski, and S. Kumar, "MEGA6: molecular evolutionary genetics analysis version 6.0," *Molecular Biology and Evolution*, vol. 30, no. 12, pp. 2725–2729, 2013.

[29] J. Yang, R. Yan, A. Roy, D. Xu, J. Poisson, and Y. Zhang, "The I-TASSER suite: protein structure and function prediction," *Nature Methods*, vol. 12, no. 1, pp. 7-8, 2015.

[30] T. A. Hall, "Bioedit: a user friendly biological sequences alignment and analysis program for windows 95/98/NT," *Nucleic Acids Symposium*, vol. 41, pp. 95–98, 1999.

[31] E. Gasteiger, C. Hoogland, A. Gattiker, M. R. Wilkins, R. D. Appel, and A. Bairoch, "Protein identification and analysis tools on the expasy server," in *The Proteomics Protocols Handbook*, J. M. Walker, Ed., pp. 501–607, Humana Press, New Jersey, NJ, USA, 2005.

[32] B. Vulliez-Le Normand, M. L. Tonkin, M. H. Lamarque et al., "Structural and functional insights into the malaria parasite moving junction complex," *PLoS Pathogens*, vol. 8, no. 6, article e1002755, 2012.

[33] N. I. Proellocks, L. M. Kats, D. A. Sheffield et al., "Characterisation of PfRON6, a plasmodium falciparum rhoptry neck protein with a novel cysteine-rich domain," *International Journal for Parasitology*, vol. 39, no. 6, pp. 683–692, 2009.

[34] G. Arévalo-Pinzón, H. Curtidor, L. C. Patiño, and M. A. Patarroyo, "PvRON2, a new *Plasmodium vivax* rhoptry neck antigen," *Malaria Journal*, vol. 10, article 60, 2011.

[35] M. Aikawa, L. H. Miller, J. Johnson, and J. Rabbege, "Erythrocyte entry by malarial parasites. A moving junction between erythrocyte and parasite," *Journal of Cell Biology*, vol. 77, no. 1, pp. 72–82, 1978.

[36] J. F. Dubremetz, N. Garcia-Réguet, V. Conseil, and M. N. Fourmaux, "Apical organelles and host-cell invasion by Apicomplexa," *International Journal for Parasitology*, vol. 28, no. 7, pp. 1007–1013, 1998.

[37] P. J. Bradley, C. Ward, S. J. Cheng et al., "Proteomic analysis of rhoptry organelles reveals many novel constituents for host-parasite interactions in *Toxoplasma gondii*," *Journal of Biological Chemistry*, vol. 280, no. 40, pp. 34245–34258, 2005.

[38] B. Shen and L. D. Sibley, "The moving junction, a key portal to host cell invasion by apicomplexan parasites," *Current Opinion in Microbiology*, vol. 15, no. 4, pp. 449–455, 2012.

[39] J. Cao, O. Kaneko, A. Thongkukiatkul et al., "Rhoptry neck protein RON2 forms a complex with microneme protein AMA1 in Plasmodium falciparum merozoites," *Parasitology International*, vol. 58, no. 1, pp. 29–35, 2009.

[40] Y. Huang, L. He, H. Jifang, P. He, J. He et al., "Characterization and annotation of *Babesia orientalis* apicoplast genome," *Parasites & Vectors*, vol. 8, p. 543, 2015.

[41] V. K. Goel, X. Li, H. Chen, S. C. Liu, A. H. Chishti, and S. S. Oh, "Band 3 is a host receptor binding merozoite surface protein 1 during the *Plasmodium falciparum* invasion of erythrocytes," *Proceedings of the National Academy of Sciences of the United States of America*, vol. 100, no. 9, pp. 5164–5169, 2003.

[42] X. Li, H. Chen, T. H. Oo et al., "A co-ligand complex anchors plasmodium falciparum merozoites to the erythrocyte invasion receptor band 3," *Journal of Biological Chemistry*, vol. 279, no. 7, pp. 5765–5771, 2004.

17

Multifunctional Thioredoxin-Like Protein from the Gastrointestinal Parasitic Nematodes *Strongyloides ratti* and *Trichuris suis* Affects Mucosal Homeostasis

Dana Ditgen,[1,2] **Emmanuela M. Anandarajah,**[1,2] **Jan Hansmann,**[3] **Dominic Winter,**[4] **Guido Schramm,**[5] **Klaus D. Erttmann,**[2] **Eva Liebau,**[1] **and Norbert W. Brattig**[2]

[1]*Department of Molecular Physiology, Westfälische Wilhelms-University, Münster, Germany*
[2]*Department of Molecular Medicine, Bernhard Nocht Institute for Tropical Medicine, Hamburg, Germany*
[3]*Department of Tissue Engineering and Regenerative Medicine (TERM), University of Würzburg, Germany*
[4]*Institute for Biochemistry and Molecular Biology, University of Bonn, Bonn, Germany*
[5]*Ovamed GmbH, Hamburg, Germany*

Correspondence should be addressed to Norbert W. Brattig; nbrattig@bni-hamburg.de

Academic Editor: Ana Maria Jansen

The cellular redox state is important for the regulation of multiple functions and is essential for the maintenance of cellular homeostasis and antioxidant defense. In the excretory/secretory (E/S) products of *Strongyloides ratti* and *Trichuris suis* sequences for thioredoxin (Trx) and Trx-like protein (Trx-lp) were identified. To characterize the antioxidant Trx-lp and its interaction with the parasite's mucosal habitat, *S. ratti* and *T. suis* Trx-lps were cloned and recombinantly expressed. The primary antioxidative activity was assured by reduction of insulin and IgM. Further analysis applying an *in vitro* mucosal 3D-cell culture model revealed that the secreted Trx-lps were able to bind to monocytic and intestinal epithelial cells and induce the time-dependent release of cytokines such as TNF-α, IL-22, and TSLP. In addition, the redox proteins also possessed chemotactic activity for monocytic THP-1 cells and fostered epithelial wound healing activity. These results confirm that the parasite-secreted Trx-lps are multifunctional proteins that can affect the host intestinal mucosa.

1. Introduction

Parasitic intestinal nematodes are widespread, affecting human and vertebrates. Worldwide, more than one-third of mankind is infected with helminths [1] of which 100–200 million people are infected with *Strongyloides* [2, 3] and approximately 800 million with *Trichuris* [4]. The investigated nematodes *Strongyloides ratti* and *Trichuris suis* are very closely related to their human-pathogenic homologues *Strongyloides stercoralis* and *Trichuris trichiura* [5, 6].

In contrast to immune responses to microbes with mainly inflammation, the immune responses to helminths are mostly less intense and highly regulated [7]. Modulation of the host's immune response reported from *T. suis* ova can be beneficial for an attenuation of inflammatory bowel diseases (IBD) such as Crohn's disease and ulcerative colitis [8, 9]. Helminths release multiple excretory/secretory (E/S) products which enable them to establish, survive, and reproduce in their hosts successfully [10, 11]. In case of *S. ratti* and *T. suis*, these E/S products include antioxidative proteins such as thioredoxin (Trx), heat shock proteins, and numerous proteases as well as protease inhibitors, galectins, and orthologous of host cytokines [10, 12–16]. Trx has also been reported in E/S products of multiple helminths [17–20]. Recently, these E/S proteins have also been detected in extracellular vesicles from helminths [21].

Trx or the Trx system in general is widespread from archaea to human consisting of Trx, the Trx reductase, and NADPH [22]. Hereby, Trx is reduced by the Trx reductase in an NADPH-dependent manner [23]. In general, Trx superfamily members regulate thiol-based redox control, operating as protein disulfide oxidoreductases, and protect cytosolic proteins against aggregation in the cell [24]. Its redox-regulating activity is important for DNA replication, maintenance of the cellular redox state, and, therefore, the cellular homeostasis and antioxidant defense [22, 25]. Furthermore, Trx is part of multiple cellular pathways [26] and capable of regulating transcription factor activities, inhibition of apoptosis, protection from high-energy oxygen radicals, and regeneration of denatured proteins and is critical for signal transduction through thiol redox control as well as more specific processes like presenting antigens [22, 23, 26–28]. Without a signal peptide, Trx is secreted by a nonclassical secretory pathway by various cells [29, 30].

The numerous extracellular activities of Trx include anti-inflammatory and antiapoptotic, and thus cytoprotective effects [31–33]. Of interest, multifunctional prokaryotic Trx, which displays unrelated properties, that is, antioxidant activity and promotion of DNA replication, has been described as moonlighting protein [34–36]. In the E/S products of Strongyloides and of multiple other helminths numerous multifunctional proteins have been detected like the moonlighting enzymes enolase and glyceraldehyde-3-phosphate dehydrogenase [10, 13, 37–39].

While Trx is well characterized, less is known about the functions of Trx-lp [26]. The Trx-lp, a member of the Trx superfamily, is a fusion protein composed of the classical Trx domain (WCGPC) at the N-terminus and a C-terminal proteasome-interacting thioredoxin (PITH) domain, formerly known as DUF1000 (protein families database, http://pfam.xfam.org/family/PF06201). It is larger than the classical Trx (12 kDa), which is highly conserved in all species [23, 25]. Proteins of the Trx superfamily have been reported in various protozoan parasites including Plasmodium, Trypanosoma, and Toxoplasma [40–43] and in the trematode Clonorchis sinensis [44]. Besides thiol-based redox control, eukaryotic Trx-lps are also involved in signaling processes as cofactors of certain enzymes, regulating specific signal proteins [45, 46]. For example, the human Trx-related protein (TRP32), known as TXNL-1, protects the cell against glucose deprivation-induced cytotoxicity and is involved in activation of antiapoptotic Akt/PI3K signaling as well as PTEN (phosphatase and tensin homologue deleted on chromosome ten) inhibition [47, 48]. Another example is the thioredoxin domain containing 17 (TXNDC17), also known as Trx-related protein of 14 kDa (TRP14), which is STAT-3-dependent and responsible for the drug resistance in human colorectal cancer cells. TRP14 also shows, like Trx1, S-nitrosylase activity and furthermore is able to control the TNF-α/NF-κB signaling pathway [49–51]. In addition, PTEN is also an interaction partner of human Trx and among others Trx controls the TNF-α/NF-κB signaling pathway as well [52, 53]. The novel thioredoxin-related transmembrane protein TMX4 is a type I transmembrane protein with its Trx-like domain inside the ER which possibly plays a role in the correct folding of proteins inside the ER due to its reductase function [54].

Since Trx have been reported to act as chemoattractant for leukocytes and to induce cytokines [31] we wanted to examine if SrTrx-lp has similar impact on monocytic cells.

In the present study we cloned and characterized two Trx-lps and investigated some functional activities including their chemotactic activity, their ability to promote wound healing processes in the intestinal epithelial cell (IEC) Caco-2 model, and their involvement in cytokine release in a three-dimensional- (3D-) cell culture model.

2. Material and Methods

2.1. Parasites. The *S. ratti* life cycle was maintained in our laboratory as reported [13, 15]. Animal experiments were approved by and conducted in accordance with guidelines of the Animal Protection Board of the City of Hamburg (G21131/591-00.33). The life cycle was maintained using Wistar rats by serial passage and the developmental stages isolated as described [14]. *T. suis* stages were obtained from Ovamed (Hamburg, Germany).

2.2. Preparation of Somatic Extracts. *S. ratti* and *T. suis* extracts were prepared from freshly harvested life stages as described before [13, 15].

2.3. DNA Sequencing and Bioinformatic Analysis. PCR products and plasmids were sequenced by the dideoxy termination method of Sanger performed by eurofinsgenomics.eu. For homology searches the NCBI Blast Program was used (http://www.ncbi.nlm.nih.gov/). Further, for bioinformatics analyses the Expert Protein Analyses System (ExPASy) proteomics server of the Swiss Institute of Bioinformatics (http://expasy.org/tools/) was used. To obtain the conserved domains of the Trx-lps the protein families database (Pfam) of the USA server (http://pfam.xfam.org/family/PF06201) was used which represents proteins by multiple sequence alignments and hidden Markov models (HMMs). Multiple sequence alignments were performed by the program CLUSTAL_W2 (http://www.ebi.ac.uk/Tools/msa/clustalw2/) from the European Bioinformatics Institute which is part of the European Molecular Biology Laboratory (EMBL-EBI).

2.4. Mass Spectrometry. SrTrx-lp and TsTrx-lp SDS-PAGE bands were excised, cut into small cubes, and transferred to microtubes and in gel digestion was performed as described elsewhere [57]. Briefly, gel pieces were destained using 30% acetonitrile (ACN), 0.07 M NH_4HCO_3, reduced with 20 mM dithiothreitol and alkylated by 1% acrylamide, and dehydrated using 100% ACN [57]. ACN was removed and the gel pieces were dried using a vacuum centrifuge and rehydrated in 0.1 M NH_4HCO_3 containing 0.5 μg of trypsin (Promega, Mannheim, Germany). A sufficient volume of 0.1 M NH_4HCO_3 was added to cover the gel pieces completely and digestion was performed at 37°C overnight. The peptide containing supernatant was transferred to new microtubes and the gel pieces were extracted with 50% ACN, 0.1% trifluoroacetic acid followed by 0.1 M NH_4HCO_3 and ACN. Samples were dried in the vacuum centrifuge, resuspended in 5% ACN and 5% formic acid, desalted using C_{18} StageTips [58], dried again, and resuspended in 5% ACN and 5% formic acid. For reversed phase chromatography in house manufactured analytical columns were used. Using 100 μm inner diameter fused silica capillaries, spray tips were generated with a P2000 laser puller (Sutter Instruments, Novato, CA, USA) and packed with 5 μm ReproSil-Pur 120 C_{18}-AQ particles (Dr. Maisch, Ammerbuch-Entringen, Germany). Peptides were loaded directly on the analytical column using a nanoflow UHPLC system (EASY-nLC 1000, Thermo Fisher Scientific, Bremen, Germany) at a flow rate of 1 μL/min solvent C (water with 0.1% formic acid). Peptides were eluted applying a 60 min linear gradient from 100% solvent A (water with 5% DMSO [59], 0.1% formic acid), to 65% solvent A, 35% solvent B (ACN with 5% DMSO, 0.1% formic acid) at a flow rate of 400 nL/min. Eluting peptides were ionized in the positive ion mode at 1.6 kV in the nanospray ion source of an Orbitrap Velos mass spectrometer (Thermo Fisher Scientific, Bremen, Germany). Survey scans (m/z 400 to 1200) were performed in the Orbitrap analyzer at a resolution of 30,000 followed by fragmentation of the 10 most abundant ions in the linear ion trap by collision induced dissociation. Dynamic exclusion was set to 30 sec with an exclusion list size of 500. Thermo *.raw files were analyzed using Maxquant (version 1.5.2.8) using the following settings: protein N-terminal acetylation and oxidation of methionine were set as variable modifications and propionamide at cysteine was set as fixed modification; enzyme specificity was set to trypsin and up to two missed cleavage sites were allowed. Data were searched against a database consisting of all *S. ratti* and *T. suis* entries from Uniprot/TrEMBL (version from 12/01/2014, 12,462 entries) as well as common contaminations. The false discovery rate was set to 1%.

2.5. Cloning, Expression, and Purification of Recombinant Trx-lps. *S. ratti* and *T. suis* RNA were isolated from adult parasitic females as described before [15] and the cDNA was synthesized by using the First Strand cDNA Kit from New England BioLabs® Inc. according to the manufacturer's instructions. Forward and reverse primers were generated using the online tool provided by Clontech (http://bioinfo.clontech.com/infusion/) (TsTrx-lp: forward: AAGGTCGTCATATGATGGCT ATAAAGGAGATAA; reverse: TCCTCGAGAATTCCTAATGAGCTTCTCCCT-T; SrTrx-lp: forward: AAGGTCGTCATATGATGGCTA-TAAAGGAGATAA; reverse: TCCTCGAGAATTCCTAAT-GAGCTTCTCCCTT). Fragments were amplified by PCR using the InFusion® HD Cloning Kit from Clontech according to the manufacturer's instructions and the Phusion High-Fidelity DNA-Polymerase from Thermo Scientific (Waltham, USA). The Trx-lp PCR fragments from *S. ratti* and *T. suis* were cloned into pJC45 vector [60] and IBA 3 plus vector, transformed into *Escherichia coli* Stellar cells (Clontech, USA) and sequenced (eurofins MWG).

The *S. ratti* and *T. suis* Trx-lps were expressed in lipopolysaccharide- (LPS-) free *E. coli* strain ClearColi® BL21 (DE3) (Lucigen Simplifying Genomics), which do not trigger the endotoxic response in human cells, in Luria-Bertani medium containing 100 μg/mL ampicillin. The expression of the His-tag fusion proteins was induced by isopropyl-β-D-thiogalactopyranoside (IPTG, final concentration 1 mM) and the expression of the Strep-tag fusion proteins by anhydrotetracycline (AHT, final concentration 200 μg/L), for 5 h

at 37°C. The bacterial cells were collected by centrifugation (6,000 ×g) for 15 min and kept at −20°C until use. Recombinant proteins were purified by using Ni^{2+} affinity chromatography (Qiagen, Hilden, Germany) or Strep-Tactin® Superflow Plus (Qiagen, Germany) according to the manufacturer's instructions. The imidazole or desthiobiotin was removed by dialysis overnight using phosphate-buffered saline (PBS, pH 7.4). Even though the endotoxin-free *E. coli* strain was used the LPS inhibitor polymyxin B (30 μg/mL) was added to all buffers used. Sodium dodecyl sulfate polyacrylamide gel electrophoresis (SDS-PAGE) was applied to verify expression and purity of the proteins, which were visualized by Coomassie brilliant blue G-250 staining. The protein concentration was quantified by Bradford assay. Furthermore, the elutions were analyzed by semidry Western blot. After SDS-PAGE and the following transfer onto nitrocellulose membranes, the membranes were incubated with the anti-his6-peroxidase (2) (mouse monoclonal; 1 : 5000; Roche life science, Mannheim, Germany) overnight at 4°C.

2.6. Functional Activity Assays

2.6.1. Insulin Reduction. According to the method of Holmgren [61] (1979) as well as Luthman and Holmgren [62] (1982), disulfide reduction activity was measured by reduction of insulin [61, 62]. In this test, the turbidity of the sample was measured, which is caused by the precipitating reduced insulin. The resulting decrease in absorbance was measured at 650 nm. During the reaction, the SrTrx-lp was repeatedly regenerated by DTT. Here, the regeneration of active Trx-lp is faster than the direct reduction of insulin by DTT. Initially, 1.6 mM insulin (bovine pancreas, Sigma-Aldrich, Hamburg, Germany) was prepared by a suspension of 50 mg of insulin in 2.5 mL 100 mM potassium phosphate buffer (pH 6.5) for the reaction approach. Here, the pH was first adjusted to 3 with 1 M HCl solution to completely dissolve the protein and the pH was adjusted to 6.5 with 1 M NaOH. The solution was supplemented with dH_2O to a volume of 5 mL. Thereafter, a master mix of 825 μL 1.6 mM insulin (160 μM final volume) and 4675 μL PE (100 mM potassium phosphate, 2 mM EDTA, pH 6.5) buffer was prepared. SrTrx-lp was tested at a concentration of 1 μM (30 μg/mL), 2.5 μM (75 μg/mL), and 5 μM (150 μg/mL). In an interval of 1 min over a period of 40 min, the reduction of insulin by SrTrx-lp was measured. As a negative control, the same reaction approach was used without redox regulatory protein. The amount of SrTrx-lp was replaced by PE-buffer. The relative specific enzymatic activity was calculated by the following formula: ΔA_{650} × 1000/mg protein concentration in the reaction mix.

2.6.2. IgM Reduction. According to the method of Wollman et al. (1988), the Trx-lp from either *S. ratti* or *T. suis* was reduced by 100 mM DTT for 1 h at room temperature (RT) and dialyzed against 80 mM HEPES and 10 mM EDTA buffer for 1 h at 4°C to remove DTT [63]. The dialysis buffer was also used as reaction buffer. The buffer was mixed with 1.7 μM IgM (PierceTM Mouse IgM Isotype Control, Thermo Scientific, Czech Republic) and 0.5 μL, 1 μL, and 5 μL of the reduced

Trx-lp solution for overnight reaction at RT. For protein size determination SDS-PAGE analysis was performed under nonreducing conditions (5–12% acrylamide gradient). Silver nitrate staining was used to visualize proteins [63].

2.7. Cells

2.7.1. Preparation of Peripheral Blood Cells.
In agreement with institutional guidelines healthy volunteers served as source for peripheral blood mononuclear cells (MNC) and polymorphonuclear cells (PMN) purified from venous blood samples (collected in sodium citrate tubes). First, erythrocytes were sedimented from anticoagulated blood samples by addition of equal amounts of 6% hydroxyethyl starch (HEAS-steril®, Fresenius, Friedberg, Germany). MNCs were separated from PMN as reported before by density centrifugation using a two-level density gradient consisting of Mono-Poly Resolving Media (1.114 g/ML; MP Biomedicals, Stockholm, Sweden) and Lymphoflot (1.077 g/mL; Bio-Rad, Dreieich, Germany) [14]. Both the MNC interphase and the PMN interphase were collected and the rest discarded. The cells were washed carefully with PBS, followed by a centrifugation step at 1,800 rpm for 10 min. This step was optionally repeated one more time, if too many platelets were present. While the MNCs were added to the THP-1 media, the PMNs were resuspended in HBSS both at a concentration of 5×10^5 cells/mL and stored on ice until further use.

2.7.2. Three-Dimensional Coculture.
To analyze the immunological effect of SrTrx-lp and TsTrx-lp, the recombinant proteins were used as stimuli in a 3D-coculture model, composed of human intestinal epithelial and dendritic cells (DCs), derived from monocytic THP-1 cells, grown on a collagen scaffold that mimics the in vivo natural microenvironment [64].

The human intestinal epithelial cells, Caco-2 cells, were grown in DMEM media (with 10 % FCS, 1% nonessential amino acids, 1% Pen/Strep; Liefer-Co) until denseness of 70–80% was reached and seeded on 12-well plates in ThinCerts™ TC inserts (Greiner BioOne) followed by the addition of 200 μL collagen (University Hospital Würzburg) to each insert. Prior to adding the Caco-2 cells, the collagen was incubated 1 h at 37°C for gelation. To detach the Caco-2 cells from the flask the cells were trypsinized prior to transfer 10^5 cells/well into the collagen-layered inserts and incubated for 2 h at 37°C and 5% CO_2 to let them adhere on the collagen. Afterwards, wells were floated with DMEM media. The cells were grown for at least 14 days until a monolayer was formed. For differentiation to DCs, THP-1 cells were washed twice in PBS and seeded in serum-free RPMI 1640 media supplemented with IL-4 (1000 IU/mL; Peprotech, Hamburg, Germany) and GM-CSF (1000 IU/mL; Peprotech) and were grown for 7–10 days [65]. Subsequently the generation of mature DCs was verified by staining 10^5 washed cultured cells with phycoerythrin- (PE-) conjugated monoclonal anti-CD86 (B7-2) antibodies (mouse anti-human CD86-PE-conjugated antibody; Becton-Dickinson Bioscience, San Diego, USA, and a PE-conjugated isotype control; PharMingen, Leiden, Netherlands) analyzed by flow cytometry (CellQuestPro; BD) (data not shown) [66]. After proper development of both cell types, the Caco-2-collagen inserts were transferred to the wells with grown DCs, which were floated with DMEM media (10% FCS, 1% nonessential amino acids, 1% Pen/Strep).

The Trx-lps were added as stimuli (5 μg, 10 μg, and 25 μg/mL), while the UFM-1 activating protein UBA-5 (25 μg/mL) from the nonparasitic nematode Caenorhabditis elegans served as negative control. UBA-5 was cloned and expressed as published by our group [67]. Further controls were performed with the bacterial cell wall components LPS (1 μg/mL; Sigma-Aldrich, Taufkirchen, Germany) and lipoteichoic acid (LTA, 0.1 μg/mL; Sigma-Aldrich, Taufkirchen) to analyze potential endotoxin contaminations and to compare both responses. Worm extract from T. suis served as positive controls for a T_H2 response. The supernatants were taken after 24 h, 48 h, and 72 h and stored at −20°C until further use.

2.8. Cytokine Enzyme-Linked Immunosorbent Assay (ELISA).
For detection of the cytokines TNF-α, IL-10, IL-22, and TSLP in cell supernatants, human ELISA Ready-SET-Go! kits from eBioscience (San Diego, USA) were used according to the manufacturer's instructions. Here, IL-10 was detected with a sensitivity of 2 pg/mL, IL-22 and TSLP with a sensitivity of 8 pg/mL, and TNF-α with a sensitivity of 4 pg/mL.

2.9. Flow Cytometry.
To measure the binding affinity of the S. ratti and T. suis Trx-lps to certain cell types, the purified proteins were labeled using the Alexa Fluor® 647 Protein Labeling Kit Microscale (A30009) from Invitrogen (Oregon, USA) according to the manufacturer's instructions. The binding affinity for both Trx-lps to monocytes, lymphocytes, and granulocytes from peripheral blood, as well as to the cell lines THP-1 cells (undifferentiated and differentiated) and Caco-2 cells, were tested. Approximately 2×10^5 cells were used per reaction. The fluorescently labeled proteins were tested in four different concentrations (0.1 μg and 0.2 μg [data unpublished] and 0.4 μg and 0.6 μg). BSA labeled with Alexa Fluor® 647 was used as negative control. Each sample, which consisted of SrTrx-lp or TsTrx-lp and the cell type to be tested, was brought to a volume of 200 μL with PBS and incubated for 30 min. All experimental setups were prepared in duplicate to test various temperatures. Incubation took place at RT (data not shown) and 37°C. After incubation, samples were washed twice, resuspended in 150 μL PBS, and analyzed by flow cytometry on a FACScalibur cytometer (BD Biosciences), with 10,000 events collected from the gated populations. For further characterization of the binding specificity, cells were preincubated with 0.1 μg and 0.2 μg (data not shown) or 0.4 μg and 0.6 μg of unlabeled protein for 30 min prior to the addition of the corresponding labeled proteins. The data were analyzed with CellQuestPro.

2.10. Chemotaxis Assay.
To evaluate the chemotactic activity of human monocytic THP-1 cells, Boyden chambers were used as described previously [68, 69]. DTT (100 mM) reduced Trx-lps from S. ratti and T. suis were tested at concentrations of 3 ng, 30 ng, 300 ng, and 1 μg each in 100 μL. The assay was performed with negative controls (random migration)

such as chemotaxis buffer (PBS containing $CaCl_2$, $MgCl_2$, and BSA) and THP-1 media (RPMI containing HEPES and 10% FCS) and as positive control LPS at 100 ng, since LPS induces migration of monocytic cells [70]. THP-1 cells (2×10^5) were allowed to migrate through polyvinyl-pyrrolidone-free polycarbonate filters (pore size: 3 μm; Nuclepore, Tübingen, Germany) within 90 min at 37°C and 5% CO_2. Afterwards, migrated cells were counted by using an inverted Zeiss microscope (Axiovert 25). Triplicates were performed in three independent experiments.

2.11. Wound Healing. To monitor epithelial cell migration of Caco-2 cells and the ability of Trx-lps to improve the wound healing process, we used the CytoSelect 24-Well Wound Healing Assay (Cell Biolabs, Inc.) according to the manufacturer's instructions. By means of the CytoSelect wound healing inserts a 0.9 mm wound field was generated. 500 μL of a Caco-2 cell suspension (containing 0.5×10^6 cells) was added to each well after the inserts had firm contact with the bottom of the wells. After overnight incubation, a monolayer was formed, the inserts were removed, the cells were washed, and the different stimuli were added. We used both Trx-lps, from *S. ratti* and *T. suis*, in concentrations of 3 ng, 30 ng, 300 ng, 1 μg, 10 μg, and 25 μg per 500 μL. As a positive control the human epidermal growth factor (EGF; 0.5 ng, 5 ng, 10 ng, 15 ng, and 25 ng) was included in order to get the proper concentration for maximal wound healing effects. As negative control cell media and LPS were added. An inverted digital microscope (EVOS™ FL Thermo Fisher Scientific) by Advanced Microscopy Group was used for observation (4x magnification). The cells were incubated for 4 days, whereby each 24 h a picture was taken and the percent closure was calculated.

2.12. Statistical Analysis. Statistical differences between groups were analyzed with the *t*-test for independent samples or the Mann–Whitney *U* test. $P < 0.05$ was taken as moderate evidence of significance and $P < 0.01$ as strong evidence of significance.

3. Results

3.1. Identification of Full-Length cDNAs Encoding the S. ratti and T. suis Trx-lps, Cloning, and Sequence Analyses. SrTrx-lp is represented by the cluster SR00399 [13] and was abundantly found in *S. ratti* E/S products of parasitic *S. ratti* females. The partial sequence was identified as the thioredoxin family protein and was used to obtain the full-length cDNA sequence by PCR. Further, the full-length cDNA sequence of the *T. suis* hypothetical protein M513 (Accession no. KFD58615.1) was cloned and identified as Trx-lp. The protein sequence of the recombinantly expressed *S. ratti* and *T. suis* Trx-lps have been verified by mass spectrometry.

Conserved domains of the Trx-lps from the intestinal helminths *S. ratti* and *T. suis* were ascertained by the protein families database (Pfam). Neither the Trx-lp from *S. ratti* nor the Trx-lp from *T. suis* contain a signal peptide. Both proteins have an N-terminal thioredoxin domain containing the active side motif CXXC (CGPC) and a C-terminal

PITH (proteasome-interacting domain of thioredoxin-like) domain.

The alignment of the amino acid sequences from different organisms revealed a relatively low degree of identity between the different species. Between the Trx-lps from *S. ratti* and *T. suis* the degree of identity (39%) was not as high as between Trx-lps from *S. ratti* and *B. malayi* (56%). A high degree of identity was revealed between both *Trichuris* spp. Trx-lps (94%), similar to the sequences of *S. ratti* and *S. stercoralis* (99.9%) (data not shown). Comparing the other aligned helminth protein sequences, the similarities to the *S. ratti* and the *T. suis* Trx-lps varied between 35% and 56%. The comparison of the redox-regulating protein between *S. ratti* and *Homo sapiens* showed 43% identity.

The aligned helminth sequences share, except for the trematode *Schistosoma mansoni*, the catalytic site sequence (CGPC) with the human Trx-lp sequence of the active site. There are always two cysteines which are separated by two amino acids, mostly glycine and proline. Instead of a glycine, the *S. mansoni* catalytic site sequence has an arginine (R) (Figure 1). The two cysteines are responsible for the redox regulation in different cellular processes. The predicted structure of SrTrx-lp is exemplarily shown in Figure 2. Both parasite Trx-lps have a Trx-like domain (left) as well as the PITH domain (right) (Figure 2; Phyre2: [61]).

3.2. Recombinant Expression and Purification of S. ratti and T. suis Trx-lp. SrTrx-lp and TsTrx-lp were recombinantly expressed in endotoxin-free *E. coli* as His-tagged proteins and as strep-tagged proteins. The amount of purified His-tagged proteins, however, was higher than the amount of purified strep-tagged proteins. Thus, after preliminary tests with strep-tagged proteins, we further worked with His-tagged proteins. Both parasite proteins were verified by Western blot using anti-strep and anti-his antibody (Figure S1) and mass spectrometry.

3.3. Functional Activity Assays

3.3.1. Reduction of Proteins

(1) Insulin Reduction. For measurement of the functional activity of SrTrx-lp using insulin, the precipitation of free insulin β-chains was measured spectrophotometrically at a wavelength of 650 nm according to Holmgren (1979) as well as Luthman and Holmgren (1982) [61, 62]. A concentration of 1 μM (30 μg/mL), 2.5 μM (75 μg/mL), and 5 μM (150 μg/mL) of the SrTrx-lp was used and the measuring time was plotted against the rate of precipitation (ΔA_{650}/min $\times 10^3$), which was about 0.064 ΔA_{650}/min at the highest concentration. SrTrx-lp reduces insulin with a relative specific activity of 1556.67 and is regenerated by DTT whereby in the negative control and the lowest concentration of SrTrx-lp only a slight precipitation of insulin could be measured (Figure 3).

(2) IgM Reduction. Pentameric IgM consists of five M immunoglobulins joint by the J chain. Its molecular weight is about 950 kDa and it contains 26 interchain disulfide

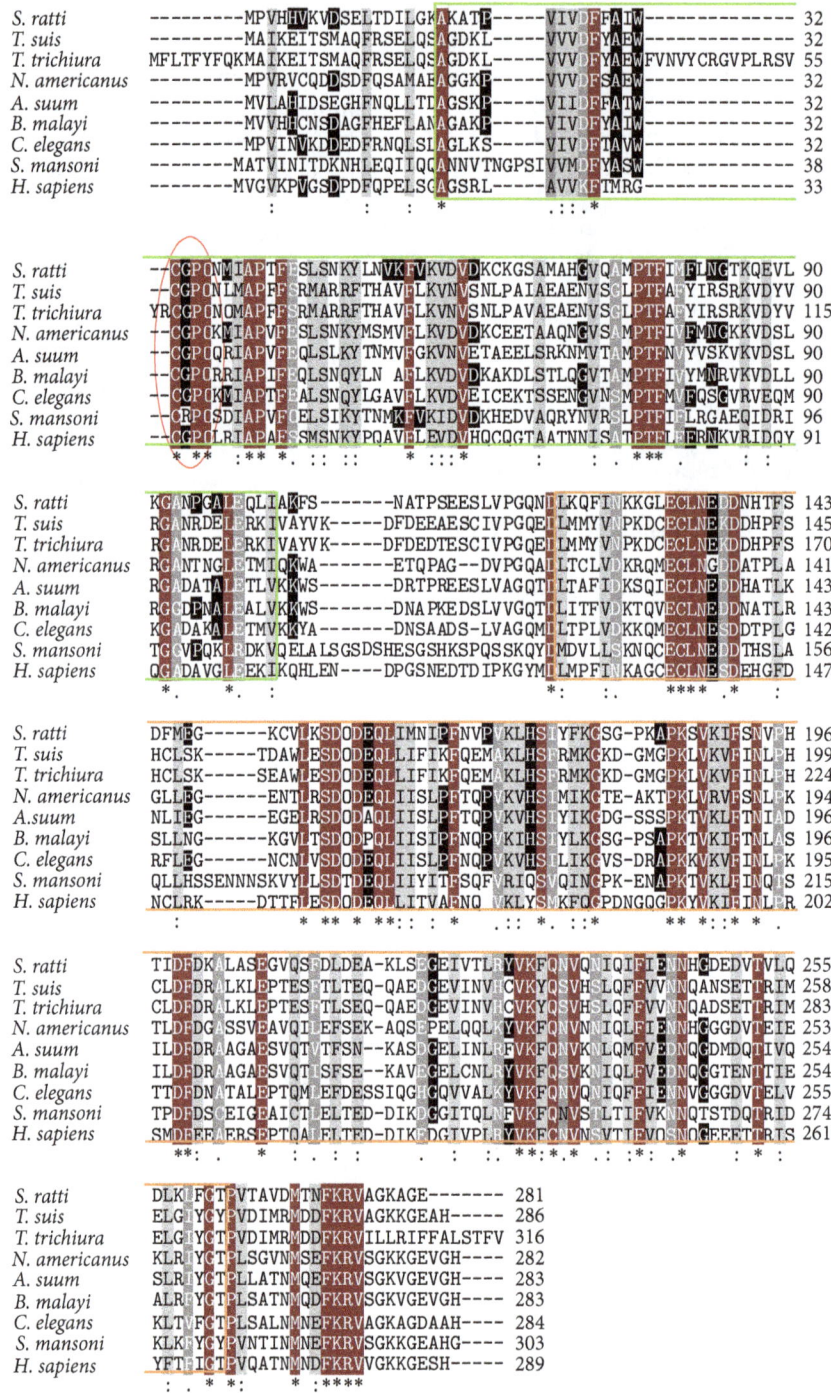

FIGURE 1: Multiple alignment of the Trx-lps from different organisms. *S. ratti* (CEF66761.1); *T. suis* (KHJ44020.1); *T. trichiura* (CDW52389.1); *Necator americanus* (XP_013304103.1); *Ascaris suum* (ERG80831.1); *Brugia malayi* (XP_001892562.1); *C. elegans* (NP_491127.1); *Schistosoma mansoni* (CD80891.1); *H. sapiens* (NP_004777.1). Green box represents the Trx-like domain; orange box represents the PITH domain; red circle shows the active site.

bridges that are potential substrates for Trx and thus for Trx-lp. Additionally to the insulin reduction activity assay, the dithiol-disulfide oxidoreductase activity of the Trx-lps was analyzed by an IgM reduction test according to Wollman et al. (1988) [63]. IgM is detectable at 250 kDa. As positive

control IgM was reduced by 100 mM DTT at which bands at about 70 kDa (heavy chain IgM) and 25 kDa (light chain IgM) occur (Figure 4, 3rd lanes). Only exposing IgM to the highest amount of SrTrx-lp, five main bands were identified (Figure 4(a), lane 7). In addition to the bands at 70 kDa and

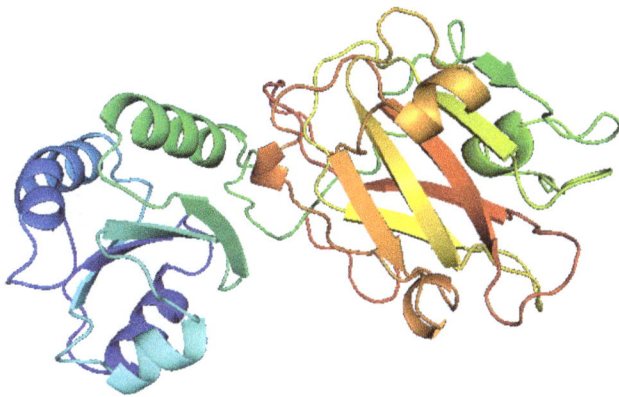

FIGURE 2: Predicted 3D-structure of parasite Trx-lp. The structure of SrTrx-lp is shown here. Both parasite Trx-lps have a Trx-like domain (left) as well as the PITH domain (right) (Phyre2: [55]).

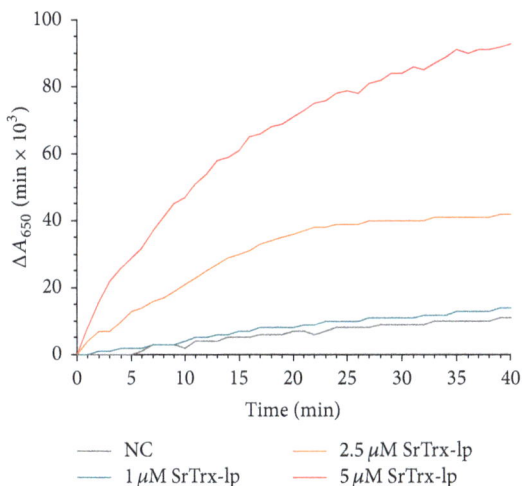

FIGURE 3: SrTrx-lp catalyzed reduction of insulin by DTT. Here, the rate of precipitation was plotted against time. While after 40 min the reduction of the SrTrx-lp was near the equilibrium, only a minor reduction of insulin was detected in the negative control (NC) without the SrTrx-lp and the lowest concentration used of SrTrx-lp ($1 \mu M$).

25 kDa, similar to the reduction of IgM with DTT, and the band at 250 kDa, now bands at about 30 kDa, representing monomeric *S. ratti* Trx-lp and 60 kDa representing dimeric *S. ratti* Trx-lp, were determined. A minor band was also seen at 45 kDa. Almost similar protein bands have been observed when *T. suis* Trx-lp was analyzed (Figure 4(b)); however, *T. suis* Trx-lp also at the low and intermediate concentration leads to the reduction of IgM. Further, bands at 140 kDa (heavy chain dimers of IgM) were predominant at all TsTrx-lp doses (Figure 4(b)).

3.4. Nematode Trx-lps Interact with Host Immune Cells

3.4.1. Binding to Mucosal and Immune Cells. The binding ability to other immune cells as well as mucosal Caco-2

cells was examined by FACS (Figure 5). Monocytes, lymphocytes, and neutrophils as well as Caco-2 cells, THP-1 cells, and THP-1-derived dendritic cells (DCs) were exposed to Alexa Flour-labeled Trx-lps. The experiments revealed considerable differential binding activities to various cells. Thus, SrTrx-lp (Figure 5(a)) as well as TsTrx-lp (Figure 5(b)) proteins strongly bound to monocytic cells shown in a dose-dependent manner for peripheral monocytes (SrTrx-lp: MFI 175–185; TsTrx-lp: MFI 19–60), THP-1 cell line (SrTrx-lp: MFI 36–108; TsTrx-lp: MFI 38–133), and generated DCs (SrTrx-lp: MFI 85–170). SrTrx-lp and at lower degree TsTrx-lp also bound to Caco-2 cells (SrTrx-lp: MFI 45–52; TsTrx-lp: MFI 14–42) and with limited affinity to neutrophilic granulocytes (SrTrx-lp: MFI 15-16; TsTrx-lp: MFI 17–50) and lymphocytes (SrTrx-lp: MFI 9–11; TsTrx-lp: MFI 10–20).

In order to verify the differentiation of THP-1 cells to DCs by IL-4 and GM-CSF, anti-CD86 antibodies were used. CD86 localized on the surface of differentiated DCs but not on THP-1 cells (data not shown).

3.4.2. Nematode Trx-lps-Induced Cytokine Profile of Intestinal Epithelial-Dendritic Cell 3D-Cultures. The *S. ratti* and *T. suis* Trx-lps were examined for their ability to induce the release of cytokines in human 3D-cocultures of intestinal epithelial cells (IEC) and DCs. The release of the inflammatory (TNF-α), anti-inflammatory (IL-10), and $T_H 2$-related cytokines (IL-22, TSLP) was analyzed. In preliminary experiments, the optimized concentrations of LPS and LTA were determined as $0.5 \mu g/mL$ and $0.1 \mu g/mL$ (data not shown). $200 \mu g/mL$ *T. suis* extract was used as a positive control and cell culture medium was used as a negative control (Figure 6(a)). The Trx-lps were tested at concentrations of 3 ng, 30 ng, 300 ng, $1 \mu g$, $10 \mu g$, and $25 \mu g$ (each per mL). The reduced state (reduction via DTT) and the oxidized state (freshly purified protein, only partly reduced, see IgM reduction) of the Trx-lps made no difference in the cytokine response (data not shown). This observation indicated that the immune responses the proteins triggered are probably active site-independent. $10 \mu g$ and $25 \mu g$ of both helminthic Trx-lps are the most representative concentrations inducing the highest cytokine release.

Cocultured cells exposed to *T. suis* (Ts) extract showed in particular an enhanced production of IL-10 and IL-22 after 48 h and an even higher release of IL-10 after 72 h, while the proinflammatory cytokine TNF-α was downregulated (Figure 6(a)). SrTrx- as well as TsTrx-lp induced initially a slightly pronounced release of proinflammatory TNF-α after 24 h ($P < 0.01$), followed by an increased production of IL-22 and TSLP after 48 h of incubation ($P < 0.01$). In response to the exposure of the cocultures to Trx-lps in particular the $T_H 2$-associated cytokine IL-22 was produced after 48 h and 72 h ($P < 0.01$). At a concentration of $25 \mu g$ of TsTrx-lp, the TNF-α release increased after 48 h and even dominated the IL-22 production. After 72 h, the IL-22 and TSLP production was dominating the overall TNF-α production. $10 \mu g/mL$ of Trx-lps appears to be slightly more potent with respect to cytokine release than $25 \mu g$ of protein with statistical significance only between the IL-22-inducing SrTrx-lp concentrations after 48 h ($P < 0.01$) (Figure 6(b)).

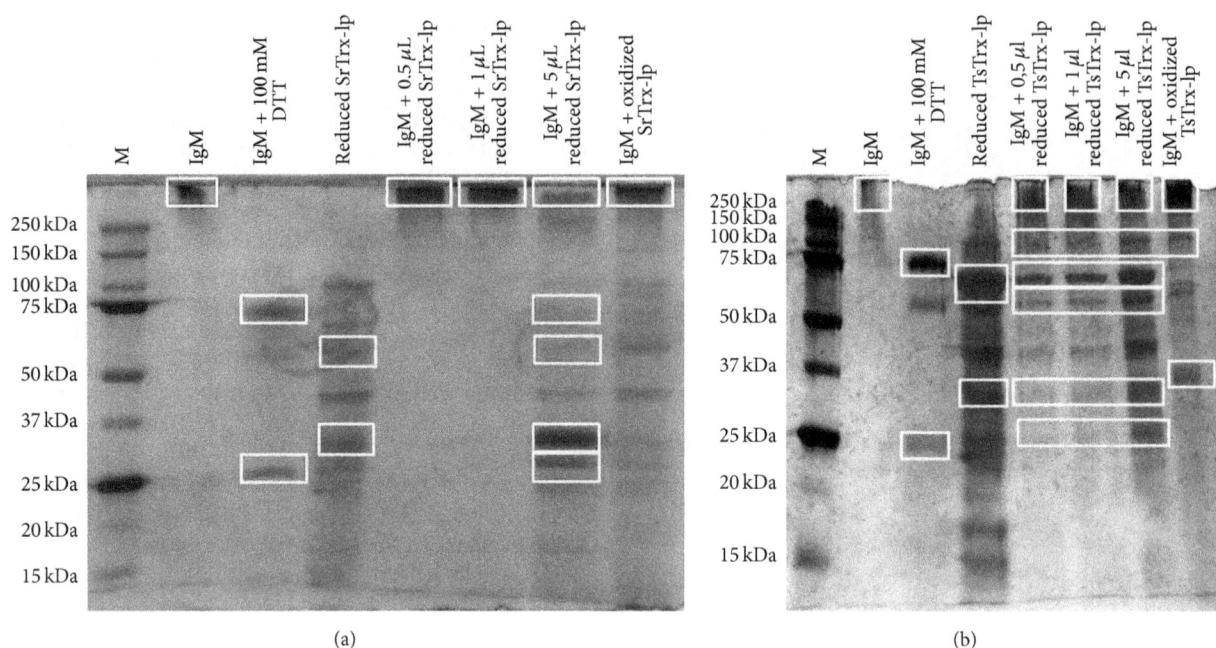

(a)   (b)

FIGURE 4: IgM reduction by the Trx-lps from *S. ratti* (a) and *T. suis* (b). Prior to incubation, the Trx-lps from both organisms were reduced by DTT. IgM was split in its chains (25 kDa, 70 kDa, and 950 kDa).

3.5. SrTrx/TsTrx-lp Displayed Chemotactic Activity for Monocytes. Human Trx is chemotactic for monocytes besides neutrophils and T lymphocytes [31]. Therefore, we investigated the chemotactic activity of the parasite Trx-lps for monocytic THP-1 cells by using Boyden chambers. Different Trx-lp concentrations (3 ng, 30 ng, 300 ng, and 1 μg; each per 100 μL) from both studied parasites were added to the lower compartment of the chambers. In the negative control, a few cells migrated through the membrane, while the cell migration using LPS as stimulant was significantly increased. Among the different applied Trx-lp concentrations the highest migration rate was detected at 3 ng. The overall cell migration was higher in case of *S. ratti* Trx-lp than after stimulation with the TsTrx-lp and half bell-shaped dose-response curve reported for chemokines is more pronounced in case of the TsTrx-lp (Figure 7).

3.6. Trx-lps Promoted Wound Healing. As an important functional activity it was investigated whether the Trx-lps from both nematode parasites express wound healing activity. Therefore, the effect of different concentrations of Trx-lps on epithelial cell (Caco-2) wound closure (Figure 8, data, and Figure 9, microscopic photography) was analyzed. Compared to the untreated cells, where the wound-like area narrowed 10–15% every day, the stimulated cells showed almost twice as much growth. 300 ng/500 μL of both parasite Trx-lps are the most potent concentration for promoting the wound healing process as well as 10 ng of EGF, which was included as positive control, while 3 ng and 30 ng and concentrations upon 1 μg (each per 500 μL) have a more moderate effect on wound healing. The wound healing process was highly significantly promoted by EGF and TsTrx-lp ($^{**}P < 0.01$) as well as significantly promoted by SrTrx-lp ($^{*}P < 0.05$).

4. Discussion

Trx is a physiologically important multifunctional protein and prokaryotic Trx has been described as so-called moonlighting protein [34, 35]. The multiple biological functions comprise features as growth factor and antioxidant, as inhibitor of apoptosis and transcriptional factor, and as chemokine [22, 23, 25–28]. Very little is known about Trx-lps, in particular about those from helminths and their potential role in parasite-host interaction.

There is only one publication about an endoplasmic reticulum located Trx transmembrane related protein from the trematode *Clonorchis sinensis*, containing a Trx domain with the active site motif Cys-Pro-Ala-Cys (CPAC). This redox molecule is suggested to serve as protection against host- and parasite-generated ROS [44].

Contrariwise, the *S. ratti* Trx-lp has the catalytic domain sequence of the uniformly small (12 kDa) ubiquitous Trx proteins (WCGPC) but has a size of approximately 30 kDa. Comparably, the *T. suis* Trx-lp has a size of approximately 33 kDa and the same catalytic domain sequence as the classic Trx.

In the present study, Trx-lp from two parasitic nematodes, *S. ratti* and *T. suis*, were cloned, expressed, and characterized for the first time. In case of both helminths the protein was present in the E/S products of the parasites [13, Brattig et al., unpublished]. The molecular mass (30–33 kDa) as well as the proteins structure suggested similar functions to those of the human Trx-related protein (TRP32), also known as TXNL-1, which protects the cell against glucose deprivation-induced cytotoxicity and is involved in antiapoptotic signaling [47, 48, 71]. Like SrTrx- and TsTrx-lp, TRP32 consists of an N-terminal Trx and a C-terminal PITH domain as well [44].

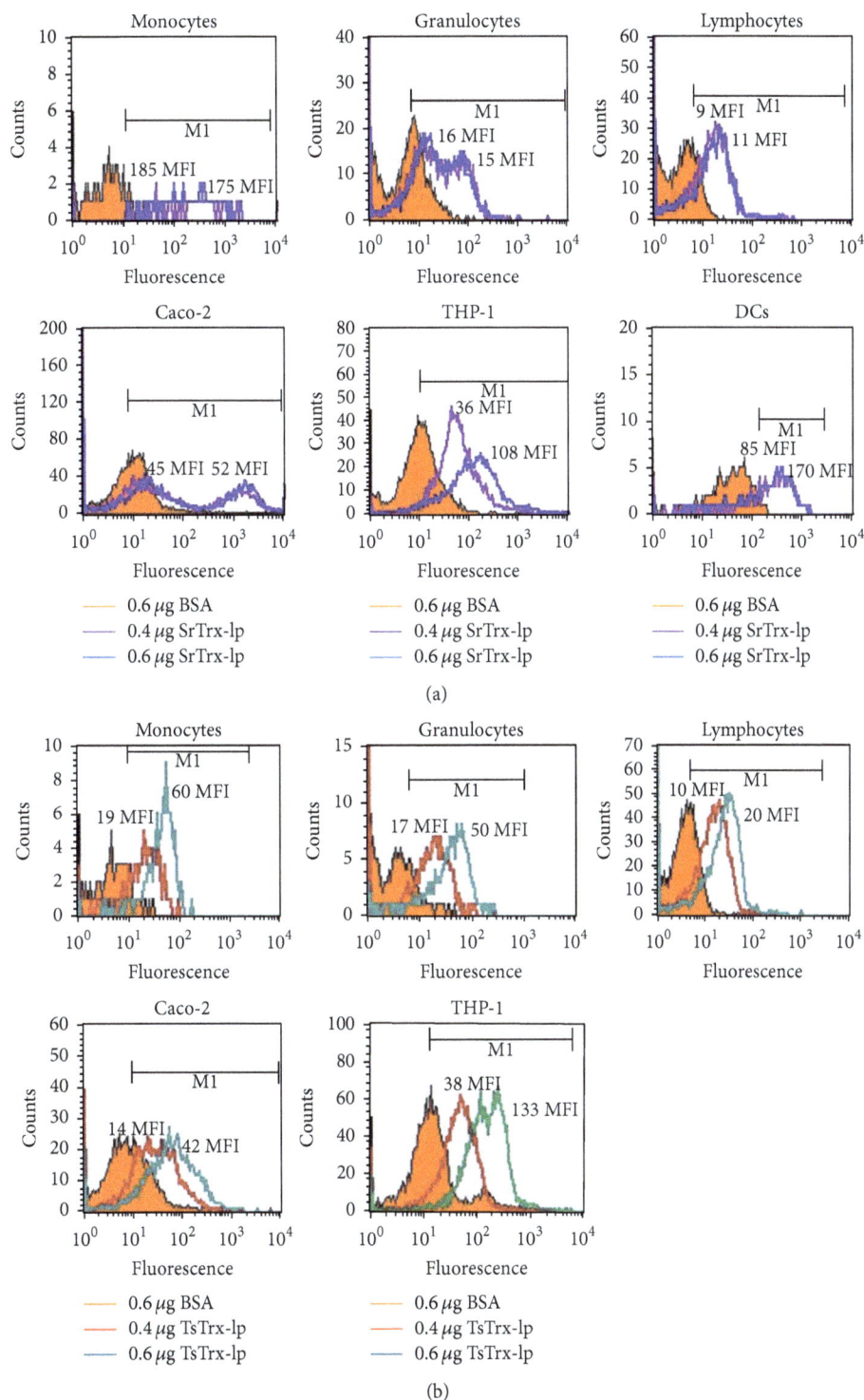

FIGURE 5: Binding of the SrTrx-lp (a) and TsTrx-lp (b) to different cell types. 2×10^5 cells were incubated at 37°C for 30 min with Alexa Flour®-labeled SrTrx-lp or TsTrx-lp. Here, peripheral blood cells (monocytes, granulocytes, and lymphocytes) as well as cell culture cells (Caco-2 cells, THP-1 cells, and THP-1-derived DCs) were tested with 0.4 μg (purple (a), red (b) line) and 0.6 μg (blue (a), green (b) line) of labeled protein determining the median fluorescent intensity (MFI). The intensity of surface fluorescence (FI, x-axis) is plotted against cell counts. (The counts in the figures represent the median fluorescence index values.) Representative results of five independent experiments are shown.

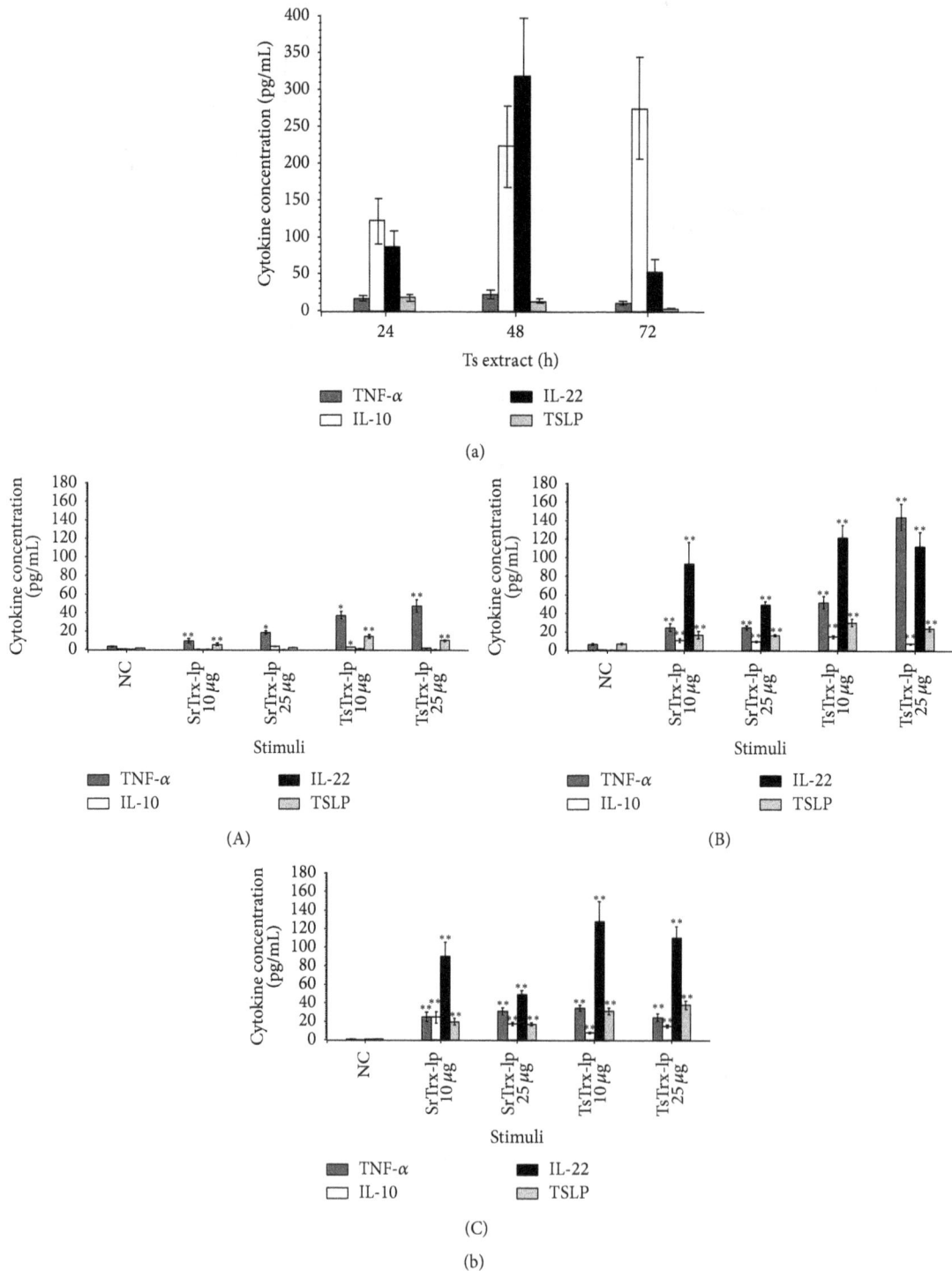

FIGURE 6: (a) Exposure of 3D-cocultures to *Trichuris suis* (Ts) extract. Culture supernatants were harvested after 24 h, 48 h, and 72 h. The release of inflammatory (TNF-α), anti-inflammatory (IL-10), and T_H2-related cytokines (IL-22, TSLP) was analyzed in a 3D-cell culture model. Representative results of at least three independent experiments are shown as median. (b) Exposure of 3D-cocultures to SrTrx- and TsTrx-lp or medium (NC). Culture supernatants were harvested after 24 h (A), 48 h (B), and 72 h (C). The release of inflammatory (TNF-α), anti-inflammatory (IL-10), and T_H2-related cytokines (IL-22, TSLP) was analyzed in a 3D-cell culture model. Representative results of at least three independent experiments are shown. Significant increase of all measured cytokines compared to NC ($^{**}P < 0.01$). $^{*}P < 0.05$; $^{**}P < 0.01$.

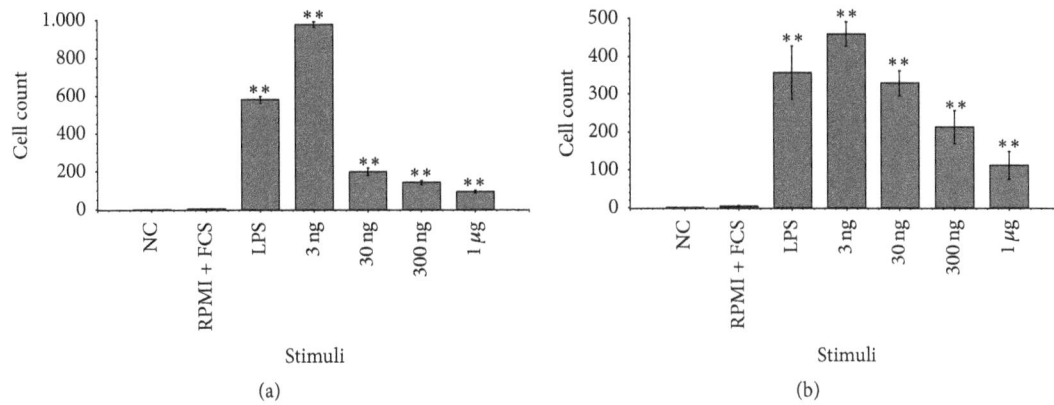

FIGURE 7: Chemotactic activity of the Trx-lp from *S. ratti* (a) and *T. suis* (b) for monocytic THP-1 cells. The chemotactic activity of both proteins for THP-1 cells was investigated by using Boyden chambers. Different concentrations of the Trx-lps were added to the lower compartment of the chemotactic chambers. Protein concentrations per 100 μL of 3 ng, 30 ng, 300 ng, and 1 μg showed SrTrx-lp and TsTrx-lp have the greatest chemotactic activity at 3 ng. Chemotaxis buffer (NC) and THP-1 media (RPMI + FCS) were included as negative control (random cell migration), while LPS was used as positive control. All used Trx-lp concentrations led to significant higher cell migration than the negative controls ($^{**}P < 0.01$).

Trx-lps are known to have several binding partners and substrates they associate with by means of their Trx domain, which exerts redox-active functions. The C-terminal PITH domain is able to interact with the 26S proteasome by the substrate-recruiting factor of the 26S proteasome eEF1A1 [72, 73].

Similar to Trx, Trx-lps of eukaryotic cells are also multifunctional and involved in different cellular processes including cofactor functions or the regulation of specific signaling proteins [46] which may indicate possible moonlighting properties that have to be demonstrated in the future [34–36]. Comparisons of Trx-like homologues by multiple sequence alignments revealed a high sequence similarity between Trx-lps from *T. suis* and from *T. trichiura* (94% identity). *Strongyloides* species are all very closely related [11, 74]. Apart from this, the protein alignment showed a relatively low degree of similarity (35%–56%) between different nematodes, either parasitic or nonparasitic. Except for *S. mansoni* all other species had the strongly conserved N-terminal Trx catalytic site sequence (CGPC). At the C-terminus all Trx-lps possess the PITH domain. Like Trx, the analyzed parasite proteins have no signal peptide and are released from cells by nonclassical protein export [29, 75].

Trx-lp has also various roles in several human cellular and extracellular processes, since reactive oxygen species (ROS) occur in the normally functioning metabolism [76]. The dithiol-disulphide oxidoreductase activity of both recombinant *S. ratti* and *T. suis* Trx-lps was either analyzed by insulin reduction according to Holmgren (1979) or IgM reduction according to Wollman et al. (1988) [61, 63]. Reduced Trx reacted very quickly with insulin and the reduced insulin was precipitated. The relative specific activity of Trx from *E. coli* amounts to a value of 4930 units [61]. Findings that measured relative specific activity of the SrTrx-lp has an activity of about 1557 units show that it has a comparable activity to classical Trx. The oxidoreductase activity was further analyzed by the reduction of murine IgM. Wollman et al. (1988) have already

shown that recombinant human Trx is able to reduce the disulfide bridges of murine IgM [63]. Therefore, we suggested Trx domain containing Trx-lps may also have the ability to reduce IgM. We could show that indeed both Trx-lps reduced the S-S bonds of IgM. Since all TsTrx-lp used doses resulted in the formation of the same bands in SDS-PAGE, this Trx-lp appears to be more active than the *S. ratti* Trx-lp. Even at the lowest concentration minor protein bands were visible at 25 kDa and 70 kDa. The more intensive they were the higher the added concentration of TsTrx-lp was. A reduction of IgM by not fully removed DTT can be excluded since then the strength of the formed bands would be the same in each approach. Although even at the lowest concentration bands have been formed, they were more intensive at the highest concentration. Furthermore, in the IgM reduction assay of SrTrx-lp no bands were existent at the lowest and the intermediate concentration of the added protein.

Through those activity assays it could be demonstrated that the recombinantly expressed Trx-lps have redox functions and are able to act as classical Trx. In further analysis, we could demonstrate multifunctional activities of the helminth proteins. For Trx it has been reported to be released by monocytes [77] and also to be chemotactic for monocytes, neutrophils, and T lymphocytes [31]. Accordingly, we have observed that *S. ratti* and *T. suis* Trx-lps exhibit chemotactic activity for monocytes and have the ability to interact with them. An attraction of monocytic cells to a nematode-dwelling site could subsequently lead to an activation of the cells leading to a consecutive generation of wound healing fostering cytokines like IL-22 and immunoregulatory interleukins [78–80]. Both parasite Trx-lps bound to monocytic cells, to the THP-1 cells, and to peripheral monocytes although in some FACS analysis there were only limited counting events. Accordingly, SrTrx-lp was shown to bind to DCs. Of interest, the parasite redox-regulating proteins also bound to Caco-2 cells and more weakly to lymphocytes and granulocytes. Thus, Trx-lps seem to interact with intestinal

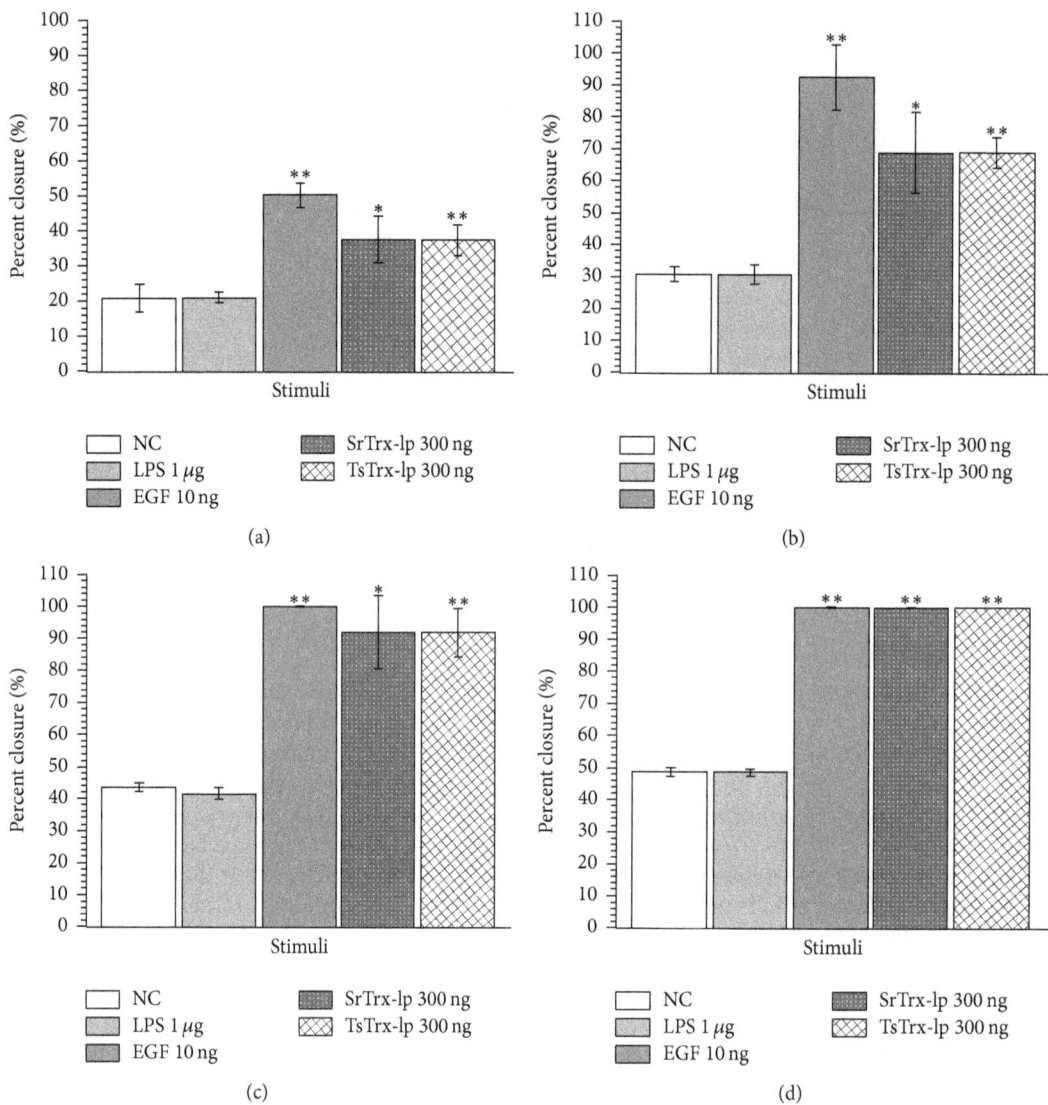

FIGURE 8: Percentage closure of the Caco-2 wound gap after 24 h (a), 48 h (b), 72 h (c), and 96 h (d). In general, the gap was narrowed by approx. 10–15% every day adding no stimulus. As a negative control, cells were observed without any stimulus (NC) and with LPS (1 μg). As a positive control, epidermal growth factor (EGF) was used, whereat 10 ng fostered wound healing the best. SrTrx-lp and TsTrx-lp were tested at different concentrations (3 ng, 30 ng, 300 ng, 1 μg, 10 μg, and 25 μg–each per 500 μL), here at 300 ng represented as the best wound healing promoting concentration. The wound healing process was highly significantly promoted by EGF and TsTrx-lp ($^{**}P < 0.01$) as well as significantly promoted by SrTrx-lp ($^{*}P < 0.05$).

epithelial cells, the first-line host cells that get exposed to E/S products released by the colonizing parasitic females, and also with second-line cells, the monocyte-derived DCs.

Of interest, Trx has been reported to possess immunological activities. Thus, it has been attributed to an anti-inflammatory role besides suppression of apoptosis and fostering cell growth [32, 81–83]. Trx can interact with immune cells and facilitates the production of TNF-α [31, 84] by monocytic lineage, but it is also able to counteract the production of inflammatory cytokines such as TNF-α [85, 86]. In the present study, 3D-coculturing of the intestinal epithelial Caco-2 cells and THP-1-derived DCs was performed. Hereby, parasite Trx-lps induced the release of proinflammatory TNF-α in the first day of the culture and at high concentration

after 48 h followed by a prevailing generation of the $T_H 2$-related cytokine IL-22 besides lower levels of TSLP and IL-10. IL-22 may be predominantly released by activated DCs in the cell cultures after 2-3 days [78, 80, 87].

IL-22, particularly produced by immune cells present beneath the epithelium, as the innate lymphoid cells [78, 80, 88], acts through signal transducer and activator of transcription (STAT-3) and is important in maintaining the homeostasis of the gut and therefore serves the protection from intestinal inflammation. An important source of IL-22 in acute colitis is TLR-stimulated CD11c^{+} DCs which are located in the surficial mucosal epithelium of the gut and are getting activated by invading pathogens like bacteria or parasites. These cells initiate, via IL-22 and thus STAT-3,

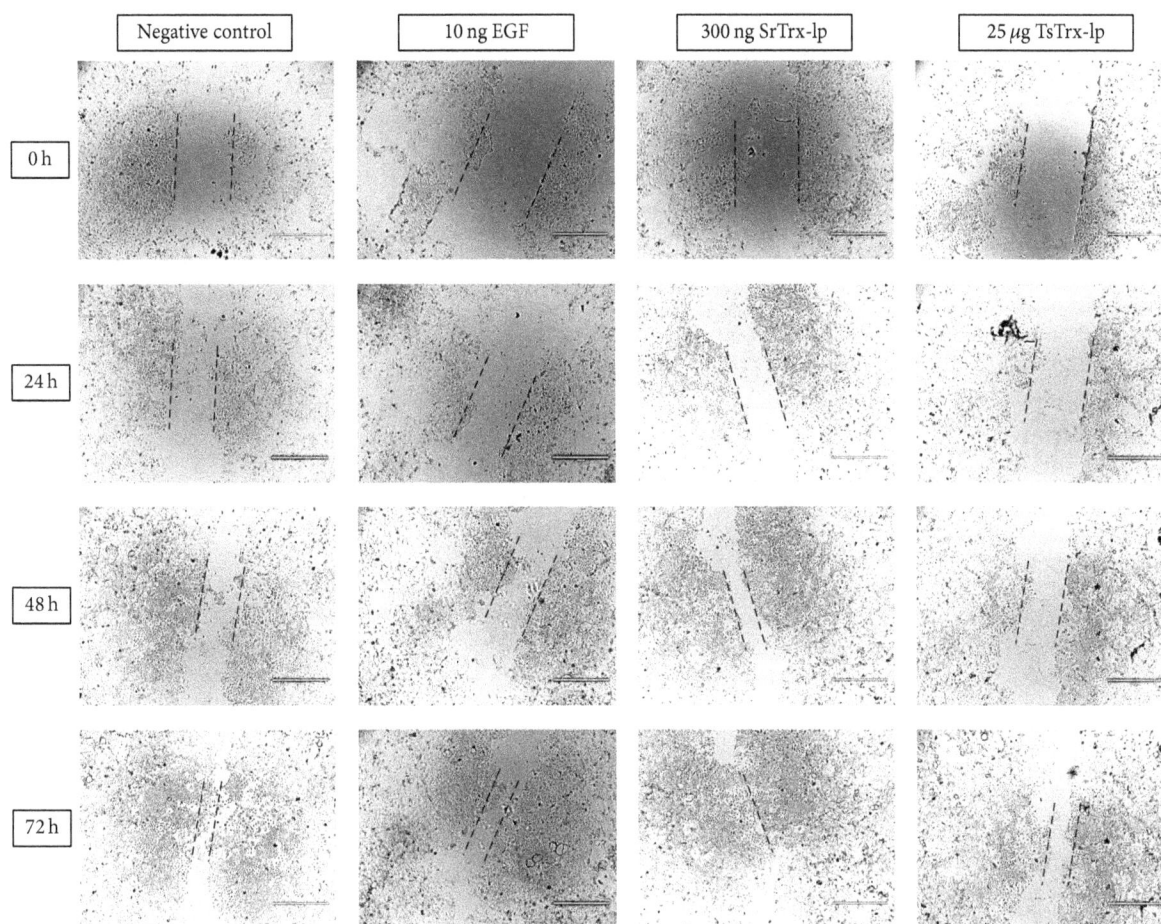

FIGURE 9: Wound healing assay with Caco-2 cells and the *S. ratti* as well as the *T. suis* Trx-lp. The CytoSelect™ 24-Well Wound healing assay was performed. Here, two examples are described as representatives for different tested concentrations. Protein concentrations of 300 ng SrTrx-lp are given as example for best concentration for wound healing promotion and 25 μg of TsTrx-lp (each per 500 μL) indicated the less wound healing-promoting concentration. The wound healing process of a 0.9 mm wound field generated was observed over 96 h, whereby each 24 h a picture was taken. The size of the scale bar is 1000 μm and the dashed black lines indicate wound-like area [56].

processes that are important for a proper stress response, mucosal wound healing, and apoptosis pathway [78, 79, 89]. IL-22 may profoundly increase the proliferation and turnover of IECs and the production of mucus and antimicrobial peptides [90]. Accordingly, the release of proteins from intestinal nematodes like Trx-lps may contribute to preserve or restore the integrity of the intestinal barrier.

Thus, there are three possible pathways for helminthic Trx-lps to act: firstly, secreted Trx/Trx-lp protects the parasite against high ROS production initiated by the host's first-line immune response via cells of the monocyte-macrophage linage. Trx may be important for redox control at wound margins, since much ROS emergence was proven there [91, 92]. ROS as second messenger ameliorates wound healing processes [93]. Therefore, among others, it serves the migration of cells and closure of wounds. Then, antioxidant molecules are probably important to maintain the balance in order to prevent stress-induced cell death. Secondly, secreted Trx-lp stimulates mucosal DCs to generate high levels of IL-22 which promotes epithelial cell proliferation and the

preservation or restitution of the integrity of the intestinal barrier. In the present study we had shown that 300 ng of parasite Trx-lps promoted the wound healing process of epithelial Caco-2 cells. A third possible function of Trx-lp secreted by the parasite may be to mimic antioxidant molecules of the host and may lead to interference reactions in the host's antioxidant metabolism concerning the substrates and binding molecules. Thus, recent reports indicated that distinct molecules secreted by helminth parasites can foster wound healing [94] and modulate the host's immune response [95].

5. Conclusion

In summary, we identified and characterized the secreted Trx-lps from *S. ratti* and *T. suis*. Both multifunctional proteins expressed antioxidative activity and the capability to interact with the host's mucosal cells, indicated by chemotactic activity for monocytic cells, binding to host's epithelial cells as well as to immune cells, by the release of cytokines. In particular, the promoting wound healing effect indicates the involvement of Trx-lp in many pathways that are initiated in the local parasite-host interaction

Disclosure

Eva Liebau and Norbert W. Brattig shared senior authorship. Nucleotide sequences for *Strongyloides ratti* thioredoxin-like protein (SrTrx-lp) and *Trichuris suis* thioredoxin-like protein (TsTrx-lp) have been deposited in the GenBank Database under the accession KX119168 for SrTrx-lp and KFD58615.1 for TsTrx-lp (originally known as hypothetical protein M13).

Competing Interests

The authors have no conflict of interests to declare.

Acknowledgments

The authors gratefully acknowledge a twelve-month scholarship of Dana Ditgen and Emmanuela M. Anandarajah by Ovamed. The doctoral student Emmanuela M. Anandarajah is supported by the Evangelisches Studienwerk Villigst. They thank F. Geisinger and L. Feige for technical and experimental assistance. The veterinary team of the Bernhard Nocht Institute for Tropical Medicine is acknowledged. Data from this work form a major part of the doctoral theses of Emmanuela M. Anandarajah and Dana Ditgen in the Department of Molecular Physiology, Westfälische Wilhelms-University, Münster, Germany.

References

[1] M. E. Viney and J. B. Lok, "Strongyloides spp," *WormBook*, pp. 1–15, 2007.

[2] A. Olsen, L. van Lieshout, H. Marti et al., "Strongyloidiasis—the most neglected of the neglected tropical diseases?" *Transactions of the Royal Society of Tropical Medicine & Hygiene*, vol. 103, no. 10, pp. 967–972, 2009.

[3] M. E. Viney, "The biology of *Strongyloides spp.*," *WormBook*, pp. 1–17, 2015.

[4] J. Bethony, S. Brooker, M. Albonico et al., "Soil-transmitted helminth infections: ascariasis, trichuriasis, and hookworm," *The Lancet*, vol. 367, no. 9521, pp. 1521–1532, 2006.

[5] C. Cutillas, R. Callejón, M. de Rojas et al., "*Trichuris suis* and *Trichuris trichiura* are different nematode species," *Acta Tropica*, vol. 111, no. 3, pp. 299–307, 2009.

[6] L. Nemetschke, A. G. Eberhardt, M. E. Viney, and A. Streit, "A genetic map of the animal-parasitic nematode *Strongyloides ratti*," *Molecular and Biochemical Parasitology*, vol. 169, no. 2, pp. 124–127, 2010.

[7] J. A. Jackson, I. M. Friberg, S. Little, and J. E. Bradley, "Review series on helminths, immune modulation and the hygiene hypothesis: immunity against helminths and immunological phenomena in modern human populations: Coevolutionary legacies?" *Immunology*, vol. 126, no. 3, pp. 18–27, 2009.

[8] D. E. Elliott and J. V. Weinstock, "Helminthic therapy: using worms to treat immune-mediated disease," *Advances in Experimental Medicine and Biology*, vol. 666, pp. 157–166, 2009.

[9] J. V. Weinstock, "Autoimmunity: the worm returns," *Nature*, vol. 491, no. 7423, pp. 183–185, 2012.

[10] J. P. Hewitson, J. R. Grainger, and R. M. Maizels, "Helminth immunoregulation: the role of parasite secreted proteins in modulating host immunity," *Molecular and Biochemical Parasitology*, vol. 167, no. 1, pp. 1–11, 2009.

[11] V. L. Hunt, I. J. Tsai, A. Coghlan et al., "The genomic basis of parasitism in the *Strongyloides* clade of nematodes," *Nature Genetics*, vol. 48, no. 3, pp. 299–307, 2016.

[12] Y. Tazir, V. Steisslinger, H. Soblik et al., "Molecular and functional characterisation of the heat shock protein 10 of *Strongyloides ratti*," *Molecular and Biochemical Parasitology*, vol. 168, no. 2, pp. 149–157, 2009.

[13] H. Soblik, A. E. Younis, M. Mitreva et al., "Life cycle stage-resolved proteomic analysis of the excretome/secretome from *Strongyloides ratti*—identification of stage-specific proteases," *Molecular and Cellular Proteomics*, vol. 10, no. 12, 2011.

[14] A. E. Younis, H. Soblik, I. Ajonina-Ekoti et al., "Characterization of a secreted macrophage migration inhibitory factor homologue of the parasitic nematode *Strongyloides* acting at the parasite-host cell interface," *Microbes and Infection*, vol. 14, no. 3, pp. 279–289, 2012.

[15] A. E. Younis, F. Geisinger, I. Ajonina-Ekoti et al., "Stage-specific excretory-secretory small heat shock proteins from the parasitic nematode *Strongyloides ratti*—putative links to host's intestinal mucosal defense system," *The FEBS Journal*, vol. 278, no. 18, pp. 3319–3336, 2011.

[16] D. Ditgen, E. M. Anandarajah, K. A. Meissner, N. Brattig, C. Wrenger, and E. Liebau, "Harnessing the helminth secretome for therapeutic immunomodulators," *BioMed Research International*, vol. 2014, Article ID 964350, 14 pages, 2014.

[17] H. Craig, J. M. Wastling, and D. P. Knox, "A preliminary proteomic survey of the in vitro excretory/secretory products of fourth-stage larval and adult *Teladorsagia circumcincta*," *Parasitology*, vol. 132, no. 4, pp. 535–543, 2006.

[18] V. G. Virginio, K. M. Monteiro, F. Drumond et al., "Excretory/secretory products from in vitro-cultured *Echinococcus granulosus* protoscoleces," *Molecular and Biochemical Parasitology*, vol. 183, no. 1, pp. 15–22, 2012.

[19] Y. Hu, L. Huang, Y. Huang et al., "Molecular cloning, expression, and immunolocalization of protein disulfide isomerase in excretory-secretory products from *Clonorchis sinensis*," *Parasitology Research*, vol. 111, no. 3, pp. 983–989, 2012.

[20] X. Cao, Z. Fu, M. Zhang et al., "iTRAQ-based comparative proteomic analysis of excretory-secretory proteins of schistosomula and adult worms of *Schistosoma japonicum*," *Journal of Proteomics*, vol. 138, pp. 30–39, 2016.

[21] A. Marcilla, M. Trelis, A. Cortés et al., "Extracellular vesicles from parasitic helminths contain specific excretory/secretory proteins and are internalized in intestinal host cells," *PLoS ONE*, vol. 7, no. 9, Article ID e45974, 2012.

[22] E. S. J. Arnér and A. Holmgren, "Physiological functions of thioredoxin and thioredoxin reductase," *European Journal of Biochemistry*, vol. 267, no. 20, pp. 6102–6109, 2000.

[23] K. Kunchithapautham, B. Padmavathi, R. B. Narayanan, P. Kaliraj, and A. L. Scott, "Thioredoxin from *Brugia malayi*: defining a 16-kilodalton class of thioredoxins from nematodes," *Infection and Immunity*, vol. 71, no. 7, pp. 4119–4126, 2003.

[24] J. C. Fierro-González, M. González-Barrios, A. Miranda-Vizuete, and P. Swoboda, "The thioredoxin TRX-1 regulates adult lifespan extension induced by dietary restriction in

Caenorhabditis elegans," *Biochemical and Biophysical Research Communications*, vol. 406, no. 3, pp. 478–482, 2011.

[25] I. M. Sotirchos, A. L. Hudson, J. Ellis, and M. W. Davey, "Thioredoxins of a parasitic nematode: comparison of the 16- and 12-kDA thioredoxins from *Haemonchus contortus*," *Free Radical Biology and Medicine*, vol. 44, no. 12, pp. 2026–2033, 2008.

[26] L. S. Nakao, R. A. Everley, S. M. Marino et al., "Mechanism-based proteomic screening identifies targets of thioredoxin-like proteins," *The Journal of Biological Chemistry*, vol. 290, no. 9, pp. 5685–5695, 2015.

[27] A. Holmgren and M. Bjornstedt, "Thioredoxin and thioredoxin reductase," *Methods in Enzymology*, vol. 252, pp. 199–208, 1995.

[28] W. H. Watson, X. Yang, Y. E. Choi, D. P. Jones, and J. P. Kehrer, "Thioredoxin and its role in toxicology," *Toxicological Sciences*, vol. 78, no. 1, pp. 3–14, 2004.

[29] W. Nickel, "The mystery of nonclassical protein secretion. A current view on cargo proteins and potential export routes," *European Journal of Biochemistry*, vol. 270, no. 10, pp. 2109–2119, 2003.

[30] A. Rubartelli, A. Bajetto, G. Allavena, E. Wollman, and R. Sitia, "Secretion of thioredoxin by normal and neoplastic cells through a leaderless secretory pathway," *The Journal of Biological Chemistry*, vol. 267, no. 34, pp. 24161–24164, 1992.

[31] R. Bertini, O. M. Z. Howard, H.-F. Dong et al., "Thioredoxin, a redox enzyme released in infection and inflammation, is a unique chemoattractant for neutrophils, monocytes, and T cells," *The Journal of Experimental Medicine*, vol. 189, no. 11, pp. 1783–1789, 1999.

[32] K. Kasuno, K. Shirakawa, H. Yoshida et al., "Renal redox dysregulation in AKI: application for oxidative stress marker of AKI," *American Journal of Physiology—Renal Physiology*, vol. 307, no. 12, pp. F1342–F1351, 2014.

[33] N. Kondo, H. Nakamura, H. Masutani, and J. Yodoi, "Redox regulation of human thioredoxin network," *Antioxidants & Redox Signaling*, vol. 8, no. 9-10, pp. 1881–1890, 2006.

[34] D. H. E. W. Huberts and I. J. van der Klei, "Moonlighting proteins: an intriguing mode of multitasking," *Biochimica et Biophysica Acta—Molecular Cell Research*, vol. 1803, no. 4, pp. 520–525, 2010.

[35] C. J. Jeffery, "Moonlighting proteins," *Trends in Biochemical Sciences*, vol. 24, no. 1, pp. 8–11, 1999.

[36] C. J. Jeffery, "Why study moonlighting proteins?" *Frontiers in Genetics*, vol. 6, article 211, 2015.

[37] A. Jolodar, P. Fischer, S. Bergmann, D. W. Büttner, S. Hammerschmidt, and N. W. Brattig, "Molecular cloning of an α-enolase from the human filarial parasite *Onchocerca volvulus* that binds human plasminogen," *Biochimica et Biophysica Acta—Gene Structure and Expression*, vol. 1627, no. 2-3, pp. 111–120, 2003.

[38] K. R. Lorenzatto, K. M. Monteiro, R. Paredes et al., "Fructose-bisphosphate aldolase and enolase from *Echinococcus granulosus*: genes, expression patterns and protein interactions of two potential moonlighting proteins," *Gene*, vol. 506, no. 1, pp. 76–84, 2012.

[39] V. Steisslinger, S. Korten, N. W. Brattig, and K. D. Erttmann, "DNA vaccine encoding the moonlighting protein *Onchocerca volvulus* glyceraldehyde-3-phosphate dehydrogenase (Ov-GAPDH) leads to partial protection in a mouse model of human filariasis," *Vaccine*, vol. 33, no. 43, pp. 5861–5867, 2015.

[40] H. Lüdemann, M. Dormeyer, C. Sticherling, D. Stallmann, H. Follmann, and R. L. Krauth-Siegel, "*Trypanosoma brucei* tryparedoxin, a thioredoxin-like protein in African trypanosomes," *FEBS Letters*, vol. 431, no. 3, pp. 381–385, 1998.

[41] D. G. Arias, V. E. Marquez, M. L. Chiribao et al., "Redox metabolism in *Trypanosoma cruzi*: functional characterization of tryparedoxins revisited," *Free Radical Biology and Medicine*, vol. 63, pp. 65–77, 2013.

[42] E. Jortzik and K. Becker, "Thioredoxin and glutathione systems in *Plasmodium falciparum*," *International Journal of Medical Microbiology*, vol. 302, no. 4-5, pp. 187–194, 2012.

[43] J. Liu, L. Wetzel, Y. Zhang et al., "Novel thioredoxin-like proteins are components of a protein complex coating the cortical microtubules of *Toxoplasma gondii*," *Eukaryotic Cell*, vol. 12, no. 12, pp. 1588–1599, 2013.

[44] C. Zhou, M. Bian, H. Liao et al., "Identification and immunological characterization of thioredoxin transmembrane-related protein from *Clonorchis sinensis*," *Parasitology Research*, vol. 112, no. 4, pp. 1729–1736, 2013.

[45] J. D. Brown, A. M. Day, S. R. Taylor, L. Tomalin, B. E. Morgan, and E. A. Veal, "A peroxiredoxin promotes H_2O_2 signaling and oxidative stress resistance by oxidizing a thioredoxin family protein," *Cell Reports*, vol. 5, no. 5, pp. 1425–1435, 2013.

[46] S. Lee, S. M. Kim, and R. T. Lee, "Thioredoxin and thioredoxin target proteins: from molecular mechanisms to functional significance," *Antioxidants and Redox Signaling*, vol. 18, no. 10, pp. 1165–1207, 2013.

[47] T. Ishii, Y. Funato, and H. Miki, "Thioredoxin-related protein 32 (TRP32) specifically reduces oxidized phosphatase of regenerating liver (PRL)," *Journal of Biological Chemistry*, vol. 288, no. 10, pp. 7263–7270, 2013.

[48] X. W. Wang, Y.-C. Liou, B. Ho, and J. L. Ding, "An evolutionarily conserved 16-kDa thioredoxin-related protein is an antioxidant which regulates the NF-κB signaling pathway," *Free Radical Biology and Medicine*, vol. 42, no. 2, pp. 247–259, 2007.

[49] I. Padera, R. Sengupta, M. Cebula et al., "Thioredoxin-related protein of 14 kDa is an efficient L-cystine reductase and S-denitrosylase," *Proceedings of the National Academy of Sciences of the United States of America*, vol. 111, no. 19, pp. 6964–6969, 2014.

[50] Z. Zhang, A. Wang, H. Li, H. Zhi, and F. Lu, "STAT3-dependent TXNDC17 expression mediates Taxol resistance through inducing autophagy in human colorectal cancer cells," *Gene*, vol. 584, no. 1, pp. 75–82, 2016.

[51] W. Jeong, T.-S. Chang, E. S. Boja, H. M. Fales, and S. G. Rhee, "Roles of TRP14, a thioredoxin-related protein in tumor necrosis factor-α signaling pathways," *The Journal of Biological Chemistry*, vol. 279, no. 5, pp. 3151–3159, 2004.

[52] J. Qin, G. M. Clore, W. P. Kennedy, J. R. Huth, and A. M. Gronenborn, "Solution structure of human thioredoxin in a mixed disulfide intermediate complex with its target peptide from the transcription factor NFκB," *Structure*, vol. 3, no. 3, pp. 289–297, 1995.

[53] J.-P. Lai, J. T. Dalton, and D. L. Knoell, "Phosphatase and tensin homologue deleted on chromosome ten (PTEN) as a molecular target in lung epithelial wound repair," *British Journal of Pharmacology*, vol. 152, no. 8, pp. 1172–1184, 2007.

[54] Y. Sugiura, K. Araki, S.-I. Iemura, T. Natsume, J. Hoseki, and K. Nagata, "Novel thioredoxin-related transmembrane protein TMX4 has reductase activity," *The Journal of Biological Chemistry*, vol. 285, no. 10, pp. 7135–7142, 2010.

[55] L. A. Kelley, S. Mezulis, C. M. Yates, M. N. Wass, and M. J. E. Sternberg, "The Phyre2 web portal for protein modeling, prediction and analysis," *Nature Protocols*, vol. 10, no. 6, pp. 845–858, 2015.

[56] M. J. Smout, J. Sotillo, T. Laha et al., "Carcinogenic parasite secretes growth factor that accelerates wound healing and potentially promotes neoplasia," *PLoS Pathogens*, vol. 11, no. 10, article e1005209, 2015.

[57] D. Winter and H. Steen, "Optimization of cell lysis and protein digestion protocols for the analysis of HeLa S3 cells by LC-MS/MS," *Proteomics*, vol. 11, no. 24, pp. 4726–4730, 2011.

[58] J. Rappsilber, Y. Ishihama, and M. Mann, "Stop and Go extraction tips for matrix-assisted laser desorption/ionization, nanoelectrospray, and LC/MS sample pretreatment in proteomics," *Analytical Chemistry*, vol. 75, no. 3, pp. 663–670, 2003.

[59] H. Hahne, N. Sobotzki, T. Nyberg et al., "Proteome wide purification and identification of O-GlcNAc-modified proteins using click chemistry and mass spectrometry," *Journal of Proteome Research*, vol. 12, no. 2, pp. 927–936, 2013.

[60] J. Clos and S. Brandau, "pJC20 and pJC40—two high-copy-number vectors for T7 RNA polymerase-dependent expression of recombinant genes in *Escherichia coli*," *Protein Expression and Purification*, vol. 5, no. 2, pp. 133–137, 1994.

[61] A. Holmgren, "Thioredoxin catalyzes the reduction of insulin disulfides by dithiothreitol and dihydrolipoamide," *Journal of Biological Chemistry*, vol. 254, no. 19, pp. 9627–9632, 1979.

[62] M. Luthman and A. Holmgren, "Rat liver thioredoxin and thioredoxin reductase: purification and characterization," *Biochemistry*, vol. 21, no. 26, pp. 6628–6633, 1982.

[63] E. E. Wollman, L. d'Auriol, L. Rimsky et al., "Cloning and expression of a cDNA for human thioredoxin," *The Journal of Biological Chemistry*, vol. 263, no. 30, pp. 15506–15512, 1988.

[64] J. Pusch, M. Votteler, S. Göhler et al., "The physiological performance of a three-dimensional model that mimics the microenvironment of the small intestine," *Biomaterials*, vol. 32, no. 30, pp. 7469–7478, 2011.

[65] C. Berges, C. Naujokat, S. Tinapp et al., "A cell line model for the differentiation of human dendritic cells," *Biochemical and Biophysical Research Communications*, vol. 333, no. 3, pp. 896–907, 2005.

[66] N. W. Brattig, U. Rathjens, M. Ernst, F. Geisinger, A. Renz, and F. W. Tischendorf, "Lipopolysaccharide-like molecules derived from Wolbachia endobacteria of the filaria Onchocerca volvulus are candidate mediators in the sequence of inflammatory and antiinflammatory responses of human monocytes," *Microbes and Infection*, vol. 2, no. 10, pp. 1147–1157, 2000.

[67] P. Hertel, J. Daniel, D. Stegehake et al., "The ubiquitin-fold modifier 1 (Ufm1) cascade of caenorhabditis elegans," *Journal of Biological Chemistry*, vol. 288, no. 15, pp. 10661–10671, 2013.

[68] M. T. Rubio De Krömer, M. Krömer, K. Lüersen, and N. W. Brattig, "Detection of a chemotactic factor for neutrophils in extracts of female Onchocerca volvulus," *Acta Tropica*, vol. 71, no. 1, pp. 45–56, 1998.

[69] N. W. Brattig, D. W. Büttner, and A. Hoerauf, "Neutrophil accumulation around Onchocerca worms and chemotaxis of neutrophils are dependent on Wolbachia endobacteria," *Microbes and Infection*, vol. 3, no. 6, pp. 439–446, 2001.

[70] T. Tajima, T. Murata, K. Aritake et al., "Lipopolysaccharide induces macrophage migration via prostaglandin D_2 and prostaglandin E_2," *Journal of Pharmacology and Experimental Therapeutics*, vol. 326, no. 2, pp. 493–501, 2008.

[71] H. Wang, L. A. Vardy, C. P. Tan et al., "PCBP1 suppresses the translation of metastasis-associated PRL-3 phosphatase," *Cancer Cell*, vol. 18, no. 1, pp. 52–62, 2010.

[72] E. Herrero and M. A. De La Torre-Ruiz, "Monothiol glutaredoxins: a common domain for multiple functions," *Cellular and Molecular Life Sciences*, vol. 64, no. 12, pp. 1518–1530, 2007.

[73] R. D. Finn, P. Coggill, R. Y. Eberhardt et al., "The Pfam protein families database: towards a more sustainable future," *Nucleic Acids Research*, vol. 44, no. 1, pp. D279–D285, 2016.

[74] M. Dorris, M. E. Viney, and M. L. Blaxter, "Molecular phylogenetic analysis of the genus *Strongyloides* and related nematodes," *International Journal for Parasitology*, vol. 32, no. 12, pp. 1507–1517, 2002.

[75] W. Nickel and M. Seedorf, "Unconventional mechanisms of protein transport to the cell surface of eukaryotic cells," *Annual Review of Cell and Developmental Biology*, vol. 24, pp. 287–308, 2008.

[76] J. E. Linz, S.-Y. Hong, and L. V. Roze, "Oxidative stress-related transcription factors in the regulation of secondary metabolism," *Toxins*, vol. 5, no. 4, pp. 683–702, 2013.

[77] B. Sahaf and A. Rosén, "Secretion of 10-kDa and 12-kDa thioredoxin species from blood monocytes and transformed leukocytes," *Antioxidants & Redox Signaling*, vol. 2, no. 4, pp. 717–726, 2000.

[78] G. Pickert, C. Neufert, M. Leppkes et al., "STAT3 links IL-22 signaling in intestinal epithelial cells to mucosal wound healing," *Journal of Experimental Medicine*, vol. 206, no. 7, pp. 1465–1472, 2009.

[79] F. Schreiber, J. M. Arasteh, and T. D. Lawley, "Pathogen resistance mediated by IL-22 signaling at the epithelial-microbiota interface," *Journal of Molecular Biology*, vol. 427, no. 23, pp. 3676–3682, 2015.

[80] A. Mizoguchi, "Healing of intestinal inflammation by IL-22," *Inflammatory Bowel Diseases*, vol. 18, no. 9, pp. 1777–1784, 2012.

[81] H. Nakamura, Y. Hoshino, H. Okuyama, Y. Matsuo, and J. Yodoi, "Thioredoxin 1 delivery as new therapeutics," *Advanced Drug Delivery Reviews*, vol. 61, no. 4, pp. 303–309, 2009.

[82] R. Watanabe, H. Nakamura, H. Masutani, and J. Yodoi, "Antioxidative, anti-cancer and anti-inflammatory actions by thioredoxin 1 and thioredoxin-binding protein-2," *Pharmacology & Therapeutics*, vol. 127, no. 3, pp. 261–270, 2010.

[83] H. Tamaki, H. Nakamura, A. Nishio et al., "Human thioredoxin-1 ameliorates experimental murine colitis in association with suppressed macrophage inhibitory factor production," *Gastroenterology*, vol. 131, no. 4, pp. 1110–1121, 2006.

[84] H. Schenk, M. Vogt, W. Dröge, and K. Schulze-Osthoff, "Thioredoxin as a potent costimulus of cytokine expression," *The Journal of Immunology*, vol. 156, no. 2, pp. 765–771, 1996.

[85] T. Hoshino, H. Nakamura, M. Okamoto et al., "Redox-active protein thioredoxin prevents proinflammatory cytokine- or bleomycin-induced lung injury," *American Journal of Respiratory and Critical Care Medicine*, vol. 168, no. 9, pp. 1075–1083, 2003.

[86] H. Tian, Y. Matsuo, A. Fukunaga, R. Ono, C. Nishigori, and J. Yodoi, "Thioredoxin ameliorates cutaneous inflammation by regulating the epithelial production and release of pro-Inflammatory cytokines," *Frontiers in Immunology*, vol. 4, article 269, 2013.

[87] E. R. Mann, D. Bernardo, S. C. Ng et al., "Human gut dendritic cells drive aberrant gut-specific T-cell responses in ulcerative colitis, characterized by increased IL-4 production and loss of

IL-22 and IFNγ," *Inflammatory Bowel Diseases*, vol. 20, no. 12, pp. 2299–2307, 2014.

[88] M. A. Kinnebrew, C. G. Buffie, G. E. Diehl et al., "Interleukin 23 production by intestinal CD103 +CD11b + dendritic cells in response to bacterial flagellin enhances mucosal innate immune defense," *Immunity*, vol. 36, no. 2, pp. 276–287, 2012.

[89] M. Rimoldi, M. Chieppa, V. Salucci et al., "Intestinal immune homeostasis is regulated by the crosstalk between epithelial cells and dendritic cells," *Nature Immunology*, vol. 6, no. 5, pp. 507–514, 2005.

[90] M. J. Broadhurst, J. M. Leung, V. Kashyap et al., "IL-22⁺ CD4⁺ T cells are associated with therapeutic *Trichuris trichiura* infection in an ulcerative colitis patient," *Science Translational Medicine*, vol. 2, no. 60, article 60ra88, 2010.

[91] P. Niethammer, C. Grabher, A. T. Look, and T. J. Mitchison, "A tissue-scale gradient of hydrogen peroxide mediates rapid wound detection in zebrafish," *Nature*, vol. 459, no. 7249, pp. 996–999, 2009.

[92] B. Enyedi and P. Niethammer, "Mechanisms of epithelial wound detection," *Trends in Cell Biology*, vol. 25, no. 7, pp. 398–407, 2015.

[93] C. Dunnill, T. Patton, J. Brennan et al., "Reactive Oxygen Species (ROS) and wound healing: the functional role of ROS and emerging ROS-modulating technologies for augmentation of the healing process," *International Wound Journal*, 2015.

[94] W. C. Gause, T. A. Wynn, and J. E. Allen, "Type 2 immunity and wound healing: evolutionary refinement of adaptive immunity by helminths," *Nature Reviews Immunology*, vol. 13, no. 8, pp. 607–614, 2013.

[95] R. M. Maizels and H. J. McSorley, "Regulation of the host immune system by helminth parasites," *Journal of Allergy and Clinical Immunology*, vol. 138, no. 3, pp. 666–675, 2016.

18

Prevalence of Intestinal Parasitic Infection among Food Handlers in Northwest Iran

Davoud Balarak,[1] Mohammad Jafari Modrek,[2] Edris Bazrafshan,[1] Hossein Ansari,[3] and Ferdos Kord Mostafapour[1]

[1]Department of Environmental Health, Health Promotion Research Center, Zahedan University of Medical Sciences, Zahedan, Iran
[2]Department of Parasitology, Infectious Diseases and Tropical Medicine Research Center,
 Zahedan University of Medical Sciences, Zahedan, Iran
[3]Department of Epidemiology and Biostatistics, Health Promotion Research Center, Zahedan University of Medical Sciences,
 Zahedan, Iran

Correspondence should be addressed to Davoud Balarak; dbalarak2@gmail.com

Academic Editor: Dong H. Shin

Parasitic diseases are among the most important infectious diseases and pose health problems in many countries, most especially in developing countries. Workers at food centers could transmit parasitic infections in the absence of sanitation. This is a descriptive study conducted to determine the prevalence of intestinal parasitic infections in food clerks in the city of Tabriz in 2014. Data was recorded in the offices of the health center for all food handlers who were referred to the laboratory for demographic and stool tests to receive the health card. Parasitic infection was observed in 172 cases (3.73%) of 4612 samples. A total of 156 positive samples (90.69%) were related to protozoa and 16 (9.3%) were related to helminthes. Most of the parasitic infections were related to *Giardia* and *Entamoeba coli* and the lowest infection was related to *H. nana*. Also, there was a significant relationship between level of education and parasitic infection rate ($P = 0.0044$). But there was no significant difference between the type of infection and amount of intestinal parasites. The results show that the prevalence of intestinal parasites, especially pathogenic protozoa, is common in some food handlers. Therefore, more sanitary controls are required and increasing of education will play a crucial role in improving the health of these people.

1. Introduction

The parasitic infections are considered as one of the major health problems in the world and especially in developing countries [1, 2]. According to the World Health Organization (WHO), nearly two-thirds of the world is infected with one kind of intestinal parasite and *Ascaris and Giardia* infections have the highest rate among all kinds [3]. Due to geographical location, climate, extent of the area, and cultural and biological characteristics, there is a suitable environment for the activity of various parasites in Iran [4, 5]. The reasons for high incidence of parasites in some parts of the country are a result of specific climate of the regions, local customs, and the use of human and animal fertilizers in agriculture and olericulture [6]. Lack of clean and safe water, high population density, lack of proper disposal of waste, noncompliance with

health standards (social and individual), lack of adequate washing of vegetables, and lack of well cooked meat lead to high prevalence of intestinal parasites [7, 8]. Studies in different parts of Iran revealed that there is infection of intestinal parasites between studied groups. For this reason, several studies have been conducted on the prevalence of parasitic infections in different parts of the country that show the prevalence of parasitic infections between 2 and 61% in studied populations [9–11]. According to pervious researches, the prevalence level of parasitic diseases was as follows: Kermanshah (59.13%) [3], Mazandaran (21%) [12], Kashan (46.9%) [13], and Ardabil (27.7%) [14], while it was 13.7% and 8.4% for Semnan [15] and Ghaemshahr [16] that indicates the great prevalence of these infections from the statistical viewpoint. As a result of the high incidence of parasitic infection in the country, the identification of infection, knowing the methods of transmission, and preventing their transmission are of paramount importance and are considered as health priorities [17].

One of the most important areas of risk of intestinal parasitic infections in the community is the risky nature of some jobs. These conditions facilitate the transmission of disease through the close contact with the infectious sources that are an integral part of some of the jobs, so they provide the possibility of easy infection [18]. The production and sales of health food products are always a concern in many developed and developing countries. Today, a number of factors including existence of quality control procedures, strict supervision of production processes, reduction in the amount of contact with food, and food categories in the process of production are the most important tasks in the production of high quality products. In this case, the role of food handlers, as the final agent of product provider, is very important because they can make unsafe and hazardous foods for consumption. If these people fail to follow the principles of primary health care and personal hygiene, they are considered as one of the main sources of transmission of pathogens through food and can easily pass many infectious agents, especially intestinal parasites to their customers [18, 19]. Obviously, the study of parasitic diseases among employees of the food industry and the prevalence rate of infection in them will help to reduce and control the spread of these diseases. Since our country has several areas with different climatic diversity and social and cultural patterns, study of parasitic infections in any province or region is absolutely necessary.

Given the role of food handlers in the spread of parasitic diseases and also due to the prevalence of infection in the country and lack of awareness towards the infection level in the city of Tabriz, this study was conducted to investigate the intestinal parasitic contamination in food handlers in the city of Tabriz in 2014.

2. Materials and Methods

According to existing regulations, Health Care Card is necessary to monitor the health status of employees in the health sector of different jobs and especially jobs that were related to food. All those who were responsible for procurement, distribution, and sale at food centers have referred to health centers to extend their Health Care Cards twice per year. The food handlers should undergo the necessary tests, especially parasitic tests, to receive the Health Care Card. For each food handler, 3 stools were tested by direct method (wet method with Gram) and the results were recorded along with other personal information such as age, sex, and address in the offices for each job. According to the Office Inspection Act of Tabriz, everyone is required to obtain a Health Care Card and undergo test for parasites. In this study, the recorded data related to epidemiological intestinal parasitic infection of individuals who are working at procurement, distribution, and sale of food products in public places of Tabriz in 2014 were used. In addition, other data including details of the type of business, business location, gender, and the results of tests were obtained from these offices. Based on the present data, Tabriz city had 1465 center for sales and distribution of food in

2014 and a total of 4612 person were working at these centers. The results of tests were analyzed by Spss version 18 (IBM, USA) using the chi-square test and t-test.

3. Result

The results showed that 3966 male food handlers (85.99%) and 646 female food handlers (14.01%), who were engaged in the distribution and sale of food, have been referred to health centers to achieve the Health Care Card. Their ages were between 14 and 68 years. The education level of the food handler was observed to be under high school diploma (30.68%), high school diploma to B.S. (58.2%), and over B.S. (11.12%). Table 1 shows the number of positive cases according to age, sex, and education level. There was no significant difference between age and gender in any case. The infection was observed in 156 cases (3.38%) of males and 16 cases (0.345%) of females but statistically significant difference was not observed in this field ($P = 0.094$).

Parasitic comparison between the food suppliers is given in Table 2, indicating that the highest percentage of infection is related to restaurants and supermarkets, but in general, there was no statistically significant difference between different professions ($P = 0.112$). A total of 172 subjects (3.73%) were diagnosed with intestinal parasites and five types of intestinal parasites; *Giardia lamblia*, *Entamoeba coli*, *Ascaris*, *Entamoeba histolytica*, and *H. nana* were observed. Most of the infections were related to *Giardia* with 109 (63.33%) of 172 positive cases. *Entamoeba* coli infection was recorded as 38 positive cases (22.1%), while *Ascaris* infection was 10 cases (5.83%). Furthermore, the parasitic worm infection was related to *Ascaris* and *H. nana* and equal to 16 cases (0.38%) but protozoa infections were 156 cases (3.38%).

4. Discussion

The result of this study showed that 3.73% of the food handlers in Tabriz were infected with intestinal parasites. The parasitic infections in this study were less than the results of other studies which have been conducted in Yazd [10], Varamin and Sanandaj [20, 21], and Hamadan [22]. These studies have been conducted in the past decade and, as it was stated, show high amounts of parasitic infection, but in the present decade their prevalence in different areas has significantly declined parallel to improvement of public health. One of the most effective methods in reducing parasitic infection was strict enforcement of sanitary regulations in recent years, so that all people in Tabriz were issued a Health Care Card and parasitological tests were performed on them accurately.

The highest prevalence of infection was in supermarket and sandwich and pizza businesses which recorded 22.1, 8 and 19.2% of parasitic infections, respectively. The case study conducted in Kerman by Salary and Safizadeh is fully in accordance with the results of our study, so that the maximum amount of infection was 27.1% which was recorded for supermarket workers and clerks among the food handlers of Kerman [11]. Also, the prevalence level was reported to be 30.2% in another survey that was conducted by Rouhani et al. in Nowshahr and Chaloos on various food vendors

TABLE 1: The number of positive cases of parasites in food production and distribution centers employees in the city of Tabriz according to individual characteristics.

Characteristics	Studied people		Positive cases	
	Numbers	%	Numbers	%
Gender				
Male	3966	85.99	156	3.38
Female	646	14.01	16	0.345
Age (year)				
Under 20	418	9.05	20	0.433
20 to 40	2576	55.85	108	2.34
Over 40	1618	35.1	44	0.954
Education				
Under high school diploma	1415	30.68	68	1.75
High school diploma to B.S.	2684	58.2	76	1.64
Over B.S.	513	11.12	28	0.607

[23]. Koohsar et al. have also reported that intestinal parasitic infection in food handlers of Gorgan was equal to 6%. According to these observations, most parasitic infections were caused by butchery staffs by 25% and the highest rate of infection was due to the parasite *Giardia* [24]. The difference between the results could be due to the sample size, the studied population, social, economic, and geographical characteristics, and climate changes, including humidity and heat, direct exposure to the sun's rays, and a lack of vegetation. In another study, the reason for these differences is highlighted [25]. Kamau et al. have conducted a similar study in Kenya and they found out that *Giardia* parasite is one of 6 common types of parasites among members of restaurant staff [26].

Although there is no direct contact with food by supermarket worker as other businesses but the contamination of food are possible when they are uncovered; therefore, sandwich, pizza, and restaurant handlers have most contribution in incidence of parasitic infection after supermarkets, which is accordance with most studies in the our country. Since these groups have direct contact with food, intestinal parasitic infection is more possible; thus the staff in these sites require personal and professional hygiene in their work environment, and these health professionals are decisively required to follow up and monitor. In this study, most parasitic infections are related to *Giardia* with 109 of the 172 positive samples which assigned 63.37% of the total parasitic infection that are well correlated to the study that took place in the city of Kerman by Salary and Safizadeh but the *Giardia* parasite has been isolated only in their study which is considered as difference with our study [11]. In another study, the prevalence of intestinal parasites was recorded in food handlers involved in the procurement, distribution, and sale of food in public places in Gilan Province and it was detected that the most common parasite is *Giardia* [9]. The studies conducted in Saudi Arabia and Sudan cleared that *Giardia* and *E. histolytica* parasites were observed in food handlers and the amount of *Giardia* parasite was very high in Sudan [27, 28].

In this study, the rate of infection with worms was 16 samples (0.345%) and protozoan infection was at the rate of 156 samples (3.38%). These findings are in line with the earlier studies conducted by Kheyrandish et al. in Khorramabad, Shidfar et al. in Ilam, and Balarak et al. in Ahar [29–31]. Also, a correlation was found with a study in Ethiopia and India [32, 33]. These studies show higher prevalence of protozoan parasites than worms. The high prevalence of protozoan parasites is due to the proliferation of the spread and ability to produce large simple cysts and protozoa stability in environmental conditions [34]. In the study conducted in Nigeria and Manila, protozoan parasites like *Giardia* and *Entamoeba* coli were more observed, due to the higher proliferation of protozoa than worms [35, 36]. According to various studies, it is indicated that the prevalence of intestinal parasitic infections has recently changed compared to the past and the prevalence of infection has declined, totally. The reduction in the incidence of infections over the years can be attributed to the development of networks for the distribution of drinking water, more comprehensive monitoring of health systems, and ongoing communication with employees and stricter rules than in the past to provide health advice and provide hygiene standards, continuous testing of parasitic infections (6 months) and the availability of drugs for the treatment of infections, higher levels of life expectancy in terms of health and increasing level of individual health information and the use of less human fertilizer by farmers [19, 37].

Literacy level reduces the number of positive samples; in other words, there is a significant relationship between level of education and degree of parasitic infection ($P = 0.0044$). It could be interpreted that if the literacy rate increased, then awareness about parasitic infections will also increase. Therefore, the lower need for health advice and better compliance with sanitary regulations will be achieved, as noted in other studies [29]. People in South-East Asia have less knowledge about these infections; thus, there are more infections in those regions [38], while the infection level is less in developed countries like Italy [39].

5. Conclusion

The results show that the prevalence rate of intestinal parasites, particularly *Giardia* protozoan, in food handlers is

TABLE 2: Frequency of pathogenic intestinal parasites on the separation of producers and suppliers of food trade in Tabriz.

Pollution trade unit	Total	Without infection	Infection					Number of positive samples	% of positive samples	% of all samples
			Giardia	Ascaris	Entamoeba coli	H. nana	Entamoeba histolytica			
Supermarket	880	842	23	4	7	2	2	38	22.1 (38/172)	0.823 (38/4612)
Restaurant	821	797	14	2	5	1	2	24	13.95 (24/172)	0.52 (24/4612)
Sandwich and pizza	709	676	21	2	8	1	1	33	19.2 (33/172)	0.715 (33/4612)
Confectionery	229	221	5	—	2	—	1	8	4.65 (8/172)	0.173 (8/4612)
Bakery	274	265	5	1	2	1	—	9	5.25 (9/172)	0.195 (9/4612)
Butcher	95	91	2	—	1	—	1	4	2.32 (4/172)	0.086 (4/4612)
Dairy	403	383	13	1	5	—	1	20	11.65 (20/172)	0.433 (20/4612)
Production workshop	889	876	11	—	2	—	—	13	7.55 (13/172)	0.281 (13/4612)
Ice cream and juice	312	289	15	—	6	1	1	23	13.3 (23/172)	0.498 (23/4612)
Sum	4612	4440	109	10	38	6	9	172	172	—
% of positive samples	—	—	63.37	5.81	22.1	3.48	5.23	100	100	—
% of all samples	100	96.27	2.36	0.0021	0.0082	00.13	00.195	3.73	—	3.73

3.73%. This percentage of infection between the workers in food producing places can be considered as an important way to transmit and incidence of these infectious agents among the public. Therefore, the attention to the health status of these places and the continuous monitoring of the workers in these places can play significant role to prevent the infection. Furthermore, holding of training and educational programs leads to increase the knowledge of the workers towards these infections and the associated risk to themselves and their community which can be assumed as other effective factors to diminish the health risk problems associated with this sector.

Competing Interests

The authors declare that they have no competing interests.

References

[1] M. Malakotian, M. Hosseini, and H. Bahrami, "Survey of the parasires of vegetable in Kerman province," *Medical Journal of Hormozgan University*, vol. 13, no. 1, pp. 55–62, 2009.

[2] M. Shahnazi, M. Sharifi, Z. Kalantari, M. Alipour Heidari, and N. Agamirkarimi, "The study of consumed vegetable parasitic infections in Qazvin," *The Journal of Qazvin University of Medical Sciences*, vol. 12, no. 4, pp. 83–89, 2009.

[3] M. Vojdaani, A. Barzegar, and A. Shamsiaan, "Frequency of parasitic infections in patients referred to special clinic of Kermanshah University of Medical Sciences in years 1995–99," *Journal of Kermanshah University of Medical Sciences*, vol. 6, no. 2, pp. 31–37, 2002.

[4] T. K. Hazrati, M. Mostaghim, H. Khalkhalli, and M. A. Aghayar, "The prevalence of intestinal parasitic infection inthe students of primary schools in Nazloo region in Urmia during 2004-2005," *Urmia Medical Journal*, vol. 16, no. 4, pp. 212–217, 2004.

[5] K. Sharifi Sarasiabi, A. Madani, and S. Zare, "Prevalence of intestinal parasites in primary school publishes of Bandar Abbas," *Journal of Hormozgan University of Medical Sciences*, vol. 4, no. 5, pp. 25–30, 2002.

[6] B. Ezatpour, A. S. Chegeni, F. Abdollahpour, M. Aazami, and M. Alirezaei, "Prevalence of parasitic contamination of raw vegetables in Khorramabad, Iran," *Food Control*, vol. 34, no. 1, pp. 92–95, 2013.

[7] J. G. Damen, E. B. Banwat, D. Z. Egah, and J. A. Allanana, "Parasitic contamination of vegetables in Jos, Nigeria," *Annals of African Medicine*, vol. 6, no. 3, pp. 115–118, 2007.

[8] H. Soleimnanpoor, A. Zohor, and A. Ebrahimzadeh, "The survey of parasitic contamination of vegetables in Zabol city during 2011-2012," *Zabol University of Medical Sciences*, vol. 3, no. 2, pp. 40–47, 2013.

[9] A. S. Arani, R. Alaghehbandan, L. Akhlaghi, M. Shahi, and A. R. Lari, "Prevalence of intestinal parasites in a population in south of Tehran, Iran," *Revista do Instituto de Medicina Tropical de Sao Paulo*, vol. 50, no. 3, pp. 145–149, 2008.

[10] F. A. Dehghani and M. Azizi, "Study of the rate of contamination of intestinal parasites among workers in fast food outlets of Yazd," *Journal of Shahid Sadoughi University of Medical Sciences and Health Services*, vol. 11, no. 1, pp. 22–28, 2003.

[11] S. Salary and H. Safizadeh, "Prevalence of intestinal parasite infestation in the food suppliers of Kerman City, Iran, in 2010," *Journal of Health & Development*, vol. 1, no. 4, pp. 315–322, 2013.

[12] T. Razavyoon and J. Massoud, "Intestinal parasitic infection in Feraydoon Kenar, Mazandaran," *Journal of School of Public Health and Institute of Public Health Research*, vol. 1, no. 1, pp. 39–49, 2003.

[13] M. Arbabi and S. A. Talari, "Intestinal parasites among students of Kashan University of Medical Sciences," *Journal of Ilam University of Medical Sciences*, vol. 12, no. 44-45, pp. 24–33, 2005.

[14] A. Daryani and G. H. Ettehad, "Prevalence of intestinal infestation among primary school students in Ardabil, 2003," *Journal of Ardabil University of Medical Sciences*, vol. 5, no. 3, pp. 229–234, 2005.

[15] N. E. Atash, R. Ghorbani, S. Peyvandi, and S. Imani, "Prevalence of oxyuriasis and some related factors in kindergarten and primary school children in urban areas of Semnan province," *Journal of Semnan University of Medical Sciences*, vol. 9, no. 1, pp. 67–74, 2007.

[16] S. Ranjbar-Bahadori, A. Dastorian, and B. Heidari, "Prevalence of intestinal parasites in Ghaemshahr in 2004," *Medical Science Journal of Islamic Azad Univesity—Tehran Medical Branch*, vol. 15, no. 3, pp. 151–155, 2005.

[17] A. Siyadatpanah, F. Tabatabaei, A. E. Zeydi et al., "Parasitic contamination of raw vegetables in Amol, North of Iran," *Archives of Clinical Infectious Diseases*, vol. 8, no. 2, pp. 159–183, 2013.

[18] P. Ayeh-Kumi, S. Quarcoo, G. Kwakye-Nuako, J. Kretchy, A. Osafo-Kantanka, and S. Mortu, "Prevalence of intestinal parasitic infections among food vendors in Accra, Ghana," *The Journal of Tropical Medicine and Parasitology*, vol. 32, no. 1, pp. 1–8, 2009.

[19] M. Zarezadeh and M. Malakotian, "Prevalence of bacteria (Salmonella, Shigella) and intestinal parasites among food handlers in Kerman, Iran, in 1390," *Pajoohandeh Journal*, vol. 19, no. 1, pp. 55–59, 2014.

[20] Z. Aminzadeh, S. Afrasiabian, and L. Gachkar, "Intestinal parasitism in food-sellers in Sanandaj, 1997," *Pejouhandeh*, vol. 6, no. 5, pp. 449–452, 2002.

[21] Z. Aminzadeh, F. Shaker, M. Nazari, and L. Gachkar, "Prevalence of intestinal parasite in food handlers in Varamin 2002," *Journal of Paramedical Sciences*, vol. 1, no. 3, pp. 157–162, 2002.

[22] M. Fallah, S. Sadeghian, H. Taherkhani, F. Habibi, and Z. Heidar Barghi, "Study of parasitic and bacterial infections in the food-handling personnel, Ramadan, Iran," *Journal of Research in Health Sciences*, vol. 4, no. 1, pp. 3–10, 2011.

[23] S. Rouhani, K. Reshad, and A. Athari, "Surveying the prevalence of intestinal parasites' infection in food handlers in Nowshahr and Chalous," *Pajouhesh dar Pezeshki*, vol. 24, no. 1, pp. 15–20, 2000.

[24] F. Koohsar, A. Amini, A. Ayatollahi, G. Noshak, H. Hedayat-Mofidi, and M. Namjoo, "The prevalence of intestinal parasitic infections in food handlers inGorgan, Iran," *Med Lab J*, vol. 6, no. 1, pp. 26–34, 2012.

[25] G. Molavi, J. Masoud, I. Moubedi, and G. Hassanpour, "Prevalence of intestinal parasites in Esfahan municipal workers," *Journal of School of Public Health and Institute of Public Health Research*, vol. 5, no. 3, pp. 43–50, 2007.

[26] P. Kamau, P. Aloo-Obudho, E. Kabiru et al., "Prevalence of intestinal parasitic infections in certified food-handlers working in food establishments in the City of Nairobi, Kenya," *Journal of Biomedical Research*, vol. 26, no. 2, pp. 84–89, 2012.

[27] D. Zaglool, Y. Khodari, R. Othman, and M. Farooq, "Prevalence of intestinal parasites and bacteria among food handlers in a tertiary care hospital," *Nigerian Medical Journal*, vol. 52, no. 4, pp. 266–270, 2011.

[28] M. A. Babiker, M. S. M. Ali, and E. S. Ahmed, "Frequency of intestinal parasites among food-handlers in Khartoum, Sudan," *Eastern Mediterranean Health Journal*, vol. 15, no. 5, pp. 1098–1104, 2009.

[29] F. Kheyrandish, E. Badparva, and M. Tarahi, "Prevalence of intestinal parasites in Khorramabad bakeries' workers in 2001," *Yafteh*, vol. 5, no. 17, pp. 45–50, 2004.

[30] F. Shidfar, M. Aghilinegad, and R. Nasrifar, "Intestinal Parasitological infection of employee in food manufacture anddistribution centers of Ilam University of Medical Sciences," *Iran Occupational Health*, vol. 2, no. 1, pp. 24–27, 2005.

[31] D. Balarak, Y. Mahdavi, M. J. Modrek, S. Sadeghi, and A. Ali Joghataei, "Prevalence of parasitic contamination of raw vegetables in Ahar, Iran," *International Journal of Analytical, Pharmaceutical and Biomedical Sciences*, vol. 5, no. 1, pp. 28–32, 2016.

[32] G. Andargie, A. Kassu, F. Moges, M. Tiruneh, and K. Huruy, "Prevalence of bacteria and intestinal parasites among food-handlers in Gondar Town, Northwest Ethiopia," *Journal of Health, Population and Nutrition*, vol. 26, no. 4, pp. 451–455, 2008.

[33] S. Khurana, N. Taneja, R. Thapar, M. Sharma, and N. Malla, "Intestinal bacterial and parasitic infections among food handlers in a tertiary care hospital of North India," *Tropical Gastroenterology*, vol. 29, no. 4, pp. 207–209, 2008.

[34] B. Abera, F. Biadegelgen, and B. Bezabih, "Prevalence of *Salmonella typhi* and intestinal parasites among food handlers in Bahir Dar Town, Northwest Ethiopia," *Ethiopian Journal of Health Development*, vol. 24, no. 1, pp. 46–50, 2010.

[35] O. A. Morenikeji, N. C. Azubike, and A. O. Ige, "Prevalence of intestinal and vector-borne urinary parasites in communities in south-west Nigeria," *Journal of Vector Borne Diseases*, vol. 46, no. 2, pp. 164–167, 2009.

[36] D. G. Esparar, V. Y. Belizario, and J. Relos, "Prevalence of intestinal parasitic infections among food-handlers of a tertiary hospital in Manila using direct fecal smear andformalin ether concentration technique," *The Philippine Journal of Microbiology and Infectious Diseases*, vol. 33, no. 3, pp. 99–103, 2004.

[37] D. Balarak, M. Ebrahimi, M. J. Modrek, E. Bazrafshan, A. H. Mahvi, and Y. Mahdavi, "Investigation of parasitic contaminations of vegetables sold in markets in the city of Tabriz in 2014," *Global Journal of Health Science*, vol. 8, no. 10, pp. 178–184, 2016.

[38] M. M. Zain and N. N. Naing, "Sociodemographic characteristics of food handlers and their knowledge, attitude and practice towards food sanitation: a preliminary report," *Southeast Asian Journal of Tropical Medicine and Public Health*, vol. 33, no. 2, pp. 410–417, 2002.

[39] I. F. Angelillo, N. M. A. Viggiani, L. Rizzo, and A. Bianco, "Food handlers and foodborne diseases: knowledge, attitudes, and reported behavior in Italy," *Journal of Food Protection*, vol. 63, no. 3, pp. 381–385, 2000.

19

Vectors and Spatial Patterns of *Angiostrongylus cantonensis* in Selected Rice-Farming Villages of Muñoz, Nueva Ecija, Philippines

Ma. Angelica A. Tujan,[1] **Ian Kendrich C. Fontanilla,**[2] **and Vachel Gay V. Paller**[1]

[1]*Animal Biology Division, Institute of Biological Sciences, College of Arts and Sciences, University of the Philippines Los Baños, 4031 Laguna, Philippines*
[2]*DNA Barcoding Laboratory, Institute of Biology, College of Science, University of the Philippines Diliman, 1001 Quezon City, Philippines*

Correspondence should be addressed to Ma. Angelica A. Tujan; tujangelica@gmail.com

Academic Editor: Dave Chadee

In the Philippines, rats and snails abound in agricultural areas as pests and source of food for some of the local people which poses risks of parasite transmission to humans such as *Angiostrongylus cantonensis*. This study was conducted to determine the extent of *A. cantonensis* infection among rats and snails collected from rice-farming villages of Muñoz, Nueva Ecija. A total of 209 rats, 781 freshwater snails, and 120 terrestrial snails were collected for the study. Heart and lungs of rats and snail tissues were examined and subjected to artificial digestion for parasite collection. Adult worms from rats were identified using SSU rDNA gene. Seven nematode sequences obtained matched *A. cantonensis*. Results revealed that 31% of the rats examined were positive with *A. cantonensis*. *Rattus norvegicus* and *R. tanezumi* showed prevalence of 46% and 29%, respectively. Furthermore, only *Pomacea canaliculata* (2%) and *Melanoides maculata* (1%) were found to be positive for *A. cantonensis* among the snails collected. Analysis of host distribution showed overlapping habitats of rats and snails as well as residential and agricultural areas indicating risks to public health. This study presents a possible route of human infection for *A. cantonensis* through handling and consumption of *P. canaliculata* and *M. maculata* or crops contaminated by these snails.

1. Introduction

Angiostrongylus cantonensis or the rat lungworm is a zoonotic helminth responsible for the disease called angiostrongylosis. Its life cycle involves rodents as definitive hosts and mollusks as intermediate hosts. It can also infect other animals, for example, shrimps and frogs, without further development and still be infective when ingested. Humans, however, are dead-end hosts for *A. cantonensis* and can be infected through ingestion of infected mollusks, things contaminated by infected mollusks, for example, soil and vegetables [1, 2], and ingestion of paratenic hosts. As a result, *A. cantonensis* is the major cause of eosinophilic meningitis in humans particularly in Indo-Pacific regions where it is endemic. The animal-human environmental interface of *A. cantonensis* is difficult to assess and one of the reasons is that its hosts are easily affected by changes in the environment [3]. Changes in ecology and environment may also result in changes in the epidemiology of this parasite. Thus, it is important to assess the possible transmission route of the parasite due to its risks to both veterinary and public health.

Nueva Ecija is the rice granary of the Philippines and one of its towns, Muñoz, was observed to have *A. cantonensis* [4, 5]. The intensity and molecular biology of the observed parasites were not determined in previous studies. However, it is important to present stronger evidence regarding the presence of *A. cantonensis* as it could be mistaken for other species of *Angiostrongylus*. Furthermore, the intermediate host remains unknown in the region. This is an important key in assessing the infection of *A. cantonensis* particularly in humans because these hosts harbor the infective stage of the parasite. Thus, this study was conducted to determine the extent of infection of *A. cantonensis* among rats and snails collected from Muñoz.

2. Materials and Methods

2.1. Study Site. Muñoz is located in Nueva Ecija, Philippines, comprising of 37 villages with a total population of 85,461. It is globally positioned at 15.71°N latitude and 120.90° longitude E and has a total land area of 16,305 hectares which is mainly for agricultural utilization (9,819 hectares) followed by residential zone (2,847 ha) [6]. Five rice-farming villages were randomly selected for the study during June 2014-October 2015.

2.2. Harvest of Angiostrongylus cantonensis in Rats. Single live capture traps were used for the collection of rats. Rats were euthanized and dissected for the presence of *A. cantonensis*. The heart and lungs of the rat were examined for adult worms. Collected worms were preserved in 100% ethanol. Adult *A. cantonensis* were identified with the female having a barber pole appearance and male having a copulatory bursa [7]. Additionally, the organs were artificially digested with pepsin-HCl solution in a hot plate with magnetic stirrer at 37°C ± 2°C for one hour. It was filtered and placed in a petri dish for microscopic examination of larvae.

2.3. DNA Extraction, PCR Amplification, and Sequencing of Adult Worms Recovered from Rats. The identification of obtained worms was performed using the standard molecular barcode of soil nematodes, the $5'$ end of the small subunit ribosomal RNA gene (SSU rDNA) [8–10]. Total genomic DNA was extracted from adult worms using PureLink® Genomic DNA (Life Technologies) kit. The SSU rDNA gene was amplified through polymerase chain reaction (PCR) using the following primers: SSU_F_07 (sense) $5'$-AAA-GATTAAGCCATGCATG-$3'$ and SSU_R_09 (anti-sense) $5'$-AGCTGGAATTACCGCGGCTG-$3'$ [10]. A total of 50 μL of PCR mix was prepared consisting of 5 μL PCR buffer with 1.5 mM MgCl$_2$, 1.0 μL 10 mM dNTP, 2.5 μL 10 μM of each primer, 10 μL Q buffer (Qiagen, Netherlands), 0.25 μL 1.25 T Taq (Roche™, USA), and 4 μL DNA sample. The amplification was performed using Labnet MultiGene™ thermocycler with PCR conditions of 94°C for three minutes, 43 cycles of 94°C for 30 seconds, 45°C for 30 seconds, and 65°C for one minute, the final extension at 72°C for five minutes. PCR products were visualized in 1% agarose gel with ethidium bromide under ultraviolet illumination. Qiagen™ Gel Extraction Kit (USA) was used to extract the PCR products from the gel. The purified PCR products were sent to 1st Base, Malaysia, for sequencing of the antisense strands. DNA sequences were assembled using STADEN package version 1.5.3 [11] and aligned using BioEdit Sequence Alignment Editor 7.0.9.0 [12]. The species with the closest SSU rDNA sequence from GenBank for each nematode sequence was determined using the nucleotide Basic Local Alignment

Search Tool (BLAST, http://blast.ncbi.nlm.nih.gov/Blast.cgi/, [13]).

2.4. Collection of Potential Intermediate Snail Hosts. Freshwater snails were handpicked from rivers, rice fields, irrigations, and around houses while terrestrial snails were handpicked along walls and trees. All collected snails were placed in labeled containers and transferred to the laboratory for identification and parasite examination. Each snail was chopped into small pieces and placed in a petri dish containing Ash's digestive fluid [14]. It was left overnight at 37°C and live larvae were observed under a microscope. *A. cantonensis* third stage larvae (L3) were identified with their two chitinous rods at the anterior end and a slightly curved and pointed tail [7].

2.5. Survey on Knowledge and Practices of Locals on Parasites from Rats and Snails. One hundred respondents in the study site were interviewed and a questionnaire was given out to each respondent. The respondents include farmers and housewives which are people that have high chances of getting *A. cantonensis*.

2.6. Data Analyses. Prevalence and mean intensity were computed for both rats and snails. Index of discrepancy was used to determine the distribution pattern of *A. cantonensis* in both rat species [15]. Furthermore, chi-square test of independence and Fisher's exact test were used to compare the prevalence between rat species whereas Mann Whitney U test was used to compare intensities between rat species. All statistical analyses were performed at 95% confidence level using several statistical analysis software programs including Quantitative Parasitology version 3.0 and Predictive Analysis Software version 18.0 both for Windows.

3. Results

A total of 209 rats, 781 freshwater snails, and 120 terrestrial snails were collected and examined for *A. cantonensis* infection. The rats were identified as *R. norvegicus* (n = 24) and *R. tanezumi* (n = 185). The freshwater snails were grouped according to their species: *Indoplanorbis exustus* (n = 14), *Jagora asperata* (n = 7), *Melanoides maculata* (n = 89), *Pomacea canaliculata* (n = 200), *Radix quadrasi* (n = 138), *Tarebia granifera* (n = 108), *Vivipara angularis* (n = 41), and *Vivipara carinata* (n = 184). Moreover, all terrestrial snails collected were identified as *Achatina fulica* (n = 120).

3.1. SSU rDNA Gene Sequences Detected in Adult Worms from Rats. Seven representative adult female worms were successfully subjected to DNA extraction, PCR amplification, and sequencing. Floyd et al. [8] proposed that two sequences belong to the same species when they are 99.5–100% identical for the 450 bp of the $5'$ end of the SSU rDNA gene. Thus, *A. cantonensis* was identified with certainty to species level based on GenBank BLAST results (Table 1).

3.2. Angiostrongylus cantonensis Infection in Rats. A total of 64 rats (31%) were found to be infected with *A. cantonensis*

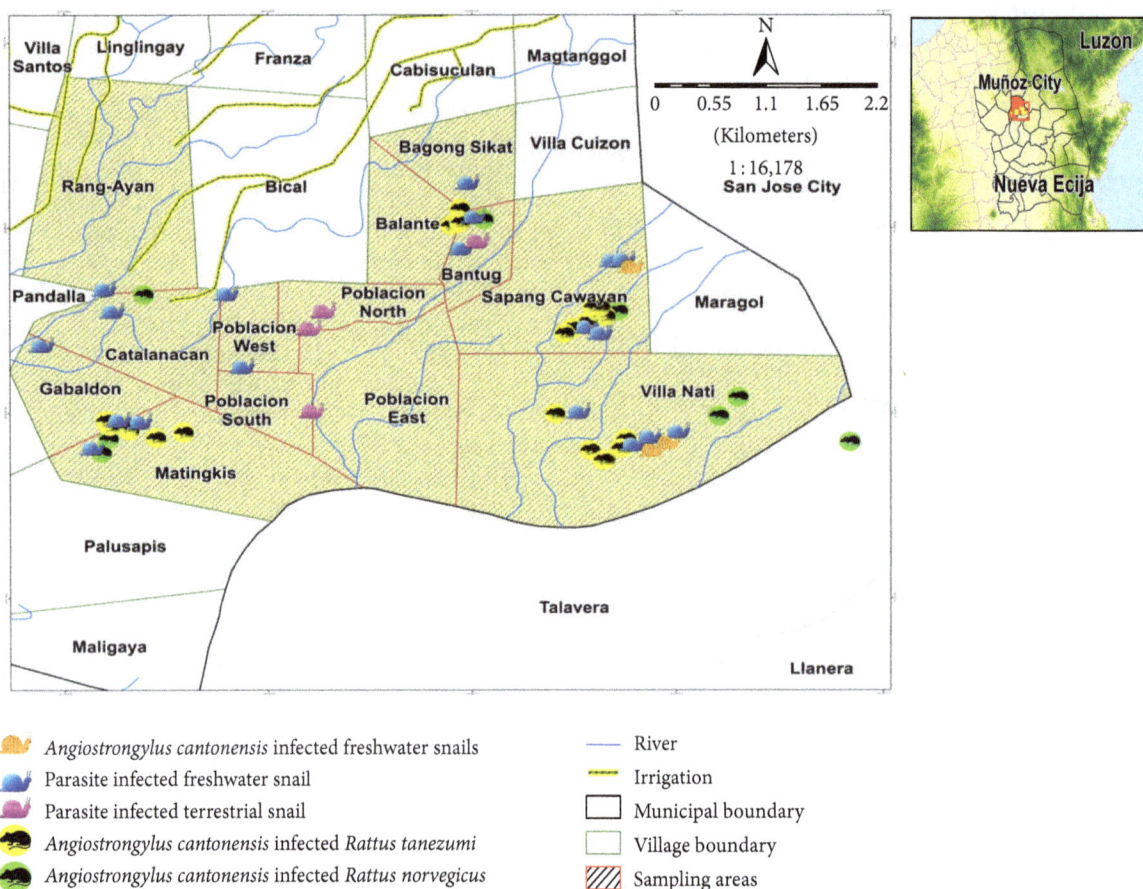

FIGURE 1: Map of Muñoz, Nueva Ecija, showing the distribution of *Angiostrongylus cantonensis* infected rats and snails.

TABLE 1: Adult worm sequences obtained in the study and their closest match in GenBank.

Sequence	Closest match	Identity
RN (4BDYKU3S01R)	*Angiostrongylus cantonensis*	100%
RT1 (4BEG550U016)	*Angiostrongylus cantonensis*	99.8%
RT2 (4BER379J016)	*Angiostrongylus cantonensis*	100%
RT3 (4BEW135B016)	*Angiostrongylus cantonensis*	100%
RT4 (4BEYNXB6016)	*Angiostrongylus cantonensis*	99.8%
RT5 (4FJJJW6R013)	*Angiostrongylus cantonensis*	99.8%
RT6 (4FJNEHJH013)	*Angiostrongylus cantonensis*	99.8%

(Table 2). The rats belonging to *R. norvegicus* showed 46% (11/24) prevalence for *A. cantonensis* infection while *R. tanezumi* showed 29% (53/185) prevalence. Statistical analysis revealed no significant differences between the rat species ($\chi^2 = 0.086$, $P > 0.05$). Furthermore, intensity of *A. cantonensis* for *R. tanezumi* (62 parasite/rat) is higher than *R. norvegicus* (43 parasite/rat) and also showed no significant differences ($U = 286.5$; $P > 0.05$).

3.3. Angiostrongylus cantonensis from Snails. A total of 84 snails (9%) were found infected with nematode larvae

(Table 3). However, only *M. maculata* (1/89) and *P. canaliculata* (3/200) harbored the *A. cantonensis* L3. These infected snails were collected from rice fields and irrigation.

3.4. Distribution of Rats and Snails in Muñoz. The distribution map revealed that the sampling points for infected rats and snails overlap implying potential hosts occurring in the same area (Figure 1). The map also showed that the selected villages have rats and snails harboring *A. cantonensis* and other parasites. Two villages, namely, Sapang Cawayan and Villa Nati have the known definitive and intermediate hosts of *A. cantonensis*. However, there are also villages which have *A. cantonensis* infected rats but no *A. cantonensis* infected snails.

3.5. Knowledge and Practices of Locals on Parasites from Rats and Snails. One hundred respondents from Muñoz consisting of adult males ($n = 72$) and females ($n = 28$) were interviewed for the study (Table 4). According to the survey, most of the respondents eat rat. The local people believe that it cures skin diseases. Apparently, most of the locals have eaten rat meat once or twice especially when they were still young. Some of them also made their children eat rat meat primarily because of its presumed medicinal property. Meanwhile, most of the locals usually eat *R. quadrasi*, *M. maculata*, and *Vivipara* species and a few of them eat

TABLE 2: Rats examined for *Angiostrongylus cantonensis* infection in Muñoz, Nueva Ecija.

Rats	Sites					Total	Prevalence (%)
	Bantug	Catalanacan	Matingkis	Sapang Cawayan	Villa Nati		
Rattus norvegicus	8 (4)	3 (1)	5 (2)	4 (1)	4 (3)	**24 (11)**	46
Rattus tanezumi	34 (13)	31 (1)	28 (11)	54 (14)	38 (14)	**185 (53)**	29
Total	**42 (17)**	**34 (2)**	**33 (13)**	**58 (15)**	**42 (17)**	209 (64)	31

P. canaliculata. In addition, locals are not aware that ingestion of snails with parasites can cause diseases in humans.

4. Discussion

Transmission of *Angiostrongylus cantonensis* is very complex and difficult to assess. It involves parasitizing both warm- and cold-blooded animals which do not necessarily share the same spatial distribution and ecological requirements [16]. Furthermore, its larval stages are easily affected by different environmental factors such as temperature, oxygen, and pH as well as other factors such as host range and spatial and temporal variations. However, there is a need to further investigate its transmission in order to understand how to control and prevent its associated disease.

The importance of fast and accurate diagnostic tools for diseases caused by nematodes lies on the knowledge of their prevalence and geographical distribution [17] which is specifically true for *A. cantonensis* having nearly all mollusks as its intermediate host. Moreover, there are two other rat lungworms that have similar life cycle with *A. cantonensis*, *A. malaysiensis,* and *A. mackerrasae*. Most diagnoses of these nematodes are based on morphological characteristics; however, specific identification of these nematodes is unfeasible because of the similar descriptions on the size and body shapes among its species [17]. Hence, misidentification and misdiagnoses of these nematodes may result in the under-estimation of their infections. That is why most works on nematodes have chosen rDNA regions because they are useful genetic markers for studies on diagnosis, systematics, and molecular evolution [17]. Thus, utilization of genetic markers is widely used in differentiating closely related species like in the case of *Angiostrongylus*.

In this study, the sampling sites were mainly comprised of agricultural areas where *R. tanezumi* is abundant. Moreover, these sites are located in rice fields and water reservoirs which are perfect habitat for snails. *R. tanezumi* primarily consume insects, snails, slugs, and other invertebrates found in their habitats and they specifically survive on frogs and snails during non-rice periods [5]. This may have resulted in the increased contact of *R. tanezumi* with possible intermediate hosts of *A. cantonensis* leading to the transmission of the parasite. On the other hand, *R. norvegicus* is usually located in cities and towns and is less commonly found in culti-vated areas [18]. Nevertheless, other factors may have also contributed to the prevalence and intensity of *A. cantonensis* in both rat species such as the exposure of these rats to the intermediate hosts, behavior, and feeding habits of rats, as well as their genetic traits.

The low prevalence of *A. cantonensis* in freshwater snails in the study may has resulted from a number of factors such as the seasonal infection of rats with *A. cantonensis*. Based on a study by Antolin et al. [4], female *R. tanezumi* in PhilRice farms from Nueva Ecija were found to be infected with *A. cantonensis* during June to September. This could mean that the transmission of the parasite between its hosts may have happened during these months. In connection to this, snails for the study were collected during other months of the year. Furthermore, ecological characteristics of *P. canaliculata* and *M. maculata* such as their benthic life cycle may also be accounted for the low prevalence of *A. cantonensis* infection. According to Lv et al. [16], only a few species of freshwater snails naturally transmit *A. cantonensis* because rat feces containing its L1 are diluted in freshwater bodies. Previous studies suggest that *A. cantonensis* infection in terrestrial snails and slugs are higher than freshwater snails [19–21]. In the present study, no *A. cantonensis* L3 was observed in *A. fulica*. The reason is unclear; however, possibilities exist such as effects of seasonal variations and habitat of hosts on the transmission of parasite. Based on observations, *A. fulica* becomes active when it starts to rain. In the study of Salibay and Luyon [22], fewer rats were caught in rainy days. This indicates that there may have been less contact between rats and *A. fulica*. Studies in other countries have shown that *A. fulica* are naturally infected with the parasite [8, 17, 19, 23–25]. In the Philippines, few reports regarding *A. cantonensis* infection in snails can be found. *A. fulica* and *Laevicaulis altae* from Metro Manila were revealed to be infected with *A. cantonensis* [9, 10, 26] but other than the snail, no other reports have been made. This study, however, is the first record of *A. cantonensis* infected *P. canaliculata* and *M. maculata* in the Philippines.

Maps may be used to predict the probability of animals occurring in an area [27–29] and in this case the rats and snails. Rodents exhibit territorial behavior [30]; however, both rat species have overlapping sampling points in the study. This may be due to residential areas being adjacent to agricultural lands. *R. norvegicus* is a major urban pest worldwide but it was also reported as a field pest in some parts of the Philippines. They are also often found close to water sources like rivers and irrigations [18]. Furthermore, Salibay and Luyon [22] revealed in their study that *R. norvegicus* were mostly observed in areas where *R. tanezumi* were also caught. The unusual change in habitat of *R. norvegicus* from residential to agricultural is probably due to habitat alterations. Moreover, food consumption of rats can also be accounted for the captured rat distribution. When food is not available in their habitat, rats would either resort to

TABLE 3: Snails examined for *Angiostrongylus cantonensis* infection in Muñoz, Nueva Ecija.

(a)

Snails	Sites						Total	*Angiostrongylus cantonensis*	Infected with other nematodes
	Bantug	Catalanacan	Matingkis	Sapang Cawayan	Villa Nati				
Freshwater									
Indoplanorbis exustus	1	1	10	0	2		14	–	–
Jagora asperata	2	1	0	4	0		7	–	–
Melanoides maculata	20	18	12	21 (1)	18		89	+	–
Pomacea canaliculata	25	37	37	76 (1)	25 (2)		200	+	+
Radix quadrasi	22	24	39	20	33		138	–	+
Tarebia granifera	13	7	0	88	0		108	–	+
Vivipara angularis	1	0	25	7	8		41	–	+
Vivipara carinata	31	29	15	66	43		184	–	+
Total	**115**	**117**	**138**	**282**	**129**		**781**		

(b)

Snails	Sites						Total	*Angiostrongylus cantonensis*	Infected with other nematodes
	Bagong Sicat	Poblacion East	Poblacion North	Poblacion South					
Terrestrial									
Achatina fulica	30	30	30	30			120	–	+

TABLE 4: Response of the local people in Muñoz, Nueva Ecija regarding awareness on parasitic diseases from rats and snails.

Question	Response (%)
Eat rats	
Yes	60
No	40
Difference of rat species	
Yes	98
No	2
Eat snails	
Yes	84
No	16
Snail species as food	
Birabid (R. quadrasi)	41
Susong pilipit/palipit (M. maculata)	71
Susong papa (Vivipara sp.)	69
Golden apple snail (P. canaliculata)	15
Source of snail	
River	52
Irrigation	53
Rice field	40
Fish Pond	20
Small creek	16
Other animals from the field as source of food	
Frog	53
Igat (eel)	35
Tulya (shellfish)	77
Fish	11
Talangka (crab)	20
Awareness of disease associated with eating snails	
Yes	5
No	95

cannibalism [5] or look for other sources of food. Meanwhile, snails were collected in rice fields, waterways, and near houses which can be easily accessed by rats. The locations of A. cantonensis infected snails coincide with those of A. cantonensis infected rats. Hence, transmission of parasites between rats and snails has been occurring in the study sites.

Eosinophilic meningitis is caused by several helminth and nonhelminth parasites [31] but the most common cause is A. cantonensis [32] which is linked to the introduction, farming, and consumption of some snail species. Meningitis was recorded as one of the leading causes of child mortality in the Philippines from 2001 to 2006 and 2008 to 2010 [33]. The analysis of cerebrospinal fluid (CSF) as well as blood culture aids in their proper classification of CNS infections; however, it may not always be followed because of the limitations on the part of the patient. Thus, the final diagnosis is dependent on the assessment and opinion of the attending physician [34]. Even though the cases of meningitis were not solely based on A. cantonensis infection alone, the prevalence of the parasite is of concern and even more if diagnosis of the parasite in humans is neglected most of the times. In conclusion,

human cases of angiostrongylosis have not been recorded in Muñoz but this could be due to the misdiagnoses and lack of readily available diagnostic tools. The presence of the parasite in potential hosts could not eliminate the possibility of its transmission to humans. Hence, public education regarding zoonotic parasites should be implemented. Proper handling of its intermediate hosts and crops that may be contaminated by its hosts should be practiced. Moreover, P. canaliculata and M. maculata should be further examined since it is in these populations that larvae of A. cantonensis were observed. Other regions in the Philippines should also be evaluated particularly in those areas where rice planting is the main source of livelihood.

Competing Interests

The authors declare that there is no conflict of interests regarding the publication of this paper.

Acknowledgments

This study was funded by the Department of Science and Technology, Accelerated Science and Technology Human Resource Development Program (DOST-ASTHDRP), Philippine Rice Research Institute (PhilRice), and the University of the Philippines, Office of International Linkages (UP-OIL). The authors would also like to acknowledge Dr. Clarisse Yvonne Domingo of the Central Luzon State University (CLSU), Professor Ryan Emmanuel de Chavez and Leendel Punzalan of the University of the Philippines Los Baños (UPLB), and Ronniel Pedales of the University of the Philippines Diliman (UPD) for their contribution in this study.

References

[1] M. M. Kliks and N. E. Palumbo, "Eosinophilic meningitis beyond the Pacific Basin: the global dispersal of a peridomestic zoonosis caused by *Angiostrongylus cantonensis*, the nematode lungworm of rats," *Social Science and Medicine*, vol. 34, no. 2, pp. 199–212, 1992.

[2] W. C. Marquardt, R. S. Demaree, and R. B. Grieve, *Parasitology and Vector Biology*, Academic Press, San Diego, Calif, USA, 3rd edition, 2000.

[3] S. R. Chen, S. Lv, L. P. Wang et al., "The first outbreak of angiostrongyliasis in Dali," *Parasitic Infectious Diseases*, vol. 6, pp. 137–138, 2008.

[4] M. M. Antolin, R. C. Joshi, L. S. Sebastian, L. V. Marquez, and U. G. Duque, "Endo- and ectoparasites of the Philippine rice field rat, *Rattus tanezumi* Temminck, on PhilRice farms," *International Rice Research Notes*, vol. 31, no. 1, pp. 26–27, 2009.

[5] G. R. Quick, "Rodent and rice," in *Report and Proceeding of an Expert Panel Meeting on Rice Rodent Control*, International Rice Research Institute, Los Banos, Calif, USA, September 1990.

[6] Science City of Muñoz, Science City of Muñoz, February 2014, http://www.sciencecityofmunoz.ph/.

[7] A. Vitta, R. Polseela, S. Nateeworanart, and M. Tattiyapong, "Survey of Angiostrongylus cantonensis in rats and giant African land snails in Phitsanulok province, Thailand," *Asian Pacific Journal of Tropical Medicine*, vol. 4, no. 8, pp. 597–599, 2011.

[8] R. Floyd, E. Abebe, A. Papert, and M. Blaxter, "Molecular barcodes for soil nematode identification," *Molecular Ecology*, vol. 11, no. 4, pp. 839–850, 2002.

[9] I. K. C. Fontanilla and C. M. Wade, "The small subunit (SSU) ribosomal (r) RNA gene as a genetic marker for identifying infective 3rd juvenile stage *Angiostrongylus cantonensis*," *Acta Tropica*, vol. 105, no. 2, pp. 181–186, 2008.

[10] D. M. A. Constantino-Santos, Z. U. Basiao, C. M. Wade, B. S. Santos, and I. K. C. Fontanilla, "Identification of *Angiostrongylus cantonensis* and other nematodes using the SSU rDNA in *Achatina fulica* populations of Metro Manila," *Tropical Biomedicine*, vol. 31, no. 2, pp. 327–335, 2014.

[11] R. Staden, K. F. Beal, and J. K. Bonfield, "The staden package, 1998," *Bioinformatics Methods and Protocols*, vol. 132, pp. 115–130, 1994.

[12] T. Hall, *Bioedit: A User-Friendly Biological Sequence Alignment Editor and Analysis Program for Windows 95/98/NT*, vol. 41 of *Nucleic Acids Symposium Series*, 1999.

[13] S. F. Altschul, W. Gish, W. Miller, E. W. Myers, and D. J. Lipman, "Basic local alignment search tool," *Journal of Molecular Biology*, vol. 215, no. 3, pp. 403–410, 1990.

[14] L. R. Ash, "Diagnostic morphology of the third-stage larvae of *Angiostrongylus cantonensis*, *Angiostrongylus vasorum*, *Aelurostrongylus abstrusus*, and *Anafilaroides rostratus* (Nematoda: Metastrongyloidea)," *Journal of Parasitology*, vol. 56, no. 2, pp. 249–253, 1970.

[15] R. Poulin, "The disparity between observed and uniform distributions: a new look at parasite aggregation," *International Journal for Parasitology*, vol. 23, no. 7, pp. 937–944, 1993.

[16] S. Lv, Y. Zhang, P. Steinmann et al., "The emergence of angiostrongyliasis in the People's Republic of China: the interplay between invasive snails, climate change and transmission dynamics," *Freshwater Biology*, vol. 56, no. 4, pp. 717–734, 2011.

[17] R. L. Caldeira, C. L. G. F. Mendonça, C. O. Goveia et al., "First record of molluscs naturally infected with *Angiostrongylus cantonensis* (Chen, 1935) (Nematoda: Metastrongylidae) in Brazil," *Memorias do Instituto Oswaldo Cruz*, vol. 102, no. 7, pp. 887–889, 2007.

[18] K. P. Aplin, P. R. Brown, J. Jacob, C. Krebs, and G. Singleton, *Field Methods for Rodent Studies in Asia and the Indo-Pacific*, vol. 100 of *ACIAR Monograph*, 2003.

[19] X. Hu, J. Du, C. Tong et al., "Epidemic status of *Angiostrongylus cantonensis* in Hainan island, China," *Asian Pacific Journal of Tropical Medicine*, vol. 4, no. 4, pp. 275–277, 2011.

[20] H. M. Zhang, Y. G. Tan, X. M. Li et al., "Survey on the infectious focus of *Angiostrongylus cantonensis* in Guangxi," *Journal of Tropical Diseases and Parasitology*, vol. 5, pp. 79–80, 2007.

[21] Z. H. Deng, Q. M. Zhang, R. X. Lin et al., "Survey on the natural focus of angiostrongyliasis in Guandong province," *South China Journal of Preventive Medicine*, vol. 34, pp. 42–45, 2008.

[22] C. Salibay and H. A. V. Luyon, "Distribution of native and non-native rats (Rattus spp.) along elevational gradient in a Tropical Rainforest of Southern Luzon, Philippines," *Ecotropica*, vol. 14, pp. 129–136, 2008.

[23] J. H. Cross, "*Angiostrongylus (Parastrongylus) cantonensis* in the Western Hemisphere," *Southeast Asian Journal of Tropical Medicine and Public Health*, vol. 35, supplement 1, pp. 107–111, 2004.

[24] J. L. Teem, Y. Qvarnstrom, H. S. Bishop et al., "The occurrence of the rat lungworm, *Angiostrongylus cantonensis*, in nonindigenous snails in the Gulf of Mexico region of the United States," *Hawai'i Journal of Medicine & Public Health*, vol. 72, no. 6, supplement 2, pp. 11–14, 2013.

[25] S. C. Thiengo, A. Maldonado, E. M. Mota et al., "The giant African snail *Achatina fulica* as natural intermediate host of *Angiostrongylus cantonensis* in Pernambuco, northeast Brazil," *Acta Tropica*, vol. 115, no. 3, pp. 194–199, 2010.

[26] N. P. Salazar and B. D. Cabrera, "*Angiostrongylus cantonensis* in rodent and molluscan hosts in Manila and suburb," *Acta Medica Philippina*, vol. 6, pp. 20–25, 1969.

[27] J. A. L. Jeffery, P. A. Ryan, S. A. Lyons, P. T. Thomas, and B. H. Kay, "Spatial distribution of vectors of Ross River virus and Barmah Forest virus on Russell Island, Moreton Bay, Queensland," *Australian Journal of Entomology*, vol. 41, no. 4, pp. 329–338, 2002.

[28] C. Nansen, J. F. Campbell, T. W. Phillips, and M. A. Mullen, "The impact of spatial structure on the accuracy of contour maps of small data sets," *Journal of Economic Entomology*, vol. 96, no. 6, pp. 1617–1625, 2003.

[29] P. A. Ryan, S. A. Lyons, D. Alsemgeest, P. Thomas P, and B. H. Kay, "Spatial statistical analysis of adult mosquito (Diptera: Culicidae) counts: an example using light trap data, in Redland Shrine, Southeastern Queensland, Australia," *Journal of Medical Entomology*, vol. 41, no. 6, pp. 1143–1156, 2004.

[30] J. L. Barrett and P. Prociv, "Neuro-angiostrongyliosis in wild black and grey-headed flying foxes (*Pteropus* spp.)," *Australian Veterinary Journal*, vol. 80, pp. 554–558, 2002.

[31] C. Graeff-Teixeira, A. C. A. Da Silva, and K. Yoshimura, "Update on eosinophilic meningoencephalitis and its clinical relevance," *Clinical Microbiology Reviews*, vol. 22, no. 2, pp. 322–348, 2009.

[32] Q.-P. Wang, D.-H. Lai, X.-Q. Zhu, X.-G. Chen, and Z.-R. Lun, "Human angiostrongyliasis," *The Lancet Infectious Diseases*, vol. 8, no. 10, pp. 621–630, 2008.

[33] Department of Health, Leading Causes of Child Mortality, 2013, http://www.doh.gov.ph.

[34] M. O. Niñal, J. R. Navarro, A. T. Gepte, J. Rimando-Magalong, G. M. Samonte, and L. Villar, "Retrospective surveillance of CNS infections, bacterial meningitis and meningococcl disease cases admitted at benguet general hospital," National Epidemiology Center, Department of Health, Philippines, 2005, http://www.doh.gov.ph/.

20

Prevalence of Helminths in Dogs and Owners' Awareness of Zoonotic Diseases in Mampong, Ashanti, Ghana

Papa Kofi Amissah-Reynolds,[1] **Isaac Monney,**[2]
Lucy Mawusi Adowah,[1] **and Samuel Opoku Agyemang**[1]

[1]Department of Science Education, University of Education, Winneba, P.O. Box M40, Mampong, Ashanti, Ghana
[2]Department of Environmental Health and Sanitation Education, University of Education, Winneba, P.O. Box M40, Mampong, Ashanti, Ghana

Correspondence should be addressed to Papa Kofi Amissah-Reynolds; kofireynolds@gmail.com

Academic Editor: Bernard Marchand

Dogs are popular pets that live closely with humans. However, this cohabitation allows for the transmission of zoonotic parasites to humans. In Ghana, very little is known about zoonotic parasites in dogs. We examined excrements of 154 dogs for intestinal helminthes using saturated sodium chloride as a floatation medium and further interviewed 100 dog owners regarding knowledge on zoonosis and pet management practices. Thirteen parasite species were identified, with an overall prevalence of 52.6%. Nematodes were more common than cestodes, with *Toxocara canis* being the most prevalent helminth (18.8%). Age ($p = 0.011$; $\chi^2 = 9.034$) and location ($p = 0.02$; $\chi^2 = 12.323$) of dogs were significant risk factors of helminthic infections, while mode of housing, function, and gender of dogs were not. Knowledge on zoonosis and pet management practices were poor, including irregular deworming and feeding of animals off the bare ground. Dogs may play an active role in the transmission of zoonotic diseases in the area, given the cohabitation of infected dogs with humans; irregular deworming pattern of dogs; and rampant excretion of helminth-infested dog excreta into the environment.

1. Introduction

Dogs live in close association with humans, providing them with companionship and security, among others [1]. However, these companion animals can as well transmit diseases to humans who have close contact with them [2, 3]. Infections could be transmitted to humans through contact with animal hair [4, 5], food and water contaminated with dog excreta or secretions, and/or consumption of dog meat [6]. According to the literature, dogs can host well-known zoonotic parasites, including *Toxocara canis*, *Diphyllobothrium latum*, *Ancylostoma* spp., *Uncinaria stenocephala* [7], and *Echinococcus granulosus* [8]. The presence of these parasites in dogs causes different clinical symptoms depending on the parasite species and density [9].

There are numerous reports on canine intestinal parasites worldwide. Some studies reported prevalence of between 4 and 40% [7, 10–12]. Others reported higher prevalence of over 60% [9, 13–16]. The varying prevalence reported could be due to differences in status of dog sampled, geographical location, and the diagnostic techniques used [17, 18]. Gastrointestinal parasites are more common in dogs in developing countries [19]. High prevalence and heavy infections are often reported in such countries. This is attributed to the fact that dogs in these regions are rarely treated for parasitic diseases and policies on pet ownership are usually lacking [15] or poorly enforced, thereby providing fertile grounds for zoonotic transmission of parasites.

In Ghana, very little attention has been given to parasites in dogs and only two studies have been conducted in this regard. Studies conducted by Anteson and Corkish [20] and Johnson et al. [21] identified a total of 9 species of intestinal helminths in dogs in Ghana. Anteson and Corkish highlighted the ineffectiveness of antihelminthics, while asserting the possible transmission of zoonotic parasites to children [20]. Johnson et al. [21] identified housing styles, sources of dogs, and purpose of keeping dogs as significant factors associated with infection. Data from these studies were however not population-based as they focused only on owned dogs, but not stray or unowned dogs.

To the best of our knowledge, there has been no survey on intestinal helminths in dogs in Mampong, Ashanti. Current epidemiological data is therefore needed for establishing effective control measures in animal and public health. We report for the first time the prevalence and types of helminths in dogs, deworming practices, and knowledge of pet owners on zoonotic parasites in the area.

2. Experimental Section

2.1. Study Area and Study Design. Mampong, Ashanti, is the capital of the Mampong Municipal Assembly in the Ashanti region of Ghana. Geographically, it is located on latitude 7°05′42″N and longitude 1°24′49″W, approximately 60 km northeast of the regional capital, Kumasi. It has an estimated population of 40,000 people, accounting for approximately half of the population of the entire municipality. The town lies within a wet semiequatorial forest zone and has scenic undulating land forms which range from scarps and hills to low lying tropical areas. Farming activities are predominant in the township owing to the fertile soil.

The study area was divided into three sites according to the planning of settlements. Site 1 is a poorly planned settlement with dispersed housing system and poor environmental conditions. Site 2 and Site 3 have better community setup in terms of housing and environmental conditions compared to Site 1.

Dogs were classified into three age groups as puppies (0–6 months), young dogs (>6 months to 12 months), and adults (>12 months) as described by Bone, 1988 (cited in [9]). They were further categorized into stray, semidomestic, and domestic based on a modified description from the one used by Perera et al. [19]. Domestic dogs were the ones with owners, kept under strict confinement, who do not mingle with stray or semidomestic dogs and may or may not be dewormed and/or vaccinated against rabies and other diseases. The semidomestic dogs were the ones who had owners, who mingle with stray dogs and may or may not be dewormed and/or vaccinated against rabies and other diseases. Stray dogs were the free-ranging ones that did not have owners, fed off the streets, and had no deworming and vaccination against rabies and other diseases.

Random house-to-house screening of dogs of all age groups, sexes, housing styles, and functions was conducted between March and July 2015. Stool samples of one hundred and fifty-four (154) dogs were collected in sterile containers labelled with identification data. After collection, samples were taken to the Laboratory of Veterinary Service, Kumasi, and kept frozen until use. With the informed consent of dog owners, a structured questionnaire was also used to assess the dog management practices and owners' awareness of zoonotic canine parasites.

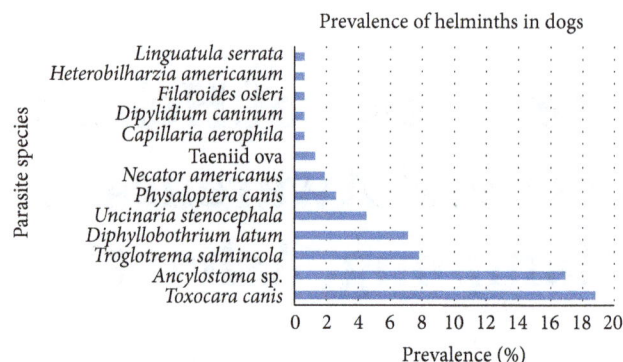

FIGURE 1: Distribution of helminths found in dog excrement (n = 154).

2.2. Laboratory Procedure. Stool samples of dogs were analysed for eggs of parasites using saturated sodium chloride solution as a floatation medium. Samples were observed under the light microscope at 10x objective. The parasites were classified according to their species based on existing keys and descriptions [22]. The results were analysed using SPSS version 17 to determine frequencies and percentages. Test for associations was conducted with the Chi-square (χ^2) test at 5% significance level.

2.3. Assessment of Pet Management Practices and Awareness of Zoonotic Diseases. Questionnaires were administered to 100 dog owners who consented to be interviewed. The questionnaires were divided into two distinct sections to capture information on reasons for keeping dog(s), number of dogs kept, knowledge of zoonosis, and pet management practices including deworming frequency, housing and feeding mode, and veterinary care.

3. Results and Discussion

3.1. Results. Out of the 154 dog excrement samples examined, approximately 53% were infected with at least one parasite. Overall, 13 parasite species were found in the dog excrement, with the top four parasites being *Toxocara canis* (18.8%), *Ancylostoma* sp. (16.9%), *Troglotrema salmincola* (7.8%), and *Diphyllobothrium latum* (7.1%) (Figure 1). Nematodes were more common than cestodes in the study dogs.

The prevalence pattern by age of the four most predominant parasites is presented in Figure 2. Only two fish parasites, namely, *Troglotrema salmincola* and *Diphyllobothrium latum*, showed an association with age. While the former showed decreasing prevalence with age of dogs, the latter showed a reverse trend.

The prevalence of helminths in dogs in relation to age and sex is shown in Table 1. Among the three age groups, the highest prevalence (86.7%) was recorded in puppies, followed by adult dogs (52.0%) and young dogs (41.5%). Male dogs recorded slightly higher prevalence (55.1%) than female dogs (48.2%), though the difference was not statistically significant

TABLE 1: Prevalence of helminths in dogs in relation to age and sex ($N = 154$).

Variable	Number examined	Number infected	Infection rate (%)	p value
Age (months)				
0–6	15	13	86.7	
>6–12	41	17	41.5	$p = 0.011; \chi^2 = 9.034$
>12	98	51	52.0	
Sex				
Male	98	54	55.1	
Female	56	27	48.2	$p = 0.410; \chi^2 = 0.678$
Total	**154**	**81**	**52.6**	

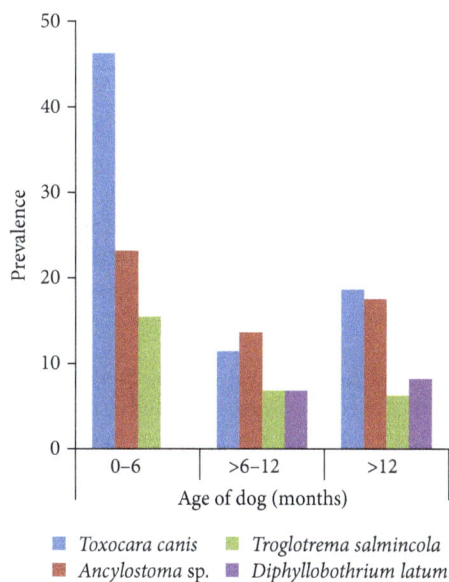

FIGURE 2: Prevalence pattern of parasites by age.

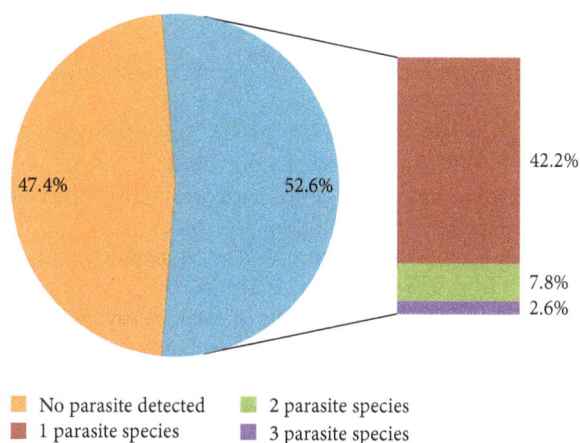

FIGURE 3: Pattern of parasitic infection among study dogs.

($p > 0.05$). Eleven (11) species of parasites were recovered from male dogs compared to six (6) from the females.

The frequency of single and mixed infections is presented in Figure 3. Single infections were more common (42.2%) than multiple infections (10.4%). The percentage of dogs harbouring mixed infections of two and three parasites was 7.8% and 2.6%, respectively. Interestingly, 13 male dogs (13.2%) harboured multiple parasites compared to 3 female dogs (5.4%). No female dog harboured more than two parasites. Multiple infections were recorded in dogs sampled from Site 1 and Site 2 only.

The prevalence of helminths in relation to the functions of the 145 owned dogs sampled is presented in Table 2. Though dogs who were kept for companionship recorded higher prevalence (69.2%) than dogs used for hunting (66.7%) and security (44.0%), the difference was not statistically significant ($p > 0.05$). Approximately 70% of the owned dogs sampled were kept for security.

Table 3 shows the prevalence of helminths in dogs in relation to their housing styles. Stray dogs recorded the highest prevalence (66.7%), followed by domestic (61.1%) and semidomestic dogs (50.4%), respectively. Nearly 90% of the dogs sampled were allowed to roam about in the community.

A statistically significant association ($p = 0.02; \chi^2 = 12.323$) was found between location and helminthic infections in dogs. The relation between location and prevalence of helminths is shown in Table 4.

None of the 100 dog owners interviewed fed their dogs with standard dog feed and close to three-quarters (73%) fed their dogs off the bare floor (Table 5). Awareness of rabies disease in dogs was comparable (62%) to that of helminth infestation in dogs (60%) among the dog owners. Most dog owners (93%) allowed their dogs to defecate anywhere without any restriction and close to 9 out of 10 dog owners had never taken their dogs to a veterinary clinic although there is one present in the town. Regular deworming of dogs is an uncommon practice among dog owners; close to half of the dog owners had never dewormed their dogs and a significant proportion (76%) had no knowledge of the transmission of zoonotic diseases to humans. About a third of dog owners (32%) kept their dogs in kennels and close to half of them (46%) cleaned the kennels once a month. The median number of dogs kept per owner was 2 (range 1–15).

3.2. Discussion. The present study reports for the first time on helminth parasites in dogs in Mampong, Ashanti, Ghana. The two previous studies on dogs in Ghana were done on owned dogs in a different location [20, 21]. All the parasites

TABLE 2: Prevalence of helminths in relation to dog function.

Function of dog	Number of dogs examined	Number of dogs infected	Infection rate (%)	p value
Companionship	39	27	69.2	
Hunting	6	4	66.7	$p = 0.59$; $\chi^2 = 5.650$
Security	100	44	44.0	
Total	**145**	**75**	**51.7**	

TABLE 3: Prevalence of helminths in relation to mode of housing.

Mode of housing	Number of dogs examined	Number of dogs infected	Infection rate (%)	p value
Domestic	18	11	61.1	
Stray	9	6	66.7	$p = 0.476$; $\chi^2 = 1.485$
Semidomestic	127	64	50.4	
Total	**154**	**81**	**52.6**	

reported in this study have been previously documented in dogs elsewhere, but with regional variation in prevalence and parasite species.

In the present study, we recorded lower overall prevalence compared to previous data in Africa [9, 13, 15, 16]. The use of single faecal floatation in the present study may have underestimated the prevalence, as a combination of methods has been reported to increase the chances of recovering more parasites [19]. Also, ecological and epidemiological differences, as well as the faecal floatation methods used [11], may account for the variations in distribution and prevalence of parasites. Single infections were more common than multiple infections and this agrees with the findings of Ugbomoiko et al. [15] and Kimura et al. [12]. In contrast, other studies [9, 16, 19] reported higher frequencies of multiple parasites compared to single parasites.

Data from the present study is consistent with previous works in Ghana and other parts of Africa which have reported *Toxocara canis*, *Ancylostoma* sp., *Dipylidium caninum*, and Taeniidae as some of the helminths parasitizing dogs [9, 13, 15, 16, 21]. *Trichuris vulpis*, *Strongyloides* sp., and *Spirocerca lupi* were absent in the present study though these parasites have also been previously reported. The absence of *Spirocerca lupi* eggs in our study may be due to the use of sodium chloride as a floatation medium [9].

Our present results agree with Ugbomoiko et al. [15] and Kimura et al. [12], who reported *Toxocara canis* as the most common helminth in dogs. *Toxocara canis* is a soil-transmitted helminth; thus, habits like feeding off floors and sleeping on bare grounds in the study dogs could account for this observation. Though the prevalence pattern of *Toxocara canis* was not age-dependent, the highest prevalence of this nematode was found in puppies. The older dogs may have developed specific immunity to *Toxocara canis* through frequent exposure at an early age. We also found two fish parasites, *Diphyllobothrium latum* and *Troglotrema salmincola*, in the dogs sampled. Our results also showed that the prevalence pattern for the fish parasites was age-dependent; *Troglotrema salmincola* decreased with age, whereas *Diphyllobothrium latum* showed a reverse trend. The age-dependent prevalence pattern of fish parasites may be due to the role of immune

responses in dogs. This role is however unclear and needs to be elucidated. These parasites have only been previously reported in dogs which feed on raw/fresh fish products. Considering that nearly all the dogs sampled fed on raw fish products or viscera, we predict that this prevalence pattern observed is largely due to the feeding habit of the dog rather than the age.

We recorded lower prevalence of *Dipylidium caninum* (0.6%) than has been previously reported elsewhere in Africa. Studies by Anteson and Corkish [20] and Zewdu et al. [9] recorded significantly higher prevalence of *Dipylidium caninum* by postmortem compared to coproscopy. Zewdu et al. [9] further indicated that necropsy provides more detailed information than coproscopy. However, we could not perform necropsy in the present study because the dogs could not be killed for such purposes.

Taeniid tapeworms are morphologically indistinguishable. Therefore, molecular analysis is needed to differentiate species of Taeniidae. *Echinococcus granulosus* is one of the Taeniid species found in dogs, which is also of zoonotic importance. However, the use of coproscopy in the present study did not allow us to detect the species of Taeniidae present.

Age, sex, location, and management practices (including housing styles and deworming practices) are some risk factors that predispose dogs to parasitism. Knowledge of risk factors of infection is vital in the development of effective control programs. Identification of risk factors is however a complex process, particularly in developing countries, given the high numbers of stray dogs with poor or no documented histories. In the present study, age of dog and location were identified as significant risk factors associated with parasitism. Perhaps, the low immunity of puppies compared to the older dogs accounts for the significantly higher prevalence of infection in the former. Though sex was not a significant risk factor, male dogs harboured more parasites species and more multiple infections compared to females. This could be attributed to the greater propensity for male dogs to roam about compared to females.

Ten (10) out of the 13 parasite species encountered were zoonotic, excluding *Physaloptera canis*, *Filaroides osleri*, and

TABLE 4: Prevalence of helminths based on location of dog.

Location	Number of dogs examined	Number of dogs infected	Infection rate (%)	p value
Site 1	48	30	62.5	
Site 2	50	32	64.0	$p = 0.02; \chi^2 = 12.323$
Site 3	56	19	33.9	
Total	**154**	**81**	**52.6**	

TABLE 5: Dog management practices and awareness of zoonotic parasites ($N = 100$).

Type of feed given to dog	
Dog feed	0%
Raw meat products and household leftovers	100%
How do you feed your dog?	
In a bowl	13%
In a bowl and/or on the floor	14%
On the floor	73%
Awareness of rabies in dogs	
Yes	62%
No	38%
Awareness of helminthic infections in dogs	
Yes	60%
No	40%
Where do(es) your dog(s) usually defecate?	
Within the house	7%
Within and/or outside the house (anywhere)	93%
Have you ever taken your dog(s) to a veterinary clinic?	
Yes	12%
No	88%
Frequency of deworming dog(s)	
Once every 3 months	16%
Once every 6 months	13%
Once a year	25%
Never	46%
Awareness of risk of zoonotic transmission of parasites	
Yes	24%
No	76%
Do you keep your dogs in kennels?	
Yes	32%
No	68%
*How often do you clean the kennels?**	
Daily	22%
Weekly	32%
Monthly	46%

*$n = 32$.

Heterobilharzia americanum. In addition, majority of the dogs were not regularly dewormed. Deworming of dogs did not appear to be associated with the reasons for keeping the dogs. For some dog owners, deworming was mostly done once a year by veterinary officers during community visits to vaccinate the pets against rabies. In areas where there are strict regulations on pet husbandry practices, dogs are generally given better care. These dogs are more regularly dewormed, mostly kept confined, or always accompanied outside and are not likely to defecate indiscriminately into the environment. These practices have the potential to limit the transfer of zoonotic agents. In comparison with these best practices, veterinary care for dogs in the study area was very poor. Except for yearly vaccination of dogs against rabies by veterinarians, awareness creation on zoonosis and proper pet management practices was virtually nonexistent. With most dogs harbouring zoonotic parasites, having close bonds with dog owners who irregularly deworm their pets, and defecating indiscriminately, public health is threatened as a result of easy transfer of zoonotic parasites into the environment.

Compared to the most recent report on canine helminths in Ghana, we observed a number of similarities. The prevalence rate of 52.6% reported in our study is comparable with the 62.6% prevalence rate reported by Johnson et al. [21] in Accra, Ghana. The highest prevalence of *Toxocara canis* reported in both studies was found in puppies. Again, more than half of the dogs sampled in the two studies were kept for security and with a similar proportion being allowed to roam about in the communities. Most dog owners kept multiple dogs at home (median: 2 and 3, resp.), but pet management practices were poor. Both studies identified zoonotic canine parasites and reported low awareness of dog owners on risk of zoonotic transmission of parasites. Based on these findings, we predict that data from other parts of the country could show a similar trend and this could have serious implications for animal and public health, with dogs playing active roles in zoonotic transmission.

4. Conclusions

The study shows that more than half (approximately 53%) of the study dogs ($N = 154$) were infected with helminthic parasites, mostly nematodes. The top four parasites were *Toxocara canis* (18.8%), *Ancylostoma* sp. (16.9%), *Troglotrema salmincola* (7.8%), and *Diphyllobothrium latum* (7.1%). Age of dogs ($p = 0.011; \chi^2 = 9.034$) and location ($p = 0.02; \chi^2 = 12.323$) were significant risk factors of helminth parasitism, while mode of housing, function, and gender of dogs were not. Only close to a quarter (24%) of dog owners had knowledge of transmission of zoonotic diseases to humans and about half (46%) have never dewormed their dogs although most of them (73%) fed their dogs directly off the ground. Dogs in the area are potential agents of zoonotic transmission given direct excretion of helminth-infested excreta into the environment, cohabitation with owners, and poor pet management practices. The indiscriminate excretion of dogs in the environment is a blot on the landscape and poses a potential pollution source for adjoining surface water resources. The lack of awareness on the transmission of zoonotic diseases from dogs to humans and lack of proper veterinary care for the dogs are a serious public health risk. Dog owners need to be educated and veterinary services should be offered on door-to-door basis instead of the conventional centralised mode.

Given the species diversity of parasites in dogs in the region, we recommend the use of broad spectrum antihelminthics in the treatment of helminthiasis. The study needs to be replicated in other parts of the country to give a holistic impression of the spatial variation of helminth infection among dogs across the country.

Authors' Contribution

All the authors conceived and designed the experiments; Papa Kofi Amissah-Reynolds, Lucy Mawusi Adowah, and Samuel Opoku Agyemang performed the experiments; Papa Kofi Amissah-Reynolds and Isaac Monney analysed the data; all the authors contributed reagents/materials/analysis tools; Papa Kofi Amissah-Reynolds and Isaac Monney wrote the paper. Isaac Monney, Lucy Mawusi Adowah, and Samuel Opoku Agyemang contributed equally to this work.

Acknowledgments

The authors acknowledge the contributions of the following staff of the Laboratory Service, Veterinary Directorate, Kumasi: Dr. Mabel Abudu, Ms. Belinda Sunu, Mrs. Irene, K. Afutu, and Mrs. Ernestina Antwi.

References

[1] M. Paul, L. King, and E. P. Carlin, "Zoonoses of people and their pets: a US perspective on significant pet-associated parasitic diseases," *Trends in Parasitology*, vol. 26, no. 4, pp. 153–154, 2010.

[2] M. R. Lappin, "Pet ownership by immunocompromised people. Bayer Zoonosis Symposium," *North American Veterinary Conference*, vol. 24, no. 5, pp. 16–25, 2002.

[3] N. Itoh, K. Kanai, Y. Hori, F. Hoshi, and S. Higuchi, "Prevalence of *Giardia intestinalis* and other zoonotic intestinal parasites in private household dogs of the Hachinohe area in Aomori prefecture, Japan in 1997, 2002 and 2007," *Journal of Veterinary Science*, vol. 10, no. 4, pp. 305–308, 2009.

[4] W. F. El-Tras, H. R. Holt, and A. A. Tayel, "Risk of *Toxocara canis* eggs in stray and domestic dog hair in Egypt," *Veterinary Parasitology*, vol. 178, no. 3-4, pp. 319–323, 2011.

[5] S. Öge, H. Öge, B. Gönenç, G. Özbakiş, and C. Yildiz, "Presence of *Toxocara* eggs on the hair of dogs and cats," *Ankara Üniversitesi Veteriner Fakültesi Dergisi*, vol. 60, pp. 171–176, 2013.

[6] J. Cui and Z. Q. Wang, "Outbreaks of human trichinellosis caused by consumption of dog meat in China," *Parasite*, vol. 8, no. 2, supplement, pp. S74–S77, 2001.

[7] D. Joffe, D. Van Niekerk, F. Gagne, J. Gilleard, S. Kutz, and R. Lobingier, "The prevalence of intestinal parasites in dogs and cats in Calgary, Alberta," *Canadian Veterinary Journal*, vol. 52, no. 12, pp. 1323–1328, 2011.

[8] J. Abdi, K. Asadolahi, M. H. Maleki, and A. A. Hafez, "Prevalence of helminthes infection of stray dogs in Ilam province," *Journal of Paramedical Sciences*, vol. 4, no. 2, pp. 47–50, 2013.

[9] E. Zewdu, Y. Semahegn, and B. Mekibib, "Prevalence of helminth parasites of dogs and owners awareness about zoonotic parasites in Ambo town, central Ethiopia," *Ethiopian Veterinary Journal*, vol. 14, no. 2, pp. 17–30, 2010.

[10] M. C. Gaunt and A. P. Carr, "A survey of intestinal parasites in dogs from Saskatoon, Saskatchewan," *Canadian Veterinary Journal*, vol. 52, no. 5, pp. 497–500, 2011.

[11] M. Mirazaei and M. Fooladi, "The prevalence of intestinal helminths in owned dogs in Kerman city, Iran," *Scientia Parasitologica*, vol. 13, no. 1, pp. 51–54, 2012.

[12] A. Kimura, Y. Morishima, S. Nagahama et al., "A coprological survey of intestinal helminthes in stray dogs captured in Osaka prefecture, Japan," *Journal of Veterinary Medical Science*, vol. 75, no. 10, pp. 1409–1411, 2013.

[13] B. M. Anene, T. O. Nnaji, and A. B. Chime, "Intestinal parasitic infections of dogs in the Nsukka area of Enugu State, Nigeria," *Preventive Veterinary Medicine*, vol. 27, no. 1-2, pp. 89–94, 1996.

[14] F. J. Martínez-Moreno, S. Hernández, E. López-Cobos, C. Becerra, I. Acosta, and A. Martínez-Moreno, "Estimation of canine intestinal parasites in Córdoba (Spain) and their risk to public health," *Veterinary Parasitology*, vol. 143, no. 1, pp. 7–13, 2007.

[15] U. S. Ugbomoiko, L. Ariza, and J. Heukelbach, "Parasites of importance for human health in Nigerian dogs: high prevalence and limited knowledge of pet owners," *BMC Veterinary Research*, vol. 4, article 49, 2008.

[16] E. T. Kutdang, D. N. Bukbuk, and J. A. A. Ajayi, "The prevalence of intestinal helminths of dogs (*Canis familaris*) in Jos, Plateau State, Nigeria," *Researcher*, vol. 2, no. 8, pp. 51–56, 2010.

[17] I. D. Robertson, P. J. Irwin, A. J. Lymbery, and R. C. A. Thompson, "The role of companion animals in the emergence of parasitic zoonoses," *International Journal for Parasitology*, vol. 30, no. 12-13, pp. 1369–1377, 2000.

[18] T. C. G. Oliveira-Sequeira, A. F. T. Amarante, T. B. Ferrari, and L. C. Nunes, "Prevalence of intestinal parasites in dogs from São Paulo State, Brazil," *Veterinary Parasitology*, vol. 103, no. 1-2, pp. 19–27, 2002.

[19] P. K. Perera, R. P. V. J. Rajapakse, and R. S. Rajakaruna, "Gastrointestinal parasites of dogs in Hantana area in the Kandy District," *Journal of the National Science Foundation of Sri Lanka*, vol. 41, no. 2, pp. 81–91, 2013.

[20] R. K. Anteson and J. D. Corkish, "An investigation of helminth parasites in well-cared for dogs in Accra," *Ghana Medical Journal*, vol. 14, no. 3, pp. 193–195, 1975.

[21] S. A. M. Johnson, D. W. Gakuya, P. G. Mbuthia, J. D. Mande, and N. Maingi, "Prevalence of gastrointestinal helminths and management practices for dogs in the Greater Accra region of Ghana," *Heliyon*, vol. 1, Article ID e00023, 2015.

[22] E. J. L. Soulsby, *Helminths, Arthropods and Protozoa of Domesticated Animals*, Bailliere Tindall, London, UK, 7th edition, 1986.

21

Profile of Geohelminth Eggs, Cysts, and Oocysts of Protozoans Contaminating the Soils of Ten Primary Schools in Dschang, West Cameroon

Vanessa Rosine Nkouayep, Blandine Ngatou Tchakounté, and Josué Wabo Poné

Research Unit of Biology and Applied Ecology, Department of Animal Biology, Faculty of Science, University of Dschang, P.O. Box 067, Dschang, Cameroon

Correspondence should be addressed to Vanessa Rosine Nkouayep; vanessa.nkouayep@gmail.com

Academic Editor: Bernard Marchand

Helminthiasis and protozoans infections have been recognized as an important public health problem. The aim of the present study was to screen soil samples collected from 10 primary schools in the city of Dschang for the presence of soil-transmitted helminth eggs, cysts, and oocysts of protozoans. A total of 400 soil samples were collected around latrines, at playgrounds, and behind classrooms in each school. These samples were examined using the sucrose flotation method. From the result obtained, an overall contamination rate of 7.75% was observed. Five genera of nematodes (*Ascaris, Trichuris, Capillaria, Cooperia,* and hookworms) were identified, while neither cysts nor oocysts of protozoans were detected. The contamination rate and the number of species found were significantly different in wet season as compared to the dry season. During the rainy season, this rate was 12.5% with all the parasitic stages identified, while, in the dry season, the soil contamination rate was 3% with the presence of only two genera (*Ascaris* and *Trichuris*). This suggests that parasite infection may occur mainly in rainy season rather than in the dry season. The most common eggs were those of *Ascaris* with 2% and 5% contamination rates in the dry and rainy seasons, respectively. Also, the soils around latrines were more contaminated (11.9%) as compared to those collected behind classrooms (7.5%) and those at playground (2.5%). It was concluded that the pupils of these schools may have played a major role in the contamination of their environment. Thus, sanitary education, enforcement of basic rules of hygiene, and deworming remain a necessity in the entire population of the study area in general and in the schools in particular in order to prevent helminth infections and to ensure effective environmental health.

1. Introduction

Soil contamination by infective forms of intestinal parasites is the most important infection risk factor for both humans and animals. These parasites have been recognized as an important public health problem, particularly in developing countries [1], where adequate water and good sanitation are lacking. The commonest and well known of these parasites are hookworms (*Necator* and *Ancylostoma*), whipworm (*Trichuris*), and the common roundworm (*Ascaris*) [2]. They are most prevalent in man and can also be found in animals. These parasites have similar epidemiological characteristics with a direct life cycle. Host contamination occurs via oral route through ingestion of infective embryonated eggs from contaminated soil, vegetables, and food products or via the percutaneous migration of infective L_3 from the environment. Recent estimation suggests that they infect over 1 billion, 770 million, and 800 million people, respectively. These parasites have been shown to negatively impact the physical fitness and cognitive performance of the pupils [3]. Mature nematode eggs, cysts, and oocysts of protozoan parasites can remain viable in the soil for a long time depending on several factors such as climatic conditions, seasonal air temperatures, humidity or desiccation of soil, and exposure to sunlight [4]. Thus, contamination of soils with infective forms of parasites may be an important source of infection and constitutes a great risk factor for human infections, especially for small children aged below 12 years because of their vulnerability to

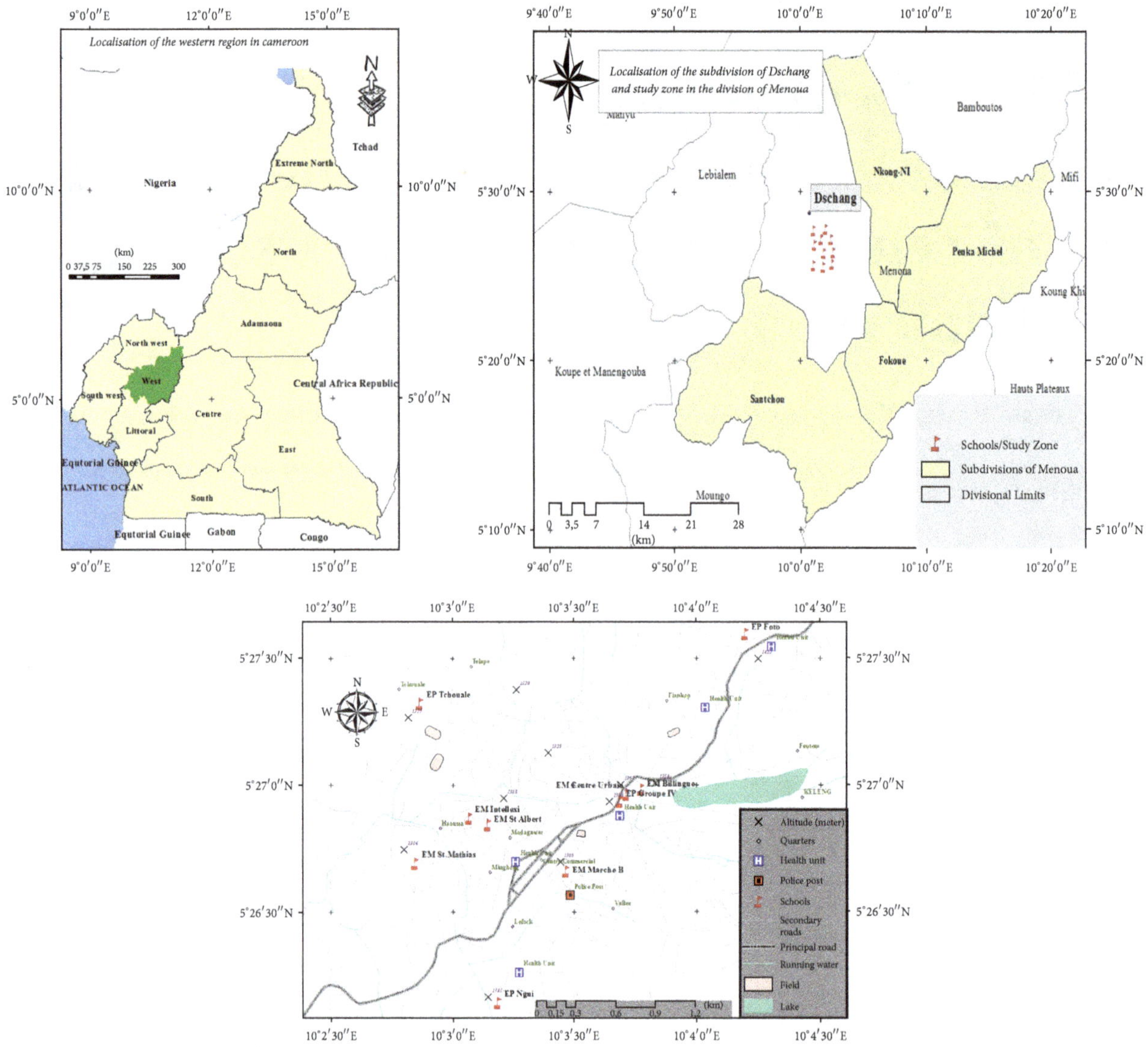

FIGURE 1: Study area and sampling points. *Source*: GPS Garmin; image satellite: QuickBird 2014, Dschang, Cameroon.

nutritional deficiency and since they usually play within the grounds [5]. Studies conducted in various cities all over the world show variable prevalence of soil contamination with different parasite genera. But, in Cameroon, to the best of our knowledge, none or few epidemiological data are available about the rate of soil contamination. In the light of that, there is a need to survey the soils around latrines, at playgrounds, and behind classrooms of school children for the presence of parasites so as to take preventive measures to avoid the impact of the disease on children. The present study was therefore conducted to determine the profile of geohelminth eggs, cysts, and oocysts of protozoans in the soils of 10 primary schools in Dschang, Cameroon.

2. Materials and Methods

2.1. Study Area.
This study was carried out in Dschang (West Cameroon) (Figure 1). Dschang is located between latitude of $5°20'$ north and longitude of $10°30'$ west and at an altitude

of about 1407 m. This city covers an area of $225\,km^2$ and has a cool, mild climate (wet tropical) of the Equatorial Guinean type characterized by two seasons: a rainy season that runs from mid-March to mid-October and a dry season from November to February. Rainfall is unimodal with an annual height of 1809 mm. High precipitation is observed in August and September. Soils encountered in this city are hydromorphic and ferralitic soils.

2.2. Study Design.
Ten schools of the city of Dschang were selected for this study based on their topographic position (plateaus, hills, and lowlands) (Figure 1). Thus, at plateaus' level, these schools were primary school of Foto (EP Foto), Tchouale primary school (EP Tchouale), and Saint-Mathias-Foréké kingdom nursery school (EM St Mathias). At the hills' level, schools were Center Urban nursery school (EM Centre Urbain), Group IV primary school (EP Groupe IV), and bilingual nursery school Dschang (EM Billingue); and at the lowlands' level, the schools were Ngui primary school

(EP Ngui), Intellexi nursery school (EM Intellexi), Market B nursery school (EM Marché B), and Saint-Albert Anglo primary school (EP St Albert).

2.3. Sample Collection. 400 soil samples were collected during two periods: August-September 2014 (rainy season) and January-February 2015 (dry season). During each season, soil samples were collected at three sites of each school: around latrines, at playgrounds, and behind classrooms. About 200 grams of soil dug at 3 cm from the ground was collected using a small hand shovel [6]. The soil samples were carried using a clean spoon and put in small plastic bags which were labeled with the date, the school name, and sample sites. The samples were kept in a bag and transported to the laboratory for parasitological analysis.

2.4. Survey Method. The samples were analyzed by the sucrose floatation technique as described by Uga et al. [7]. In the laboratory, each soil sample was sieved with a 150 μm sieve to remove large particles. 2 grams of powder was weighed and put in a test tube and mixed with 8 ml of distilled water and then centrifuged at 1000 rpm for 5 min. After centrifugation, the supernatant was discarded and the pellet rediluted with 8 ml of sucrose solution (1.20). The mixture was vigorously shaken and centrifuged at 2000 rpm for 10 min. Finally, the interface and the upper layer obtained were collected and introduced into 3 test tubes until the formation of the upper meniscus. Three cover slips were carefully placed above the tubes. After 10 minutes, cover slips were removed carefully and placed on 3 slides and observed microscopically at 10x and 40x. Since cysts are denser, the pellet of each tube was shaken and one drop removed, placed on a slide using a pipette, and covered with a cover slip. Thereafter, a drop of 1% Lugol was subsequently placed on the edge of the cover slip and the preparation observed under a microscope at 40x magnification. The identification of eggs, cysts, and oocysts was done by differential diagnosis based on morphological criteria such as size, shape, nature of the shell, the number of cores, the number of blastomeres, and karyosome [8–10].

2.5. Statistical Analysis. Statistical analyses were performed using the Statistical Package for Social Sciences 20.0 (SPSS 20.0). The Chi-square test was used to compare soil contamination percentages. The significance level for all tests was $p = 0.05$.

3. Results

3.1. Parasites Identified. Out of the 400 soil samples examined, 31 were positive to nematodes eggs given overall prevalence of 7.75%. Neither cysts nor oocysts of protozoans were identified (Table 1). Five types of nematodes were identified in the soil samples. Among these samples, 14 (3.5%) were contaminated with *Ascaris* eggs, 8 (2%) with those of *Trichuris*, 6 (1.5%) with those of *Capillaria*, 2 (0.5%) with the eggs of *Cooperia*, and 1 (0.25%) with hookworm eggs.

3.2. Seasonal Variation of Soil Contamination and Sample Site-Related Prevalence. Table 2 shows the comparison of soil

TABLE 1: Contamination rate of soil samples with identified nematodes.

Nematodes	Number of contaminated samples	Contamination rate (%)
Ascaris	14	3.5
Trichuris	8	2
Capillaria	6	1.5
Cooperia	2	0.5
Hookworms	1	0.25
Total	*31*	*7.75*

TABLE 2: Type and frequency of nematode eggs recorded in rainy and dry seasons.

Nematodes	Seasons		Total n (%)
	Rainy season n (%)	Dry season n (%)	
Ascaris	10 (5)	4 (2)	14 (3.5)
Trichuris	6 (3)	2 (1)	8 (2)
Capillaria	6 (3)	0	6 (1.5)
Cooperia	2 (1)	0	2 (0.5)
Hookworms	1 (0.5)	0	1 (0.25)
Total	*25 (12.5)*	*6 (3)*	*31 (7.75)*

contamination between the dry season and rainy season. It appears that the soil samples collected in wet season were more contaminated, 25 (12.5%), as compared to those from the dry season, 6 (3%). Irrespective of the season, the highest prevalence was seen with *Ascaris*, *Trichuris*, and *Capillaria* eggs.

Table 3 shows the prevalence of nematode eggs in soil with respect to the sample sites. We observed that soil samples collected around latrines were more contaminated by nematode eggs (11.9%) followed by those behind classrooms (7.5%) and those from playgrounds (2.5%). *Ascaris* eggs were highly present (3.5%) as compared to those of *Trichuris* (2%) and *Capillaria* (1.5%).

3.3. Soil Contamination Rate-Related School. The percentage of soil contamination related to school is shown in Figure 2. The soil samples collected in Ngui site have the highest rate (27.5%) of contamination with parasitic stages followed by those from Market B nursery school (17.5%) and those of Intellexi nursery school (7.5%). The samples from Group IV and St Albert had the same rate (10%) of contamination and those of Foto were the least contaminated (5%). The soil samples for other schools were free from contamination.

4. Discussion

A total number of 400 samples of soils obtained from 10 different schools were examined for geohelminth eggs and protozoan cysts and oocysts. We found that the contamination of soil samples was due to 5 genera of nematodes eggs: *Ascaris, Trichuris, Capillaria, Cooperia,* and hookworms. This

TABLE 3: Type and frequency of nematode eggs found in soil samples per sampling sites.

| Nematodes | Sites | | | |
	Around latrines ($n = 160$) Frequency (%)	Playgrounds ($n = 120$) Frequency (%)	Behind classrooms ($n = 120$) Frequency (%)	Total ($n = 400$) Frequency (%)
Ascaris	12 (7.5)*	0	2 (1.7)	14 (3.5)
Trichuris	6 (4)	0	2 (1.7)	8 (2)
Capillaria	1 (0.7)	3 (2.5)	2 (1.7)	6 (1.5)
Cooperia	0	0	2 (1.7)*	2 (0.5)
Hookworms	0	0	1 (0.9)*	1 (0.25)
Total	*19 (11.9)*	*3 (2.5)*	*9 (7.5)*	*31 (7.75)*

*Statistical significance ($p < 0.05$).

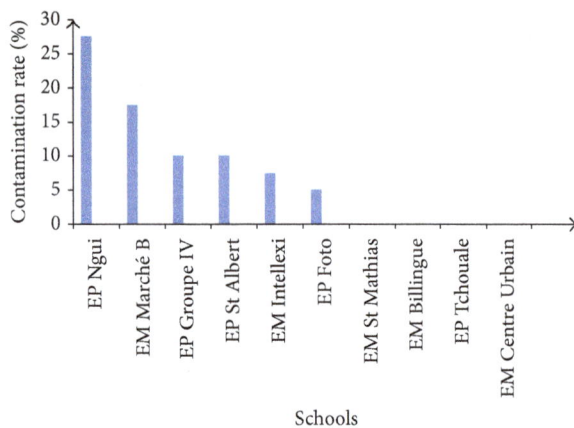

FIGURE 2: Percentages of soil contamination in different schools.

finding is similar to studies reported in Nigeria [11, 12] and in other countries of the world. For example, the study carried out by Shrestha et al. [13] on the soil-transmitted helminths in Kathmandu, Nepal, has demonstrated the presence of nematodes eggs, namely, *Ascaris*, *Trichuris*, and hookworms. Besides these, they also observed the presence of *Toxocara* eggs, *Vampirolepis nana*, and *Taenia* embryophores which were not found in this work. The absence of the embryophore of *Vampirolepis nana* in the present study could be justified by the fact that this parasite is most prevalent in hot and dry country [14], whereas Dschang has a cool climate. For the absence of *Taenia* eggs, it can be justified by the fact that, in Dschang, there is an improvement in farming conditions with livestock. Other nematodes such as *Strongyloides* and pinworms have not been identified in any of the works mentioned above. In fact, it is the larvae of *Strongyloides* that are found in nature and this work has not dwelt on the study of larvae. In general, the prevalence of nematode eggs in different works is due to their resistance to environmental conditions and poor sanitary and hygienic conditions [15]. Uga et al. [16, 17], Stojcevic et al. [18], and Tavalla et al. [19] observed in their studies in Nepal, Croatia, and Tehran, respectively, the presence of protozoan cysts (*Eimeria*, *Isospora*, *Cryptosporidium*, and *Giardia*) which were not found in this work. This could be due to (1) the

different techniques of analysis used and (2) the soil texture of Dschang. It is known that the soil texture can also affect the survival of cysts in the soil as demonstrated by Davies et al. [20].

Ascaris eggs were more frequent in soil samples (3.5%) followed by those of *Trichuris* (2%), *Capillaria* (1.5%), *Cooperia* (0.5%), and hookworms (0.25%). The high prevalence of *Ascaris* eggs in soil samples of the present study is similar to other observations reported elsewhere [11–13]. This can be explained by the fact that eggs of *Ascaris* have an inner shell layer of lipoprotein nature which makes them more resistant to harsh environmental conditions and air [21] compared to the eggs of other nematodes. Another reason is that *Ascaris* eggs can survive in adverse environmental conditions. It might also be due to the overdispersion of *Ascaris* eggs in the environment as a single female *Ascaris* lays relatively large number of eggs (200.000 eggs/day) [9]. Lower rate of soil contamination observed with *Trichuris* eggs might be due to their minimal dispersion as a single female *Trichuris* liberates relatively less numbers of eggs and also due to easy destruction of embryonated eggs by desiccation. One interesting finding in this study was the detection of *Capillaria* and *Cooperia* eggs of other animal origins which were not detected in other studies elsewhere. It might be due to the animal husbandry in the city. The absence or scarcity of hookworm eggs could be justified by their life cycle. In fact, after their release into the environment in 24 h–36 h, embryonation of the eggs takes place. The eggs hatch in the ground and the first moulting larvae (L_1) generation is then released to give the infective larvae.

Soil contamination rate found in this study (7.75%) was lower than that reported elsewhere. This rate is lower than those obtained (11.25%, 53.6%, and 32%) by Debalke et al. [22], Chukwuma et al. [11], and Odoba et al. [12], respectively, on soils in some schools in Ethiopia, in primary schools of Ebene (Nigeria), and in some primary schools in Zaria (Nigeria). The differences in these rates could be due to climatic factors, socioeconomic and topographical characteristics, and soil texture which vary within countries. Indeed, these factors have an impact on the distribution of helminths in the environment [23]. The low contamination rate observed in this work could be due to the multiple annual deworming campaigns organized since few years in schools in Cameroon. This may reflect also the improvement in the living standards, literacy rate, health awareness, use of toilet, and others. We observed that Ngui primary school had the highest contamination rate (27.5%), significantly different ($p < 0.05$) from those registered in other schools. According to Yerima and Van Ranst [24], topography influences the environmental conditions of a place. Ngui primary school is located in lowland characterized by relatively high humidity that favors the development of nematode eggs. By contrast, water flow is observed at the level of plateaus and hills that participate in soil leaching with the parasite eggs. This can partly justify the low level of contamination observed in schools located in high altitude (EM St Mathias, Bilingual EM, EM central, and EP Foto).

Around latrines, 11.9% of soil samples were more contaminated with the eggs of nematodes compared to the other two sampling points, that is, at playgrounds (2.5%) and behind classrooms (7.5%). This finding corroborates findings obtained by Chukwuma et al. [11] in Nigeria. Considering the fact that the eggs come from the school children, it clearly indicates that some school children usually defecate around/behind toilets and behind classrooms instead of using the toilet. These two points are privileged places for school children's contamination by soil-transmitted helminths.

Regarding the seasons, soil samples were relatively more contaminated during the wet season (12.5%) than during the dry season (3%). These results corroborate those obtained elsewhere [16, 18, 25, 26]. Actually, it is known that the development of the egg in the soil depends on several factors such as temperature (optimum: 20–30°C) and adequate soil moisture [27]. Thus, the observations can be explained by the fact that the rains create environmental and climatic conditions favorable for the survival of nematode eggs. However, these eggs are also resistant to extreme conditions. It is for this reason that few eggs of *Ascaris* and *Trichuris* were observed in the dry season.

5. Conclusion

The findings reported in the present study showed that the rate of soil contamination was low but it remains to be the most direct indicator of risk factor for children's health in schools of Dschang. In addition, daily awareness of the need to keep the environment healthy by adequate sanitation practices should be highlighted in order to avoid soil contamination by human intestinal parasites.

Disclosure

The senior author, Professor Josué Wabo Poné, passed away in August 2017 after this article was accepted.

Acknowledgments

The authors wish to express their sincere thanks to all the members of the Laboratory of Biology and Applied Ecology (LABEA) for their help during field and laboratory work. They also thank directors of different schools of Dschang enrolled in this study for their kind collaboration.

References

[1] G. Mordi, "The population biology and control of *Ascaris lumbricoides* in a rural community in Iran," *Transactions of the Royal Society of Tropical Medicine and Hygiene*, vol. 76, no. 2, pp. 187–197, 2009.

[2] P. J. Hotez, A. Fenwick, L. Savioli, and D. H. Molyneux, "Rescuing the bottom billion through control of neglected tropical diseases," *The Lancet*, vol. 373, no. 9674, pp. 1570–1575, 2009.

[3] E. C. Amandi and E. C. Uttah, "Environment and manage," *Journal April Science*, vol. 14, no. 2, pp. 61–64, 2010.

[4] G. W. Storey and R. A. Phillips, "The survival of parasite eggs throughout the soil profil," *Journal of Parasitology*, vol. 3, no. 85, pp. 585–590, 1985.

[5] OMS, *Deworming for health and development. Report of the third global meeting of the partners for parasite control. World Health organization*, 5p, Geneva, 2005.

[6] I. Nock, D. Duniya, and M. Galadima, "Geohelminth eggs in the soil and stool of pupils of some primary schools in Samaru, Zaria, Nigeria," *Nigerian Journal of Parasitology*, vol. 24, pp. 115–122, 2003.

[7] S. Uga, T. Matsumura, N. Aoki, and N. Kataoka, "Prevalence of Toxocara species eggs in the sandpits of public parks in Hyogo Prefecture, Japan," *Japanese Journal of Parasitology*, vol. 38, no. 5, pp. 280–284, 1993.

[8] Thienpont D., Roschette F., and Vanparijs O., *Diagnostic Des Verminoses Par Examen Coprologique*, vol. 29, Janssen Researche foundation, Beerse, Belgique, 1979.

[9] E. J. L. Soulsby, *Helminths, arthropods and protozoa of domesticated animals*, Baillere Tindal, London, 7th edition, 1982.

[10] S. Herbert, *Contribution à l'étude du parasitisme chez le Mandrill au Gabon [Ph.D. thesis]*, Faculté de Médecine de l'Université Paul-Sabatier de Toulouse, France, 2009.

[11] N. C. Chukwuma, I. M. Ekejindu, N. R. Agbakoba, D. A. Ezeagwuna, I. C. Anaghalu, and D. C. Nwosu, "The prevalence and risk factors of geohelminth infections among primary school children in Ebenebe Town," *Middle-East Journal of Scientific Research*, vol. 4, no. 3, pp. 211–215, 2009.

[12] B. Odoba, J. R. Otalu, and B. J. Balogun, "Prevalence of helminth parasites eggs in pupils and playing grounds of some selected primary schools in Zaria," *Nigeria. World Journal Life Science and Medical Research*, vol. 2, no. 5, p. 192, 2012.

[13] A. Shrestha, R. Shiba Kumar, S. Raj Basnyat et al., "Soil transmitted helminthiasis in kathmandu, Nepal," *Nepal Medical College Journal*, vol. 6, pp. 150–160, 2006.

[14] N. V. Afanwi, *The prevalence of intestinal parasites in school children in plateau state: a case [MSc thesis]*, Faculty of Medicine of University of Jos, Nigeria, 1986.

[15] S. Messaad, D. Belghyti, A. Diouane, and M. Rhajaoui, "Etude de la charge parasitaire des eaux usées de la ville de Salé," *Science Library Editions Mersenne*, vol. 5, pp. 1–11, 2013.

[16] S. Uga, K. Ono, N. Kataoka et al., "Contamination of soil with parasite eggs in Surabaya, Indonesia," *The Southeast Asian Journal of Tropical Medical and Public Health*, vol. 26, no. 4, pp. 730–734, 1995.

[17] S. Uga, W. Nagnaen, and V. Chongsuvivatwong, "Contamination of soil with parasite eggs and oocysts in Southern Thailand," *Southeast Asian Journal of Tropical Medicine and Public Health*, vol. 28, pp. 14–17, 1997.

[18] D. Stojcevic, V. Susic, and S. Lucinger, "Contamination of soil and sand with parasite elements as a risk factor for human health in public parks and playgrounds in Pula," *Veterinarski Archieves*, vol. 80, no. 6, pp. 733–742, 2010.

[19] M. Tavalla, H. Oormazdi, L. Akhlaghi et al., "Prevalence of parasites in soil samples in Tehran public places," *African Journal of Biotechnology*, vol. 11, no. 20, pp. 4575–4578, 2012.

[20] C. M. Davies, N. Altavilla, M. Krogh, C. M. Ferguson, D. A. Deere, and N. J. Ashbolt, "Environmental inactivation of Cryptosporidium oocysts in catchment soils," *Journal of Applied Microbiology*, vol. 98, no. 2, pp. 308–317, 2005.

[21] J. D. Smyth, *Animal parasitology*, Cambridge university press, Cambridge, United Kingdom, 1996.

[22] S. Debalke, A. Worku, N. Jahur, and Z. Mekonnen, "Soil transmitted helminths and associated factors among schoolchildren in government and private primary school in Jimma Town, Southwest Ethiopia," *Acta Parasitologica Globalis*, vol. 6, no. 1, pp. 29–35, 2012.

[23] C. C. Appleton and E. Gouws, "The distribution of common intestinal nematodes along an altitudinal transect in KwaZulu-Natal, south africa," *Annals of Tropical Medicine and Parasitology*, vol. 90, no. 2, pp. 181–188, 1996.

[24] P. K. Yerima and E. Van Ranst, "Introduction to soil science: soils of the tropics," in *St*, Trafford Publishing, Victoria, Canada, 6th edition, 2005.

[25] S. K. Rai, S. Uga, K. Ono, G. Rai, and T. Matsumura, "Contamination of soil with helminth parasite eggs in Nepal," *Southeast Asian Journal of Tropical Medicine and Public Health*, vol. 31, no. 2, pp. 388–393, 2000.

[26] Y. Nurdian and J. Dasar, "Soil contamination by parasite eggs in two urban villages of Jember," *Journal of Parasitology*, vol. 5, pp. 50–54, 2004.

[27] Y. Komiya and Y. Yasuraoka, "Environmental factors influencing transmission," *Medicine Parasitology*, vol. 3, no. 4, p. 114, 1966.

Permissions

All chapters in this book were first published in JPR, by Hindawi Publishing Corporation; hereby published with permission under the Creative Commons Attribution License or equivalent. Every chapter published in this book has been scrutinized by our experts. Their significance has been extensively debated. The topics covered herein carry significant findings which will fuel the growth of the discipline. They may even be implemented as practical applications or may be referred to as a beginning point for another development.

The contributors of this book come from diverse backgrounds, making this book a truly international effort. This book will bring forth new frontiers with its revolutionizing research information and detailed analysis of the nascent developments around the world.

We would like to thank all the contributing authors for lending their expertise to make the book truly unique. They have played a crucial role in the development of this book. Without their invaluable contributions this book wouldn't have been possible. They have made vital efforts to compile up to date information on the varied aspects of this subject to make this book a valuable addition to the collection of many professionals and students.

This book was conceptualized with the vision of imparting up-to-date information and advanced data in this field. To ensure the same, a matchless editorial board was set up. Every individual on the board went through rigorous rounds of assessment to prove their worth. After which they invested a large part of their time researching and compiling the most relevant data for our readers.

The editorial board has been involved in producing this book since its inception. They have spent rigorous hours researching and exploring the diverse topics which have resulted in the successful publishing of this book. They have passed on their knowledge of decades through this book. To expedite this challenging task, the publisher supported the team at every step. A small team of assistant editors was also appointed to further simplify the editing procedure and attain best results for the readers.

Apart from the editorial board, the designing team has also invested a significant amount of their time in understanding the subject and creating the most relevant covers. They scrutinized every image to scout for the most suitable representation of the subject and create an appropriate cover for the book.

The publishing team has been an ardent support to the editorial, designing and production team. Their endless efforts to recruit the best for this project, has resulted in the accomplishment of this book. They are a veteran in the field of academics and their pool of knowledge is as vast as their experience in printing. Their expertise and guidance has proved useful at every step. Their uncompromising quality standards have made this book an exceptional effort. Their encouragement from time to time has been an inspiration for everyone.

The publisher and the editorial board hope that this book will prove to be a valuable piece of knowledge for researchers, students, practitioners and scholars across the globe.

List of Contributors

Reza Shafiei and Maryam Gholami
Vector-Borne Diseases Research Center, North Khorasan University of Medical Sciences, Bojnurd, Iran

Saeed Hosseini Teshnizi
Clinical Research Development Center of Children Hospital, Hormozgan University of Medical Sciences, Bandar Abbas, Iran

Kurosh Kalantar
Department of Immunology, School of Medicine, Shiraz University of Medical Sciences, Shiraz, Iran

Golnush Mirzaee
Rehabilitation Management, University of Social Welfare and Rehabilitation Sciences, Tehran, Iran

Fatemeh Mirzaee
School of Nursing and Midwifery, Shahrekord University of Medical Sciences, Shahrekord, Iran

José Antonio Gabrie
Department of Health Sciences, Brock University, St. Catharines, ON, Canada

María Mercedes Rueda, Carol Anahelka Rodríguez and Maritza Canales
School of Microbiology, National Autonomous University of Honduras (UNAH), Tegucigalpa, Honduras

Ana Lourdes Sanchez
Department of Health Sciences, Brock University, St. Catharines, ON, Canada
Microbiology Research Institute, National Autonomous University of Honduras (UNAH), Tegucigalpa, Honduras

David Z. Munisi
Department of Global Health and Biomedical Sciences, School of Life Sciences and Bioengineering, Nelson Mandela African Institution of Science and Technology, Arusha, Tanzania
Department of Biomedical Sciences, School of Medicine and Dentistry, College of Health Sciences, University of Dodoma, Dodoma, Tanzania

Joram Buza and Emmanuel A. Mpolya
Department of Global Health and Biomedical Sciences, School of Life Sciences and Bioengineering, Nelson Mandela African Institution of Science and Technology, Arusha, Tanzania

Safari M. Kinung'hi
National Institute for Medical Research (NIMR), Mwanza Research Centre, Isamilo Road, Mwanza, Tanzania

Karshima Solomon Ngutor
Department of Animal Health, Federal College of Animal Health and Production Technology, PMB 001, Vom, Nigeria

Lawal A. Idris and Okubanjo Oluseyi Oluyinka
Department of Veterinary Parasitology and Entomology, Ahmadu Bello University, PMB 1045, Zaria, Nigeria

Ravindra Sharma, Keshaw Tiwari, Kristen Birmingham, Elan Armstrong, Andrea Montanez, Reneka Guy, Yvette Sepulveda, Veronica Mapp-Alexander and Claude DeAllie
School of Veterinary Medicine, St. George's University, West Indies, Grenada

Makarim M. Adam Suliman, Bushra M. Hamad, Musab M. Ali Albasheer and Muzamil Mahdi Abdel Hamid
Department of Parasitology and Medical Entomology, Institute of Endemic Diseases, University of Khartoum, Khartoum, Sudan

Maytha Elhadi and Mutaz Amin Mustafa
Department of Parasitology and Medical Entomology, Institute of Endemic Diseases, University of Khartoum, Khartoum, Sudan
Faculty of Medicine, University of Khartoum, Khartoum, Sudan

Maha Elobied
Faculty of Pharmacy, Al-Neelain University, Sudan

Rajiv Ravi and Zary Shariman Yahaya
School of Biological Sciences, Universiti Sains Malaysia, Minden, 11800 Penang, Malaysia

William Yavo and Abibatou Konaté
Malaria Research and Control Centre, National Institute of Public Health, BPV 47, Abidjan, Côte d'Ivoire
Faculty of Pharmacy, Department of Parasitology and Mycology, Félix Houphouët-Boigny University, BPV 34, Abidjan, Côte d'Ivoire

Denise Patricia Mawili-Mboumba, Marie L. Tshibola Mbuyi and Marielle K. Bouyou-Akotet
Faculty of Medicine, Department of Parasitology and Mycology, University des Sciences de la Santé, BP 4009, Libreville, Gabon

Fulgence Kondo Kassi and Eby I. Hervé Menan
Faculty of Pharmacy, Department of Parasitology and Mycology, Félix Houphouët-Boigny University, BPV 34, Abidjan, Côte d'Ivoire
Parasitology and Mycology Laboratory of Diagnosis and Research Centre on AIDS and Other Infectious Diseases, 01 BPV 03, Abidjan, Côte d'Ivoire

Etienne Kpongbo Angora
Faculty of Pharmacy, Department of Parasitology and Mycology, Félix Houphouët-Boigny University, BPV 34, Abidjan, Côte d'Ivoire

Miguel Romero, R. Cerritos and Cecilia Ximenez
Faculty of Experimental Medicine, Experimental Immunology Laboratory, National Autonomous University of Mexico, Dr. Balmis 148, Colonia Doctores, 06720 Mexico City, Mexico

Ángel de la Cruz Pech-Canul
Centre for Biomolecular Sciences,The University of Nottingham, University Park, University Blvd, Nottingham NG7 2RD, UK

Victor Monteón
Investigaciones Biomédicas, Universidad Autónoma de Campeche, Av. Patricio Trueba s/n, Col. Lindavista, 24039 Campeche, CAM, Mexico

Rosa-Lidia Solís-Oviedo
Centre for Biomolecular Sciences, The University of Nottingham, University Park, University Blvd, Nottingham NG7 2RD, UK
Investigaciones Biomédicas, Universidad Autónoma de Campeche, Av. Patricio Trueba s/n, Col. Lindavista, 24039 Campeche, CAM, Mexico

Catalina Tovar Acero and Dina Ricardo Caldera
Grupo de Investigación en Enfermedades Tropicales y Resistencia Bacteriana, Facultad de Ciencias de la Salud, Universidad del Sinú, Montería, Colombia

Jorge Negrete Peñata
Laboratorio de Investigaciones Biomédicas, Universidad del Sinú, Montería, Colombia

Camila González, Cielo León and Mario Ortiz
Departamento de Ciencias Biológicas, Centro de Investigaciones en Microbiología y Parasitología Tropical (CIMPAT), Universidad de los Andes, Bogotá, Colombia

Julio Chacón Pacheco
Fundación Colombia Mia, Montería, Colombia Grupo de Investigación Biodiversidad Unicordoba, Universidad de Córdoba, Montería, Colombia

Elkin Monterrosa
Area de Entomología, Laboratorio de Salud Pública de Córdoba, Montería, Colombia

Abraham Luna
Hospital San Juan de Sahagún, Sahagún, Colombia

Lyda Espitia-Pérez
Grupo de Investigación Biomédica y Biología Molecular, Facultad de Ciencias de la Salud, Universidad del Sinú, Montería, Colombia

Jorge Negrete Peñata
Laboratorio de Investigaciones Biomédicas, Universidad del Sinú, Montería, Colombia

Camila González and Cielo León and Mario Ortiz
Departamento de Ciencias Biológicas, Centro de Investigaciones en Microbiología y Parasitología Tropical (CIMPAT), Universidad de los Andes, Bogotá, Colombia

Julio Chacón Pacheco,
Fundación Colombia Mia, Montería, Colombia Grupo de Investigación Biodiversidad Unicordoba, Universidad de Córdoba, Montería, Colombia

Elkin Monterrosa
Area de Entomología, Laboratorio de Salud Pública de Córdoba, Montería, Colombia

Abraham Luna
Hospital San Juan de Sahagún, Sahagún, Colombia

Lyda Espitia-Pérez
Grupo de Investigación Biomédica y Biología Molecular, Facultad de Ciencias de la Salud, Universidad del Sinú, Montería, Colombia

J. G. Gotep, G. E. Forcados, I. A. Muraina, N. Ozele, O. O. Oladipo, M. S. Makoshi, A. L. Samuel and A. A. Atiku
Biochemistry Division, National Veterinary Research Institute, PMB 01, Vom, Nigeria

J. T. Tanko
Parasitology Division, National Veterinary Research Institute, PMB 01, Vom, Nigeria

B. B. Dogonyaro
Virology Division, National Veterinary Research Institute, PMB 01, Vom, Nigeria

O. B. Akanbi
Central Diagnostics Laboratory, National Veterinary Research Institute, PMB 01, Vom, Nigeria

H. Kinjir
Haematology Department, Federal College of Veterinary and Medical Laboratory Technology, PMB 01, Vom, Nigeria

T. E. Onyiche
Biochemistry Division, National Veterinary Research Institute, PMB 01, Vom, Nigeria
Department of Veterinary Microbiology and Parasitology, University of Maiduguri, Bama Road, Maiduguri, Borno State, Nigeria

G. O. Ochigbo
Biochemistry Division, National Veterinary Research Institute, PMB 01, Vom, Nigeria
Department of Veterinary Physiology, Pharmacology and Biochemistry, University of Ibadan, PMB 0248, Ibadan, Nigeria

O. B. Aladelokun
Biochemistry Division, National Veterinary Research Institute, PMB 01, Vom, Nigeria
Department of Biochemistry, University of Ibadan, PMB 0248, Ibadan, Nigeria

H. A. Ozoani
Biochemistry Division, National Veterinary Research Institute, PMB 01, Vom, Nigeria
Department of Medical Laboratory Science, Rivers State University of Science and Technology, PMB 5080, Port Harcourt, Nigeria

V. Z. Viyoff
Biochemistry Division, National Veterinary Research Institute, PMB 01, Vom, Nigeria
Department of Epidemiology, University of Buea, Buea, Cameroon

C. C. Dapuliga
Biochemistry Division, National Veterinary Research Institute, PMB 01, Vom, Nigeria
Microbiology Department, Kwame Nkrumah University of Science and Technology, Kumasi, Ghana

P. A. Okewole
Central Diagnostics Laboratory, National Veterinary Research Institute, PMB 01, Vom, Nigeria

D. Shamaki and M. S. Ahmed
National Veterinary Research Institute, Vom, Nigeria

C. I. Nduaka
Africa Education Initiative (NEF), 9401 Sentinel Ridge, Eagleville, PA 19403, USA

G. N. Hartmeyer, M. N. Skov and M. Kemp
Research Unit of Clinical Microbiology, Institute of Clinical Research Faculty of Health Science, University of Southern Denmark, Odense, Denmark
Department of Clinical Microbiology, Odense University Hospital, Odense, Denmark

S. V. Hoegh
Department of Clinical Microbiology, Odense University Hospital, Odense, Denmark

R. B. Dessau
Department of Clinical Microbiology, Slagelse Hospital, Slagelse, Denmark

Sivaprakasam Rajasekaran and Ramalingam Bethunaickan
Department of Immunology, National Institute for Research in Tuberculosis, Chennai, India

Rajamanickam Anuradha
International Center for Excellence in Research, National Institutes of Health, National Institute for Research in Tuberculosis, Chennai, India

Furhan T. Mhaisen
Tegnervägen 6B, 641 36 Katrineholm, Sweden

Abdul-Razzak L. Al-Rubaie
Department of Biological Control Technology, Al-Musaib Technical College, Al-Furat Al-Awsat Technical University, Al-Musaib, Iraq

Ngabu Malobi, Long Yu, Pei He, Junwei He, Yali Sun and Yuan Huang
State Key Laboratory of Agricultural Microbiology, College of Veterinary Medicine, Huazhong Agricultural University, Wuhan, Hubei 430070, China

Lan He and Junlong Zhao
State Key Laboratory of Agricultural Microbiology, College of Veterinary Medicine, Huazhong Agricultural University, Wuhan, Hubei 430070, China
Key Laboratory of Animal Epidemical Disease and Infectious Zoonoses, Ministry of Agriculture, Huazhong Agricultural University, Wuhan, China

Dana Ditgen and Emmanuela M. Anandarajah
Department of Molecular Physiology, Westfälische Wilhelms-University, Münster, Germany

Department of Molecular Medicine, Bernhard Nocht Institute for Tropical Medicine, Hamburg, Germany

Jan Hansmann
Department of Tissue Engineering and Regenerative Medicine (TERM), University of Würzburg, Germany

Dominic Winter
Institute for Biochemistry and Molecular Biology, University of Bonn, Bonn, Germany

Guido Schramm
Ovamed GmbH, Hamburg, Germany

Klaus D. Erttmann and Norbert W. Brattig
Department of Molecular Medicine, Bernhard Nocht Institute for Tropical Medicine, Hamburg, Germany

Eva Liebau
Department of Molecular Physiology, Westfälische Wilhelms-University, Münster, Germany

Davoud Balarak, Edris Bazrafshan and Ferdos Kord Mostafapour
Department of Environmental Health, Health Promotion Research Center, Zahedan University of Medical Sciences, Zahedan, Iran

Mohammad Jafari Modrek
Department of Parasitology, Infectious Diseases and Tropical Medicine Research Center, Zahedan University of Medical Sciences, Zahedan, Iran

Hossein Ansari
Department of Epidemiology and Biostatistics, Health Promotion Research Center, Zahedan University of Medical Sciences, Zahedan, Iran

Ma. Angelica A. Tujan and Vachel Gay V. Paller
Animal Biology Division, Institute of Biological Sciences, College of Arts and Sciences, University of the Philippines Los Baños, 4031 Laguna, Philippines

Ian Kendrich C. Fontanilla
DNA Barcoding Laboratory, Institute of Biology, College of Science, University of the Philippines Diliman, 1001 Quezon City, Philippines

Papa Kofi Amissah-Reynolds, Lucy Mawusi Adowah and Samuel Opoku Agyemang
Department of Science Education, University of Education, Winneba, Mampong, Ashanti, Ghana

Isaac Monney
Department of Environmental Health and Sanitation Education, University of Education, Winneba, Mampong, Ashanti, Ghana

Vanessa Rosine Nkouayep, Blandine Ngatou Tchakounté and Josué Wabo Poné
Research Unit of Biology and Applied Ecology, Department of Animal Biology, Faculty of Science, University of Dschang, Dschang, Cameroon

Index

www.ingramcontent.com/pod-product-compliance
Lightning Source LLC
Chambersburg PA
CBHW050457200326
41458CB00014B/5222